INTRODUCTION
TO BIOENGINEERING

ADVANCED SERIES IN BIOMECHANICS

Editor: Y C Fung *(University of California, San Diego)*

Vol. 1:   Selected Works on Biomechanics and Aeroelasticity
        Parts A & B
        by Y C Fung

Advanced Series in Biomechanics – Vol. 2

INTRODUCTION TO BIOENGINEERING

Editor

Y. C. Fung

*University of California
San Diego*

Authors
Shu Chien
Yuan Cheng Fung
David A. Gough
Marcos Intaglietta
Ghassan Kassab
Bernhard Palsson
Robert L. Sah
Geert Schmid-Schönbein
Lanping Amy Sung
Pin Tong
Michael R. T. Yen
Wei Huang

 World Scientific
Singapore • New Jersey • London • Hong Kong

Published by

World Scientific Publishing Co. Pte. Ltd.

P O Box 128, Farrer Road, Singapore 912805

USA office: Suite 1B, 1060 Main Street, River Edge, NJ 07661

UK office: 57 Shelton Street, Covent Garden, London WC2H 9HE

British Library Cataloguing-in-Publication Data
A catalogue record for this book is available from the British Library.

INTRODUCTION TO BIOENGINEERING
Advanced Series in Biomechanics — Vol. 2

ISBN   981-02-4023-6
ISBN   981-02-4398-7 (pbk)

Printed in Singapore by World Scientific Printers

PREFACE

This book is written for two purposes: to give beginning students a sampling of the contemporary bioengineering, and to cultivate in the minds of students a sense that engineering is invention and design, is creating things that have never been. Engineering science is the study of nature and humanity quantitatively so that engineering design is made possible. Bioengineering is engineering of living organisms. Bioengineering is very popular, and we have a large number of bright young students. Yet most of them have no idea what engineering is. They think they are majoring in bioengineering, but the University tells them to take more courses in humanities, math, physics, chemistry and biology. Later they will take engineering courses which, at this time, are usually taught very much like traditional math and science: with good lectures, logical presentations, polished derivations, lots of exercises, but no mention of invention. My feeling is that invention and design is the essence of engineering. Invention and design can be taught. A habit of inventive thinking can be cultivated. And the cultivation should begin as early as possible, by demanding the students to invent, and to develop their inventions by design. Hence in this introductory course I ask the students to invent and to design. Invent something, maybe a better mouse trap, a new gadget, a new medical device, a new material, a new drug, a new method, a new instrument, a new experiment, a new theory, a new concept, a new approach. Each new idea needs to be crystallized by a design. I give them assignments to design step by step. They will soon realize that the method must be scientific, and that the real limitation to any important invention is the lack of knowledge. By wanting to invent, they realize that they need to master the classical subjects. Thus the classical subjects become alive, and the study becomes active. Thus I lay great emphasis on the first few pages of this book. The assignments given to the students are the heart of the course. Reading and commenting on the students' papers is the job of the instructor and teaching assistants.

After initiating the student to invent and design, we show them live examples in person of their faculty members. These examples constitute the body of this book. Each example consists of a lecture and a paper. The lecture tells what a faculty member is doing. The paper is the best publication of the speaker. Therefore, this book is not a run of the mill *Introduction* that is supposed to be easy to read and undemanding. Some of the papers selected here are indeed easy to understand. Some others are quite technical and may not be easy to assimilate without proper preparation. All are "advanced" in the sense that they represent a part of the current frontiers of knowledge.

Not everybody likes the inventive approach or the subject of bioengineering. This course may persuade some of the students to choose a different career goal. Our experience in this introductory course has been very satisfying. The final reports handed in by the students are often remarkably well done. Nice, fresh ideas abound. You can see that the inventive chord in the student has been plucked.

The authors and I feel the urge to publish this book because we believe that engineering education needs a qualitative lift. We need to challenge young engineering students to think their own thoughts. We need to bring inventive thinking to the fore. We need to give engineering courses a new spirit, to teach invention, to cultivate independent thoughts, to use science as our tools and to enjoy science the same way as we enjoy humanities. We give an example here in the subject of bioengineering. Any department in any university can easily remodel our example presented here to suit its own image and ambition.

Recently, several books have appeared bearing the title *Introduction to Bioengineering*. Most of them are big tomes. I cannot see how an instructor can use them to teach a course for the beginners. This book, however, is aimed at the beginner. It has been tried out several times at UCSD, and have been found exciting. We invite other institutions to try it for their students.

Y. C. Fung
Editor

CONTENTS

Appendixes

TO THE INSTRUCTOR

This book is not an ordinary textbook. It is a cultural reader trying to tell a beginner what bioengineering is. The authors tell the story through their personal experiences. They do not claim complete balance of their views; but they are sincere, vigorous, and so committed that they are devoting their whole lives to pursue their visions. They know that the field of bioengineering is young, and is developing in many directions. They all believe that the best way to teach introductory bioengineering is to ask the beginners to think about invention and design right away, whatever they think bioengineering is. This is where this little book comes in. Each author asks certain questions, then invents and designs some method to answer their questions. Their lecture notes and papers could serve as models. We recommend a series of assignments to the students as listed in the following pages, and asking them to read the book one chapter per week. We trust that you will find this approach very enjoyable, because your students will be proud of their new found abilities.

TO THE STUDENTS

Hi,

This course is an introduction to the central topics of Bioengineering. The principles of problem definition, team design, engineering inventiveness, information access and communication, ethics and social responsibility are emphasized.

The authors of this book are faculty members of Bioengineering. Each of them presents a short lecture on some of the things they are doing, and a sample of their writing. This volume is a collection of these lecture notes and papers. By meeting your faculty members, listening to them and reading the lecture notes and papers, you will get an idea about their personal interests and achievements. Overall, you will get a feeling about Bioengineering as a discipline.

An objective of this course is to let you learn a little bit about what a bioengineer is expected to do and to behave in the industry. In the industry, you are expected to design and to work in teams with other people. Our course assignments are aimed in this direction.

Bioengineering is a young discipline. It is still developing rapidly. New knowledge is added every day. In the development of new knowledge, the application of the methods and results of classical disciplines is most important. That is why your university asks you to take courses in mathematics, physics, chemistry, humanities, etc. To help you realize that this is the case, I am introducing design as a major approach to bioengineering in this course. You are asked to think yourself. Then you are asked to think with some friends as a team, advancing your project together. At the end of the course, you gain not only some knowledge, but also some friends. That friendship may be the most valuable part of you college eduction.

My philosophy of using design as a major approach to learning science was developed over the years. In the past, students listen to lectures, read textbooks, do exercises, take exams and pass courses with only a vague idea that what is learned is useful. You have done enough of that. Now, in thinking about a design project, you may find that your success depends on certain knowledge about science and humanity. Therefore, you search for knowledge actively.

The reading material accompanying each lecture is a sample of good papers. Each paper talks about a topic of engineering science. These papers are not design reports, but usually they are the results of many successful designs.

On the following three pages are the assignments I gave the class when I was teaching. Your instructor may give you a different set.

Y. C. Fung
Editor

SAMPLE ASSIGNMENTS

Reading: Read a chapter before the class, and the author's paper after the class.

Lecture Schedule: To be presented.

Assignment No. 1   Report Due on:_____

Describe a research project you would like to do. This can be an invention of an instrument for scientific inquiries, or a device for clinical application, or a gadget for home care, or any other things that you think should qualify as Bioengineering. Or it can be a study of a scientific or technological problem concerning a phenomenon of interest to bioengineering. Note that I did not define "bioengineering." You define it yourself. I know most of us have at least a dozen projects we would like to do at any time. For the present assignment, make a choice of one which is your highest priority, one of which you would be most proud if the project were completed successfully. This assignment is to be answered intuitively, based on your "gut feeling." Explain your project as best you can. Describe what it is, why do you think it is important, and how would you approach it, all in **one page**. If the idea is your own, no reference is needed. If the whole idea or parts of it came from another person, or from a journal article, a paper, a book, then you should acknowledge the source with references. Don't ever be caught for plagiarism.

Assignment No. 2   Due date:_____

Embodiment of Your Thoughts on a Design.

Give your design a title. Supply some details with hand drawn sketches or graphics, and explain your objectives, design principles and expectations of its significance. Not to exceed 3 pages.

Important Note:
You are responsible to hand in a typed or neatly handwritten report at the due date specified. Please hand in at or immediately after class. Each page should have a page number and your full name clearly printed at the upper right corner. Use 8.5 × 11 inch sheets of plain paper. Sign and date the paper at the end. Neatness and timeliness are essential quality of a good engineer.

Assignment No. 3   Due date:_____

Report on Team Organization and Team Project.

Teamwork is encourged. Get several friends to work together. Discuss your projects to see if you can reach a consensus to concentrate on **one** project. If you can, then revise Reports to Assignments 1 and 2 into a team report, and write Reports to Assignments 4–8 together as a team. On the upper right hand corner, write down the names of all team members. If you decided to go at it alone, it is OK. In that case, turn in an improved version of Reports 1 and 2, with greater details.

Assignment No. 4   Due date:_____

Library Search.

Present references that you have found.
Not to exceed 2 pages.

Assignment No. 5   Due date:_____

Design.

Discuss the scientific basis of your design in light of library search. Discuss the feasibility and difficulties. Estimate time needed to accomplish your design. Discuss possible ethical questions involved.
Report, maximum 2 pages.

Assignment No. 6   Due date:_____

Illustrate your design with good graphs. Explain your drawings carefully. If your work is theoretical, develop your equations with sufficient details.

Assignment No. 7   Due date:_____

Discuss the feasibility of your design or research. What are the unknown factors? What may cause your design to fail? What are its strengths? What are its weaknesses? How are you going to test your design? Do you need animal experiments? Do you need human experiments? What are the governmental and university regulations with regard to these experiments?

Assignment No. 8   Due date:_____

Final Report.

Prepare a final report based on all you have written up to this point. Use the following format for the cover page:

(a) At top:
     Final report on Course BE1, Introduction to Bioengineering Instructor's name:
(b) At center: Title of your project
(c) The names of all the authors in alphabetical order
(d) At bottom: Date submitted

Beginning on page 2, write Sections in the following order:
        Introduction
        Specific Aims
        Designs and Analysis
        Discussion
        References

This final report will be retained by the department. It will not be returned to you. Please save a xerox copy for yourself.

Write the best you can. Make it a nice little paper for future reference.

Reports Expected

You are responsible to hand in a clearly written report on each item listed below at the due dates specified. Each page should have a page number and your full name clearly printed at the upper right corner. Sign and date the paper at the end.

Report 1	Proposal of a Design Project	1 page
Report 2	Embodiment of my Thoughts in a Design on...	Not to exceed 3 pages
Report 3	Scientific Basis of My/Our Design On... Team work is encourged. Get several friends to work together. Discuss your projects to see if you can reach a consensus to concentrate on **one** project. If you can, then revise Reports 1 and 2 into a team report, and write Reports 4–6 together as a team. On the upper right hand corner, write down the names of all team members.	Max. 3 pages
Report 4	Results of Library Search on My/Our Project on...	Not to exceed 2 pages
Report 5	Estimate of Time and Money Needed to Accomplish My/Our Design on... and Possible Ethical Question Involved	Max. 2 pages
Report 6	Team and Individual Final Report on the Design of... Collate your five reports together. Correct all errors and misprints. Improve the English.	A nice paper

CHAPTER 1

ROLES OF FLOW MECHANICS IN VASCULAR CELL BIOLOGY IN HEALTH AND DISEASE

SHU CHIEN

Departments of Bioengineering and Medicine, and
The Whitaker Institute for Biomedical Engineering,
University of California, San Diego

Endothelial cells (ECs) form the inner lining of blood vessels, separating the circulating blood from the remainder of the vessel wall. Endothelial cells not only serve as a permeability barrier to regulate the passage of molecules between blood and the vessel wall, but also perform many important functions, including the production, secretion, and metabolism of many biochemical substances and the modulation of contractility of the underlying smooth muscle cells. The responses of endothelial cells and smooth muscle cells to changes in pressure and flow play a significant role in regulating the functional performance of blood vessels in health and disease.

Atherosclerosis results from the accumulation of fatty materials as atheroma in the artery wall. Atherosclerotic narrowing of the vessel lumen can cause a reduction of blood flow in the organ supplied by the vessel, thus leading to clinical problems. Examples are stroke, heart attack, and walking difficulty, which result from severe reductions of blood flow to the brain, heart, and legs, respectively.

There are two major elements in the initiation of atherosclerosis (Fig. 1). The first is lipids, especially the low density lipoprotein (LDL), which can become oxidized in the vessel wall. The second is monocytes, which become transformed into macrophages (large

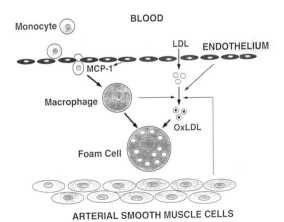

Fig. 1. Schematic drawing showing the roles of low density lipoprotein (LDL) and monocytes in the formation of foam cells.

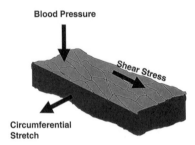

Fig. 2. The hemodynamic forces acting on the blood vessels due to blood pressure and flow.

eating cells) after entering the artery wall. The macrophages can engulf oxidized LDL to form foam cells, and the accumulation of foam cells in the vessel wall is the basis for the formation of atheroma (Steinberg, 1995).

Although every part of the arterial tree is exposed to the same LDL concentration in the circulating blood, atherosclerotic lesions show preferential distributions in branch points and curved regions (Cornhill *et al.*, 1990), suggesting that local variations in hemodynamic forces play a significant role in the focal nature of the lesions (Nerem *et al.*, 1993). Work done in our and other laboratories indicate that hemodynamic factors modulate both the transendothelial passage of large molecules such as LDL and the entry of monocytes into the artery wall. There are several components of hemodynamic forces (Fig. 2). This chapter focuses on the effects of shear stress, which is the tangential forces acting on the luminal surface of the vessel as a result of flow. Other important forces are the normal stress and circumferential stress resulting from the action of pressure. The aims of studying the role of hemodynamic forces in the regulation of EC function are to elucidate (a) the fundamental mechanism of mechano-chemical transduction, and (b) the biomechanical and molecular bases of the preferential localization of atherosclerosis in the arterial tree.

1.   Role of Hemodynamic Factors in Transendothelial Permeability of Macromolecules

At the branch points and curved regions of the arterial tree, which have a predilection for atherosclerosis, blood flow is unsteady and the shear stress shows marked spatial and temporal variations (Glagov *et al.*, 1988). Weinbaum *et al.*, (1985) proposed the hypothesis that complex flow patterns cause an accelerated EC turnover (including cell mitosis and death), such that the resulting leakiness around the ECs undergoing turnover increases the permeability of large molecules (e.g. LDL) across the endothelial layer. Our experimental studies have provided evidence that EC mitosis (Chien *et al.*, 1988; Lin *et al.*, 1989) and death (Lin *et al.*, 1990) are associated with the leakage of macromolecules such as albumin and LDL on individual cell basis. Studies performed in a number of laboratories, including our own (Schwenke and Carew, 1988; Chuang *et al.*, 1990; Truskey *et al.*, 1992) have shown that these events of accelerated EC turnover occur primarily in areas with disturbed blood flow, e.g. arterial branch points. Electron microscopic studies (Fig. 3) have identified the widening of the intercellular junctions around ECs which are dying or undergoing mitosis (Huang *et al.*, 1992; Chen *et al.*, 1996).

Fig. 3. A. Transmission electron micrographs showing a section of an endothelial cell undergoing mitosis, as indicated by the separation of the dense nucleus into two parts. The arrow points to an intercellular junction around this mitotic cell. The junction is abnormally widened as shown by the electron-dense tracer used to outline the junction. B. A higher magnification of the widened junction in A. From Chen *et al.* (1996).

The results of experimental and theoretical studies indicate the following sequence of events that contribute to the focal nature of atherosclerosis.

Local hemodynamic factors → ↑ EC turnover → ↑ Local LDL → Focal lipid
(complex flow pattern)   (mitosis & death)   permeability   accumulation

2. Effects of Shear Stress on Monocyte Chemotactic Protein-1 (MCP-1) Gene Expression

2.1. *The process of gene expression*

Proteins are encoded as genomic DNAs, which are composed of two strands of complementary nucleic acids. Each strand of the genomic DNA has a promoter region containing *cis*-elements (nucleotide sequences) which are activated following the binding of specific proteins (transcription factors) to trigger the transcription of the DNA strand into a single-stranded RNA. Such primary RNA transcripts undergo appropriate maturation processes in the nucleus to become messenger RNAs (mRNAs). The mRNA exits the nucleus to serve as a template for translation into the encoded protein in the ribosome, which manufactures proteins by using the information encoded in the RNA sequence. The synthesized proteins perform their functions either inside the cell or outside following their secretion into the extracellular fluid. Chemical and mechanical stimuli can modulate gene expression in a variety of cells. The effects of shear stress on endothelial cell gene expression have been reviewed (Davies, 1996; Chien *et al.*, 1998).

2.2. *Effects of shear stress on MCP-1 gene expression*

As mentioned above, monocytes/macrophages play a major role in atherogenesis. It has been shown that macrophages are found underneath the ECs primarily in the arterial branch points (Malinkaukas *et al.*, 1995), suggesting that local hemodynamic forces in these lesion-prone areas can enhance monocyte entry. The entry of monocytes into the blood vessel wall is induced by a number of chemotactic agents, especially the monocyte chemotactic protein-1 (MCP-1). MCP-1 is secreted by the EC, as well as other types of cells. Several chemical substances (e.g. the phorbol ester TPA) can stimulate the expression of the MCP-1 gene and the consequent secretion of MCP-1 protein by ECs to attract monocyte entry into the vessel wall.

The effects of shear stress on MCP-1 secretion by cultured human umbilical vein endothelial cells (HUVECs) have been studied by using a flow channel system (Fig. 4). The application of a shear stress at levels existing in the arterial tree (e.g. 12 dynes/cm^2) caused an increase in MCP-1 gene expression to approximately 2.5 fold in about 1.5 hours (Fig. 5) (Shyy et al., 1994). Sustained application of the shear stress was accompanied by a decrease of the gene expression to or even below the preshear background level.

Studies on the promoter region of MCP-1 showed that the TPA responsive element (TRE) is a critical cis-element in the shear stress activation of MCP-1 (Shyy et al., 1995). The transcription factor for TRE is known to be a dimer (molecule with two subunits)

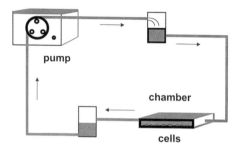

Fig. 4. A schematic drawing of the flow channel system.

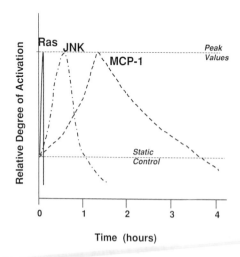

Time (hours)

Fig. 5. Time courses of the activation of signaling molecules and gene expression in endothelial cells in response to shear stress. The activation of Ras reaches a peak in less than 1 minute, followed by an increase in the kinase activity of JNK with a peak at 30 min. The induction of the MCP-1 gene occurs later than the activation of signaling molecules, with a peak at 90 min. Following prolonged shearing, the activities of the signaling molecules and the MCP-1 gene expression become lower than those in the static controls.

composed of cJun-cFos or cJun-cJun. The activation of c-Jun and c-Fos is known to be mediated by the mitogen activated protein kinase (MAPK) signaling pathways. The MAPK pathways are also involved in cell mitosis and programmed cell death (apoptosis). Therefore, the understanding of these pathways is important for the elucidation of both the chemo-attraction of monocytes by MCP-1 and the modulation of LDL permeability by acceleration of cell turnover.

3. Effects of Shear Stress on MAPK Signaling Pathways

3.1. *The MAPK signaling pathways*

Protein kinases are enzymes that can add a phosphate group to a protein (phosphorylation). The MAPK signaling pathways involve a group of protein kinases that are activated sequentially to result in a phosphorylation cascade. Thus, the phosphorylation of an upstream protein kinase will activate it so that it can serve as an enzyme to phosphorylate the next protein kinase in the pathway. There are several parallel pathways in this system (e.g. JNK and ERK shown in Fig. 6), with the molecule Ras situated as a common upstream molecule. When Ras is activated, it binds to GTP (guanosine triphosphate) instead of GDP (guanosine diphosphate). The GTP-bound Ras is the activated form that initiates the phosphorylation of the MAPK pathways, culminating in the phosphorylation of JNK and ERK and the activation of c-Jun and c-Fos, respectively. The end result is the expression of genes such as MCP-1.

3.2. *Effects of shear stress on signal transduction in endothelial cells*

The effects of shear stress on signal transduction in ECs have been studied in the flow channel system. The application of shear stress at the level found in the arteries (e.g. 12 dynes/cm^2) causes the Ras to become GTP-bound and activates both ERK and JNK (Li *et al.*, 1996). The Ras-GTP binding occurs in less than 1 min after the beginning of shearing, and this is followed by the activation of the downstream molecules (Fig. 5). The response of Ras is very transient in nature, and the early onset and short duration of its activation is in line with its upstream location relative to ERK and JNK in the signaling pathways.

Ras N17, which is a negative mutant molecule of Ras, markedly suppresses the shear-activation of TRE. Upstream to Ras, the adaptor molecules Shc, Grb2 and Sos have been

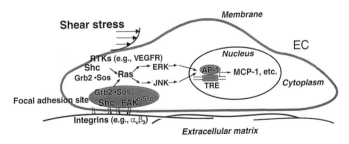

Fig. 6. Schematic diagram showing some of the signaling pathways mediating the mechano-chemical transduction in endothelial cells in response to shear stress.

found to be important (Fig. 6) because uses of their negative mutants suppress the shear-activation of TRE (Jalali *et al.*, 1997; Li *et al.*, 1997).

With the elucidation of the signaling pathway for the mechano-chemical transduction process, the next question is what are the mechano-sensors that are responsible for the initial conversion of the mechanical stimulus of shear stress into chemical changes that are transmitted through the signaling system.

4.   Mechano-sensors for Shear Stress

The experiments described in this section indicate that there are at least two sets of membrane sensors for the mechanical stimulation due to shear stress. One set is the membrane receptors located on the luminal side of the EC membrane, and the other is the integrin molecules situated on the abluminal side facing the extracellular matrix (ECM).

4.1.   *Membrane receptors — receptor tyrosine kinases*

Receptor tyrosine kinases (RTK) are membrane receptors that can undergo phosphorylation on their tyrosine residues upon appropriate stimulation. Such tyrosine phosphorylation activates the receptor such that it becomes associated with the adaptor molecules (e.g. Shc), and this in turn activates the downstream signaling pathway such as Ras-JNK. An example of receptor tyrosine kinases is the vascular endothelial growth factor receptor (VEGFR), which is known to be activated by the chemical binding of VEGF. We found that the application of shear stress causes the clustering and phosphorylation of VEGFR, as well as its binding to Shc and other adaptor molecules, and the subsequent activation of Ras and its downstream molecules (Chen *et al.*, 1999). Thus, mechanical stimulation can activate this receptor in the same manner as the chemical stimulation due to VEGF. The time courses of the responses of VEGFR to these two modes of stimuli are very similar: starting rapidly in less than 1 min and subsiding totally in 30 min. Therefore, VEGFR and possibly other membrane receptors (e.g. G-protein coupled receptors, Kuchan *et al.*, 1994) can serve as mechano-sensors to convert extracellular mechanical stimuli to intracellular chemical signals. It is likely that shear stress causes membrane perturbation, thus changing the conformation and/or interaction of membrane proteins such as VEGFR to trigger the downstream activation.

4.2.   *Membrane integrins*

Integrins are transmembrane receptors that link the intracellular cytoskeletal proteins with the proteins in the ECM to provide two-way communications between the cell and its ECM. There are more than twenty types of integrins, each of which is composed of two subunits, α and β. One of the major integrins in vascular ECs is $\alpha_v\beta_3$, which interacts specifically with the ECM proteins vitronectin and fibronectin. Shear stress causes the association of $\alpha_v\beta_3$ with Shc and the subsequent activation of the Ras pathway, only when the ECs are cultured on ECM composed of vitronectin or fibronectin (Jalali, 1998). In such systems, anti-$\alpha_v\beta_3$ antibodies markedly attenuate the shear-induced signaling. These and other findings indicate that integrins play an important role in the initiation of signaling in response to shear stress (Fig. 6). It is possible that the tension induced by shear stress on the luminal side of the EC membrane can be transmitted along the plane of the membrane and through the junctions to the integrins on the abluminal side; in this case integrins serve as a direct mechano-sensor. Alternatively, the signaling initiated by membrane receptors may be transmitted by the cytoskeleton to activate the integrins; in that case integrins may act as an amplifier for the mechano-sensing.

5. Effects of Complex Flow Pattern on Endothelial Cell Proliferation and Gene Expression

The experiments described above were performed on rectangular flow channels with a uniform height along the channel length, and the flow is laminar throughout the channel. In order to simulate the flow conditions near a branch point, we have developed a step flow channel (Chiu *et al.*, 1998), which has a lower channel height at the entrance and an expansion of the channel height thereafter (Fig. 7). Eddy flows are induced just downstream to the step, and the flow direction in the eddy region is opposite to that of the downstream forward flow region. Between the two regions with opposite flow directions, there is a narrow flow-reattachment region ("c" in Fig. 7) where shear stress is zero but the spatial gradient of shear stress is high.

The proliferation of vascular ECs in the step flow channel can be assessed by determining the incorporation of a labeled nucleotide, e.g. BrdU. Under static condition, BrdU incorporation is low and randomly distributed throughout the channel. Following 24 hours of laminar shear at 12 dynes/cm^2, BrdU incorporation is markedly enhanced in the reattachment area and its vicinity and is much lower elsewhere, including the downstream laminar flow region (Pin-Pin Hsu, unpublished results). The same distribution pattern is seen for the expression of genes related to cell cycle control for proliferation and of signaling molecules such as ERK. These results indicate that the flow pattern in the reattachment area at branch points stimulates cell proliferation. In contrast, the laminar flow region has a low cell proliferation rate and, as mentioned above, the signaling pathway and MCP-1 gene expression are down-regulated following sustained shearing. This down-regulation does not occur in the reattachment region because of the lack of sustained laminar shear. Furthermore, we have found that sustained shearing in the laminar flow region causes the activation of growth-arrest genes, which oppose cell proliferation. Taken together, these results can be interpreted to indicate that the complex flow pattern near the branch points increases EC proliferation and does not down-regulate the expression of MCP-1 gene, thus placing this region at risk for atherogenesis. In contrast, the sustained shearing in the laminar flow region

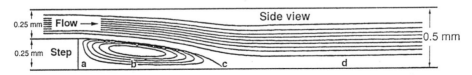

Fig. 7. A schematic drawing of the step flow channel. The expansion of the channel height beyond the step leads to secondary flow and flow reattachment

Table 1. Comparison between the straight part and branch points of the arterial tree.

	Straight Part	Branch Points
Flow Pattern	Laminar	Disturbed
Cell Turnover	Slow	Rapid
LDL Permeability	Low	High
Monocyte Infiltration	Few	Frequent
Gene Expression	Growth Suppression	Cell Proliferation
Atherosclerosis	Rare	Prevalent

induces EC cell arrest rather than proliferation and down-regulates MCP-1 expression, thus providing a protective action against atherogenesis (Table 1).

6.   Implications in Clinical Conditions

The results of the research studies outlined above have considerable implications in the prevention and treatment of cardiovascular diseases. Two examples are given below to illustrate how such knowledge is useful in relation to coronary artery disease, which is a major cause of death in the United States. The two most common modalities of treatment of coronary artery narrowing are bypass surgery and balloon angioplasty. The molecular responses of the vascular cells to mechanical forces are very relevant to both of these procedures.

6.1.   *Importance of mechanical matching between vascular bypass and native artery*

In vascular bypass surgery, the segment of the coronary artery with the atherosclerotic narrowing is removed and replaced by an artificial vascular graft or a vessel segment from the patient. One of the most commonly used bypass vessel is the saphenous vein from the patient's leg. While the saphenous vein graft has had its successes, recurrence of obstruction (restenosis) is a major complication. Veins are normally exposed to much lower pressures (< 10 mmHg) than arteries (~ 100 mmHg), and their structural features and hence mechanical properties are quite different. The insertion of a segment of saphenous vein into the coronary tree exposes the thin-walled and compliant vein to a pressure much higher than what the venous wall is used to accommodate. The mechanical mismatch causes the saphenous vein segment to bulge, and this sudden enlargement causes a geometric mismatch. Thus, the proximal junction between the coronary artery and the saphenous venous graft has a condition very similar to that in the step flow channel. Hence, the area of the saphenous vein at this junction is subjected to eddy flow, and the resulting flow reattachment area becomes vulnerable to atherogenesis because of the activation of events such as cell proliferation and MCP-1 gene expression. Therefore, it is necessary to reinforce the saphenous vein prior to grafting so that there would be a mechanical match with the coronary artery and prevents the occurrence of geometric mismatch. In the preparation of tissue-engineered vascular graft, whether cell-based and/or biomaterial-based, it is essential to match the mechanical properties of the graft with those of the native vessel.

6.2. *Use of a Ras negative mutant to prevent restenosis following balloon angioplasty*

In treating coronary artery disease by balloon angioplasty, a catheter with a balloon near its tip is advanced to the site of stenosis under X-ray visualization. Inflation of the balloon presses on the plaque and opens the obstructed vessel lumen. Although the procedure is usually successful, vessel wall thickening recurs in about 1/3 of the cases within a few months. This restenosis is principally due to the proliferation of smooth muscle cells, which is stimulated by the mechanical injury incurred by the ballooning and the chemical factors in the blood following endothelial denudation. Ras plays a pivotal role in the regulation of the intracellular signaling events for many functions in a variety of cells, including the proliferation of vascular smooth muscle cells. Therefore, we have tested the possibility of using the negative mutant RasN17 to prevent the vascular stenosis induced by balloon injury in animal experiments (Jin, 1999). In order to provide an efficient mode of transfection of the RasN17 into the vessel wall, it was packaged into nonreplicating adenovirus to produce AdRasN17. As a control, a nontherapeutic molecule LacZ was similarly packaged to produce AdLaz.

Under general anesthesia, the common carotid artery of the rat was subjected to balloon injury with a balloon catheter, similar to the procedure used clinically in balloon angioplasty. By the use of temporary clamps, the injured carotid artery was divided into a cranial and a caudal segment. AdRasN17 was introduced into one segment and AdLaz into the other. After 5 minutes, the lumen contents were rinsed out, flow was reestablished after removal of the clamps, the neck wound was closed, and the animal was allowed to recover. Six weeks later, the common carotid artery was removed for histological examination. The artery without any balloon injury had a thin wall and a wide lumen. Balloon injury (with or without AdLaz treatment) caused a marked thickening of the vessel wall and a severe narrowing of the lumen. AdRasN17 treatment greatly attenuated the wall thickening due to balloon injury, and the lumen was sufficiently wide to allow an essentially normal flow. These results suggest the potential value of using RasN17 as an agent to prevent the restenosis resulting from balloon angioplasty and illustrate how basic research can generate new approaches for clinical management of disease.

7. Summary and Conclusions

Hemodynamic forces can modulate the structure and function of endothelial cells and smooth muscle cells in blood vessels. Under normal conditions, these modulating influences allow the vascular wall to adapt to changes in pressure and flow to optimize its functional performance. In disease states, however, the abnormal responses can disturb the homeostasis and initiate or aggravate pathological processes.

Changes in flow conditions can activate EC membrane receptors such as VEGFR and integrins to initiate the signal transduction involving Ras and the downstream JNK and ERK leading to an increase in MCP1 expression. Sustained laminar flow in the straight part of the arterial tree down-regulates such activation and hence minimizes monocyte entry into the vascular wall. Laminar flow is also associated with a lower rate of cell turnover, because growth-arrest genes, rather than cell-proliferation genes, are preferentially expressed. The reduced monocyte entry and cell turnover are protective against atherogenesis.

Atherosclerosis occurs primarily in arterial branch points and curved regions. The complex flow pattern, especially the flow reattachment, in these lesion-prone regions is associated with a high cell turnover rate (accelerated cell proliferation and cell death), thus leading to a greater EC permeability to LDL due to the widening of the intercellular junction. There is also an increased tendency for monocyte accumulation due to the lack of MCP-1 down-regulation. Both the greater LDL permeability and the enhanced monocyte accumulation are important atherogenic factors.

Consideration of the results on mechano-chemical transduction can lead to the improvement of methods of prevention and treatment of coronary artery disease. In vascular bypass surgery, it is essential to consider mechanical matching between the graft and the vessel in order to prevent the creation of complex flow patterns due to geometric mismatch. Adenovirus-mediated transfer of RasN17 has the potential of being used as a prophylactic procedure against restenosis in balloon angioplasty by suppressing smooth muscle cell proliferation.

By combining mechanics with biology (from molecules to tissues), bioengineering can play a major role in enhancing our understanding of the fundamental process of mechano-chemical transduction and improving the methods of the management of important clinical conditions such as coronary artery disease.

Acknowledgments

This work was supported in part by grants HL19454, HL43026, HL44147 and HL64382 from the National Heart, Lung, and Blood Institute, a Development Award from the Whitaker Foundation, and a gift from Dr. Shi H. Huang of the Chinfon Group. The author would like to acknowledge the valuable collaboration of Drs. Fanny Almus, Benjamin Chen, Dennis Chen, H.J. Hsieh, Tony Hunter, Shila Jalali, Michael Karin, Song Li, Y. S. Julie Li, Ming-Chao Kurt Lin, Nigel Mackman, G. C. Perry, Martin Schwartz, Mohammad Sotoudeh, and Shunichi Usami, and the excellent work of Gerald Norwich, Pin-Pin Hsu, Ying-Li Hu, and Suli Yuan.

References

1. Chen, K. D., Li, Y. S., Kim, M., Li, S., Chien, S. and Shyy, J. Y. J., Mechanotransduction in response to shear stress: roles of receptor tyrosine kinases, integrins, and Shc, *J. Biol. Chem.* **274**, 18393–18400 (1999).
2. Chen, Y. L., Jan, K. M., Lin, H. S. and Chien, S., Ultrastructural studies on macromolecular permabililty in relation to endothelial cell turnover, *Atherosclerosis* **118**, 89–104 (1996).
3. Chien, S., Lin, S. J., Weinbaum, S., Lee, M. M. and Jan, K. M., The role of arterial endothelial cell mitosis in macromolecular permeability, *Adv. Exper. Med. Biol.* **242**, 99–109 (1988).
4. Chien, S., Li, S. and Shyy. J. Y. J., Effects of mechanical forces on signal transduction and gene expression in endothelial cells, *Hypertension* (Part 2), **31**, 162–169 (1998).
5. Chiu, J. J., Wang, D. L., Chien, S., Skalak, R. and Usami, S., Effects of disturbed flows on endothelial cells, *J. Biomech. Eng.* **120**, 2–8 (1998).

6. Chuang, P. T., Cheng, H. J., Lin, S. J., Jan, K. M., Lee, M. M. L. and Chien, S., Macromolecular transport across arterial and venous endothelium in rats: Studies with Evans blue-albumin and horseradish peroxidase, *Arteriosclerosis* **10**, 188–197 (1990).

7. Cornhill, J. F., Herderick, E. E. and Stary, H. C., Topography of human aortic sudanophilic lesions, *Mongraphs on Atherosclerosis* **15**, 13–19 (1990).

8. Davies, P. F., Flow-mediated endothelial mechanotransduction, *Physiol. Rev.* **75**, 519–560 (1995).

9. Frangos, J. A., Eskin, S. G., McIntire, L. V. and Ives, C. L., Flow effects on prostacyclin production by cultured human endothelial cells, *Science* **227**, 1477–1479 (1985).

10. Glagov, S., Zarins, C., Giddens, D. P. and Ku, D., Hemodynamics and atherosclerosis: Insights and perspectives gained from studies of human arteries, *Arch. Pathol. Lab. Med.* **112**, 1018–1031 (1988).

11. Kuchan, M. J., Fo, H. and Frangos, J. A., Role of G proteins in shear stress-induced nitric oxide production by endothelial cells, *Am. J. Physiol.* **267**, C753–C758 (1994).

12. Huang, A. I., Jan, K. M. and Chien, S., Role of intercellular junctions in the passage of horseradish peroxidase across aortic endothelium, *Lab. Invest.* **67**, 201–209 (1992).

13. Jalali, S., Role of c-src tyrosine kinase and integrins in shear-induced signal transduction, Ph.D. Dissertation, UCSD (1998).

14. Jalali, S., Sotoudeh, M., Yuan, S., Chien, S. and Shyy, J. Y. J., Shear stress activates p60src-Ras-MAPK signaling pathways in vascular endothelial cells, *Arteriosclerosis, Thrombosis and Vascular Biology* **18**, 227–234 (1997).

15. Jin, G., Role of Ras Signaling Pathway in Chondrocytes and Vascular Smooth Muscle Cells, Ph.D. Dissertation, UCSD (1999).

16. Li, S., Kim, M., Schlaepfer, D. D., Hunter, T., Chien, S. and Shyy, J. Y. J., The fluid shear stress induction of JNK pathway is mediated through FAK-Grb2-sos, *J. Biol. Chem.* **272**, 30455–30622 (1997).

17. Li, Y. S., Shyy, J. Y. J., Li, S., Lee, J. D., Su, B., Karin, M. and Chien, S., The Ras/JNK pathway is involved in shear-induced gene expression, *Mol. Cell. Biol.* **16**, 5947–5954 (1996).

18. Lin, S. J., Jan, K. M., Weinbaum, S. and Chien, S., Transendothelial transport of low density lipoprotein in association with cell mitosis in rat aorta, *Arteriosclerosis* **9**, 230–236 (1989).

19. Lin, S. J., Jan, K. M. and Chien, S., The role of dying endothelial cells in transendothelial macromolecular transport, *Arterioclerosis* **10**, 703–709 (1990).

20. Malinkaukas, R. A., Herrmann, R. A. and Truskey, G. A., The distribution of intimal white blood cells in the normal rabbit aorta, *Atherosclerosis* **115**, 147–163 (1995).

21. Nerem, R., Hemodynamics and the vascular endothelium, *J. Biomech. Eng.* **115**, 510–514 (1993).

22. Schwenke, D. C. and Carew, T. E., Quantification *in vivo* of increased LDL content and rate of LDL degradation in normal rabbit aorta occurring at sites susceptible to early atherosclerotic lesions, *Circulation Res.* **62**, 699–710 (1988).

23. Shyy, Y. J., Hsieh, H. J., Usami, S. and Chien, S., Fluid shear stress induces a biphasic response of human monocyte chemotactic protein 1 gene expression in vascular endothelium, *Proc. Natl. Acad. Sci. USA* **91**, 4678–4682 (1994).

24. Shyy, Y. J., Lin, M. C., Han, J., Lu, Y., Petrime, M. and Chien, S., The *cis*-acting phorbol ester "12-O-tetradecanoylphorbol 13-acetate"-responsive element is involved in shear stress-induced monocyte chemotactic protein 1 gene expression, *Proc. Natl. Acad. Sci. USA* **92**, 8069–8073 (1995).

25. Steinberg, D., Role of oxidized LDL and antioxidants in atherosclerosis, *Adv. Exp. Med. Biol.* **369**, 39–48 (1995).

26. Truskey, G. A., Roberts, W. L., Herrmann, R. A. and Malinauskas, R. A., Measurement of endothelial permeability to 125I-low density lipoproteins in rabbit arteries by use of en face preparations, *Circulation Res.* **71**, 883–897 (1992).

27. Weinbaum, S., Tzeghai, G., Ganatos, P., Pfeffer, R. and Chien, S., Effect of cell turnover and leaky junctions on arterial macromolecular transport, *Am. J. Physiol.* **248**, H945–H960 (1985).

J.-J. Chiu

D. L. Wang

Institute of Biomedical Sciences,
Academia Sinica,
Taipei 11529, Taiwan

S. Chien

R. Skalak[1]

Department of Bioengineering and Institute
for Biomedical Engineering,
University of California, San Diego,
La Jolla, CA 92093-0412

S. Usami

Institute of Biomedical Sciences,
Academia Sinica,
Taipei 11529, Taiwan;
Department of Bioengineering and Institute
for Biomedical Engineering,
University of California, San Diego,
La Jolla, CA 92093-0412

Effects of Disturbed Flow on Endothelial Cells

Atherosclerotic lesions tend to localize at curvatures and branches of the arterial system, where the local flow is often disturbed and irregular (e.g., flow separation, recirculation, complex flow patterns, and nonuniform shear stress distributions). The effects of such flow conditions on cultured human umbilical vein endothelial cells (HUVECs) were studied in vitro by using a vertical-step flow channel (VSF). Detailed shear stress distributions and flow structures have been computed by using the finite volume method in a general curvilinear coordinate system. HUVECs in the reattachment areas with low shear stresses were generally rounded in shape. In contrast, the cells under higher shear stresses were significantly elongated and aligned with the flow direction, even for those in the area with reversed flow. When HUVECs were subjected to shearing in VSF, their actin stress fibers reorganized in association with the morphological changes. The rate of DNA synthesis in the vicinity of the flow reattachment area was higher than that in the laminar flow area. These in vitro experiments have provided data for the understanding of the in vivo responses of endothelial cells under complex flow environments found in regions of prevalence of atherosclerotic lesions.

Introduction

Vascular endothelium has received much attention in the study of pathogenesis of atherosclerosis [16, 37]. As an interface between the blood and the vessel wall, the endothelium occupies a unique location directly exposed to the hemodynamic forces imposed by the flow of blood. There is ample evidence that hemodynamic forces can exert significant influences on the endothelial cells (ECs) in terms of their morphology [5, 11, 12, 21, 26, 28, 29, 30, 36], cytoskeleton organization [13, 18, 19, 34, 42, 44], ion channel activation [7, 8, 33], and gene expression [16, 28, 32, 39]. All these factors may be implicated in the early development of atherosclerosis. *In vivo*, the hemodynamics of blood flow is complex [35, 40], and the local flow patterns in the arteries are quite different from steady laminar Newtonian flow. Car and Kotha [2] have recently reported that the separation surface at a bifurcation is dependent on the Reynolds number and branching geometry, and that there is a marked difference between the separation surfaces of T and Y bifurcations, especially at higher Reynolds numbers. Flow characteristics such as boundary layer separation, eddy formation, recirculation, and secondary flow may be enhanced in regions near the arterial branches and bends, where atherosclerotic lesions are prone to develop [20].

Alignment of cells and reorientation of cytoskeletal proteins under various flow conditions have been demonstrated *in vivo* and *in vitro* [11–13, 26, 36]. *In vivo* studies [1, 22] have shown that the ECs near the apex of a flow divider, where the shear stresses on the cells are high and steady, appear to be stretched in the local flow direction. In contrast, on the lateral walls of branching sites, where the shear stresses are low and variable, the ECs tend to be round in shape.

There are a few *in vitro* studies on the effects of complex flow conditions on ECs. Dewey et al. [10] subjected vascular ECs to unsteady fluid shear stress, and emphasized that the shear stress gradient is as important as the absolute stress level in determining cell responses. Helmlinger et al. [17] have investigated the responses of cell shape, orientation, and actin microfilament localization to pulsatile flows with different types of shear stress distributions. Davies et al. [6] have shown that turbulent fluid shear stress can induce increased EC turnover. DePaola et al. [9] have examined the responses of cell shape, migration, and cell cycle to shear stress gradients. Recently, Truskey et al. [41], using a step vertical expansion flow chamber, have reported that subconfluent ECs respond to spatial gradients of wall shear stress.

The aim of the present *in vitro* study is to investigate the responses of cultured HUVECs to complex flow conditions induced in a vertical-step flow channel (VSF), in terms of alterations of cell shape, orientation, cytoskeletal organization, and DNA synthesis. In the type of expansion flow found in the VSF, eddy formation and flow separation develop well below the critical Reynolds number that leads to turbulence. These inertial effects may create flow patterns that are similar to those found in artery branch points. In the present work, confluent HUVEC monolayers were tested; the results were analyzed quantitatively and compared with the detailed flow structures, which are derived by the use of a computer program based on the finite volume method in a boundary-fitted coordinate system [23–25]. The present investigation differs from previous studies [6, 9, 10, 17, 41] in that it was performed on HUVECs, with quantitative determination of cell morphological alterations and histochemical studies of cytoskeletal reorganization.

Methods and Procedures

Flow Channel. A vertical-step flow chamber (VSF) was employed to simulate some features of the flow patterns near arterial branches and bends, e.g., flow separation and reattach-

[1] Deceased.
Contributed by the Bioengineering Division for publication in the JOURNAL OF BIOMECHANICAL ENGINEERING. Manuscript received by the Bioengineering Division September 22, 1995; revised manuscript received May 5, 1997. Associate Technical Editor: R. M. Nerem.

VERTICAL-STEP FLOW (VSF)

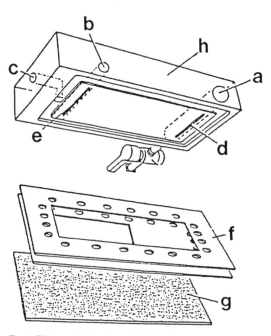

Fig. 1 Diagrams showing the parallel-plate flow chamber for vertical-step flow. The polycarbonate plate (*h*), the gasket (*f*), and the glass slide (*g*) with cell monolayer are held together by a vacuum suction applied at the perimeter of the slide via port (*c*), forming a channel with variable depth. Cultured medium enters at port (*a*) through slit (*d*) into the channel, and exits through slit (*e*) and port (*b*).

Experimental Procedures. ECs were isolated from fresh human umbilical cords by collagenase perfusion [15]. The cell pellet was resuspended in a culture medium consisting of medium M199 (Gibco, Grand Island, NY) supplemented with 20 percent fetal bovine serum (Gibco, Grand Island, NY), 30 μg/ml endothelial cell growth supplement (ECGS, Collaborative Research Inc., Bedford, MA), and 1 percent penicillin/streptomycin (Gibco, Grand Island, NY). The cell suspension was seeded onto glass slides (75 by 26 mm, Corning, Corning, NY) which had been precoated with a layer of fibronectin (5 μg/cm^2). The slide was maintained at 37°C in an incubator with a humidified mixture of 95 percent air and 5 percent CO_2, until the attainment of a confluent monolayer with a steady cell density of $1-2 \times 10^5$ cells per cm^2, which is similar to that described by Gimbrone [15] for confluent EC monolayers. The slide with the monolayer of cultured cells was mounted in the flow chamber and connected to a perfusion loop system, which was kept in a constant-temperature controlled enclosure and maintained at pH 7.4 by continuous gassing with a mixture of 5 percent CO_2 in air. The osmolality of the perfusate was checked and adjusted to 285–295 mOsm/kg H_2O during the perfusion. The chamber was placed on the stage of an inverted microscope (Diaphot, Nikon, Tokyo, Japan). A CCD video camera (WV-50, Panasonic, Tokyo, Japan) was attached to the microscope, and the video image was transmitted to a video monitor and recorder (JVC, Tokyo, Japan), enabling the recording of all results in the video field. The cells were exposed to the flow for 24 h. After flow exposure, the cells were photographed, fixed, and stained, as described in detail in the following sections.

Computational Simulation. The detailed flow patterns and shear stress distributions were computed using a numerical technique, the finite volume method, in a general curvilinear coordinate system. The calculation procedure utilized a pressure-based algorithm, which had been employed previously in simulating internal flows [23–25], such as the flows in arteries and bifurcations. Detailed procedures have been described by Lee and Chiu [23], and are summarized in the appendix. In brief, the iteration method for calculating incompressible flow was used. All variables in the program such as pressure and velocities were set to zero as the initial conditions. The boundary conditions at any wall boundaries were the nonslip conditions (no velocities). The Poiseuille velocity profile for steady flow was derived for specific flow rates as the inlet boundary conditions. For the outflow boundary conditions, the fully developed flow conditions were used, and the velocity gradients were set to zero.

ment. The resulting stagnation line and eddies may be locally similar to arterial flow separations *in vivo*. The VSF was created by combining two parallel flow channels [43] with different channel heights, as shown in Fig. 1. The glass slide and the gasket were fastened between a polycarbonate base plate and a stainless plate, using vacuum suction. In the test section, the channel width was 1 cm, the entrance height was 0.025 cm, and the main channel height was 0.05 cm. Total length and the entrance length were 4.5 cm and 1.5 cm, respectively. The Reynolds number was 123 based on the inlet channel height and 246 based on the hydraulic diameter. To visualize the experimental flow patterns, red blood cells fixed with 4 percent paraformaldehyde in PBS were used as the marker particles. The hematocrit for this purpose was about 0.1 percent and the cells were clearly visible under phase microscopy. Figure 2 shows sample experimental flow patterns in the VSF. Attention has been focused on four specific flow areas: (*a*) a stagnant flow area 20 μm from the edge of the step; (*b*) the area below the center of the recirculation eddy; (*c*) the reattachment flow area; and (*d*) the fully developed flow area. These four areas were characterized by specific flow patterns and shear stress distributions. Areas (*a*), (*b*), and (*c*) were in the disturbed flow region, while a linear laminar flow was re-established in area (*d*). The reattachment point *c* was observed experimentally to oscillate slightly (with an amplitude of about 1 percent of the eddy length) in the absence of any apparent fluctuation in overall flow. This produced a slightly oscillating shear stress in the region of the flow disturbance.

VERTICAL-STEP FLOW (VSF)

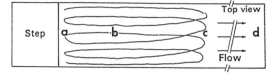

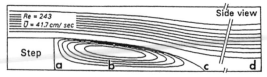

Fig. 2 Schematic diagrams of experimental flow patterns in vertical-step flow: (*a*): stagnant flow area, (*b*): center of the recirculation eddy, (*c*): reattachment flow area, (*d*): fully developed flow area

The numerical technique was validated first by comparing the computed velocity fields with the experimental flow patterns. Good agreement (within 5 percent) was obtained between the experiments and the numerical predictions of the reattachment length of the recirculation eddy.

Studies on Morphology, Cytoskeleton, and DNA Replication. The cell morphology in the preshearing condition and after 24 h of exposure to flows was examined by using phase microscopy and recorded to determine the alterations of cell shape and orientation. At the end of the flow experiments, the slide with the monolayer was rinsed with PBS and fixed in 4 percent paraformaldehyde in PBS for 30 min. After fixation, EC samples were incubated in 0.5 percent Triton X-100 (Sigma, St. Louis, MO) in PBS for 10 min, treated with 0.5 mg/ml NaBH for 10 min, and rinsed in PBS. For actin labeling, rhodamine phalloidin (Molecular Probes, Inc., Eugene, OR) was used as specified by the manufacturer's instructions. The cells were then quenched in 0.1 mM glycine for 10 min and incubated with 50 μl of rhodamine phalloidin in PBS for 20 min. The primary antibody for immunofluorescence labeling of tubulin was a mouse monoclonal antibody IgG (Sigma, St. Louis, MO) against β-tubulin. Fluorescein-conjugated goat anti-mouse IgG (Caltag, So. San Francisco, CA) was used as the secondary antibody. The general procedure for immunolabeling consisted of rehydrating specimens in PBS and overlaying the slide with 150 μl of the diluted primary antibody for 45 min. After three 5-min rinses in PBS, 150 μl of the diluted secondary antibody was overlaid for 45 min. DNA synthesis was assessed by immunocytochemical detection of bromodeoxyuridine (BrdU) (Sigma, St. Louis, MO), which is incorporated into the DNA of cells in the S phase. The labeling procedure reported in the study of Schutte et al. [38] was used. Briefly, the EC monolayer was exposed to shear flow with the medium that contained BrdU (10 mM) for 24 h. After exposure to flow, the cells were washed in PBS, fixed in 4 percent paraformaldehyde for at least 30 min, treated with 0.5 percent Triton X-100 for 10 min, and washed twice in a PBT buffer (PBS containing 1 mg/ml BSA and 0.05% v/v Tween 20, pH 7.4). The cells were then incubated in 2N HCl for 30 min at 37°C. After incubation, the cells were washed in borax buffer (0.1M sodium tetraborate, pH 8.5) and PBT, and then incubated in a medium that contained an anti-BrdU antibody (Sigma, St. Louis, MO; 1:200) for 1 h. Fluorescein-conjugated goat anti-mouse IgG (Caltag, So. San Francisco, CA) was used as the secondary antibody. The cells were then rinsed and mounted over glycerol/PBS (1:1), observed under a fluorescence microscope (Microphot-FX, Nikon, Tokyo, Japan), and photographed.

Quantitative Morphological Analysis. Photomicrographs of cell morphology in different flow areas were taken along the axisymmetric central line of the VSF. In order to determine the projected cell area, cell perimeter, and shape index (S.I.), the cell boundaries were traced manually, and all cells in each selected area were analyzed quantitatively using an NIH image software, written by Wayne Rasband of the National Institutes of Health. The accuracy of this method, as used in this study, was verified to have less than 1 percent error by calculating specific shapes with known theoretical areas. The S.I., which is defined as (4π area/perimeter^2), equals 1 for a circular cell and approaches zero for a highly elongated cell. Statistically, the Student's t-test was used to determine the significance of differences of the mean values between cells in the preshearing condition and after 24 h of flow. The level of statistical significance was selected as $p < 0.01$.

Analyses were taken from area a (0.002 cm from the step), area b (0.08 cm from the step), area c (0.17 cm from the step), and area d (2.5 cm from the step).

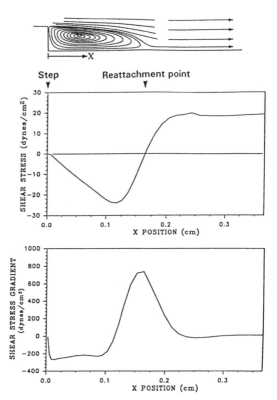

Fig. 3 Computed flow pattern (top), shear stress on the channel walls (middle), and shear stress gradient (bottom) in vertical-step flow. Direction of the main flow is left to right.

Results

Computational Results. Figure 3 shows the computed shear stress distributions on the boundaries of the VSF. It should be noted that in VSF, the shear stress varies in the flow direction primarily. Large shear stress gradients exist in the regions of flow disturbance, particularly at the sites near the step and the reattachment point. In Fig. 3, the local maximum at $X = 0.25$ cm is probably due to inertial effects. The shear stress below the center of the recirculation eddy in VSF was relatively high (24 dynes/cm^2). This high level of shear stress, which was also found by Truskey et al. [41] at slightly higher Reynolds numbers, was apparently due to the high velocity gradient of the recirculation eddy. The shear stress distribution in the fully developed flow area (d) was nearly constant (21 dynes/cm^2). These results are consistent with those obtained for a uniform Poiseuille flow, $\tau = 6Q\mu/(w \cdot h^2)$, where Q is the flow rate, μ is the dynamic viscosity, w is the channel width, and h is the channel height.

Morphological Findings. Figure 4 shows the morphological changes induced in the confluent EC monolayers by exposure in the VSF for 24 h. Prior to starting the flow (0 h), cells were uniformly distributed. After 24 h, a pronounced cell alignment with the flow direction was observed in area d in the VSF (Fig. 4). The morphological changes of the ECs in areas a and c were similar, as they both had similar relatively static flow environments; however, the cells in area b tended to be partly elongated and aligned with the local flow direction due to high shear stress level below the center of the eddy. There

VERTICAL-STEP FLOW (VSF)

100 μm

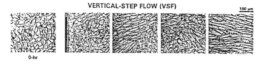

0-hr

Fig. 4 The morphological changes induced in the confluent monolayers by exposure to vertical-step flow: (a): stagnant flow area, (b): below center of the recirculation eddy, (c): reattachment flow area, (d): fully developed flow area. The direction of the main flow is left to right. 0 h sample (control) was photographed from the same experimental monolayer before shearing.

(A)

(B)

Fig. 5 Fluorescence photographs: (A) F-actin filaments and (B) microtubules after exposure to vertical-step flow for 24 h. (a): Stagnant flow area, (b): below center of the recirculation zone of a trapped eddy, (c): reattachment flow area, (d): fully developed flow area. The direction of the main flow is left to right. Solid arrows: peripheral actin filaments; open arrows: actin stress fibers; solid arrowheads: random and short actin filaments. Magnification: 400×.

were no morphological evidence of cell injury in any of the four areas.

The shape changes of the ECs after exposure to flow for 24 h were analyzed quantitatively by assessing the S.I. of the cells at 0 h and 24 h, and the results are shown in Table 1. The cells analyzed in the different flow areas were those located on the axisymmetric central regions of the VSF where there are no effects of secondary flows induced by the wall boundaries on two sides. Comparison of samples obtained from different microscopic fields of the same flow area showed no significant differences.

These results on cell shape and orientation are consistent with those reported by others on steady laminar flow [11, 12, 26] and disturbed flow [7, 9, 41]. After exposure to flow for 24 h, the S.I. of the ECs in areas a and c, where the mean shear stresses were very low (<1 dyne/cm^2) was significantly higher than those in areas b (reverse flow) and d (forward flow), where the mean shear stress was high. Most cells in area c were rounded in shape. In area d, where the cells are aligned with the flow after 24-h flow exposure, the mean value for the S.I. decreased from the static control of 0.56 ± 0.12 to 0.34 ± 0.13. The region below the VSF center of the eddy (area b) had a high shear stress level (22 dynes/cm^2), and the cells in this region were somewhat elongated in response to flow, as indicated by a decrease of the mean S.I. value to 0.44 ± 0.15.

Cytoskeletal Organization. Figure 5 shows the fluorescence photomicrographs of the organization of F-actin filaments (Fig. 5(A)) and microtubules (Fig. 5(B)) in the ECs after exposure to VSF for 24 h. The fully developed flow area d

displayed very long, well-organized, parallel actin stress fibers aligned with the flow direction in the central regions of the cells. The formation of bright F-actin bundles at cell peripheries was prominent. In the disturbed flow areas with low shear stresses (areas a and c), the actin filaments tended to localize mainly at the periphery of the cells. However, there were occasional cells that had a complete or partial loss of peripheral F-actin microfilaments, together with the formation of random and short actin bundles. It is notable that the distributions of actin filaments were similar in areas b and d of the VSF; both areas have a high shear stress but opposite directions of flow.

In general, the microtubules were distributed relatively uniformly throughout most cell regions, except for their accumulation adjacent to the nucleus. Mostly, the microtubules displayed a linear organization and an alignment with the flow direction in the fully developed flow areas with high shear stress, whereas they showed a randomly organized pattern within the cells in the disturbed flow areas with low shear stress.

DNA Synthesis. The DNA synthesis induced in the endothelial cells by exposure to shear in the VSF for 24 h has been assessed in the different flow areas by counting the number of nuclei with BrdU-incorporation. As shown in Table 1, the disturbed flow areas with low shear stress levels (areas a and c) had an enhanced DNA synthesis in comparison to the fully developed flow areas with high shear stresses (area d). The high shear, reverse flow area b in VSF had a low ratio of BrdU uptake similar to that in the high-shear, forward flow area d. Moreover, the reattachment flow area c exhibited the highest BrdU-positive stain among all areas. These results suggest that flow disturbance in low shear regions induces an increase in DNA synthesis in ECs. Area a, where the shear stress is low, also exhibits a high BrdU staining, as in the low-shear reattachment area c. It should be pointed out, however, that area a is close to the step (20 μm away) and there might have been some cell injury near the step, though such injury was not visible in morphological examination. Our results are in agreement with the concept that the temporal and spatial variations of shear stresses in the flow reattachment regions may enhance cell turnover [3, 4, 9].

Relations of Sphericity Index and DNA Synthesis to Shear Stress and Shear Stress Gradient. The results on sphericity index and BrdU uptake in the four areas of the VSF can be considered as functions of the shear stress and the shear stress gradient in these areas (Fig. 6). As mentioned above, the results obtained in area a may be subject to some error due to its closeness to the step and should be interpreted with caution. Although the data are rather limited, the results suggest that, after 24 h of shearing, the S.I. and DNA synthesis rate of HUVECs increased with decreasing shear stress magnitude and with increasing shear stress gradient.

Table 1 Regional changes in shape index and DNA replication in endothelial cell monolayer subjected to shear in a vertical step flow chamber

Area:	Control	a	b	c	d
Shear stress (dyn/cm^2)	0.0	0.5	22.0	0.4	21.0
Shear stress gradient (dyn/cm^3)	0.0	0.0	240	740	0.0
Shape index (24 hr)	0.56 ± 0.12 ($N_1 = 355$)	$0.62 \pm 0.11^*$ ($N_1 = 123$)	$0.44 \pm 0.15^*$ ($N_1 = 348$)	0.59 ± 0.14 ($N_1 = 182$)	$0.34 \pm 0.13^*$ ($N_1 = 457$)
DNA replication (% BrdU, 24 hr)	7.6 ± 1.2 ($N_2 = 6$)	$14.1 \pm 3.1^*$ ($N_2 = 4$)	6.9 ± 2.7 ($N_2 = 4$)	$21.2 \pm 3.9^*$ ($N_2 = 4$)	7.3 ± 2.2 ($N_2 = 4$)

Control values for shape index were determined from the photomicrographs taken from the same experimental monolayers before shearing. Control samples for DNA replication were kept without flow for 24 hr and then analyzed.

Column a: stagnant flow area near the step, b: area below the center of the recirculation eddies, c: reattachment flow area, d: fully developed flow area.

Values represent Mean ± S. D.

N_1: measured cell number, N_2: number of experiments. For shape index experiments, $N_2 = 2$.

* P < 0.01 compared with control values.

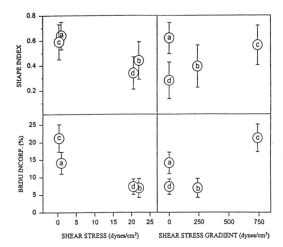

Fig. 6 The shape index of vascular endothelial cells (upper panels) and BrdU incorporation (lower panels) as functions of shear stress magnitude (left panels) and shear stress gradient (right panels) in HUVEC subjected to shear in VSF for 24 h. (*a*): Stagnant flow area, (*b*): below center of the recirculation zone of a trapped eddy, (*c*): reattachment flow area, (*d*): fully developed flow area. The data in area *a* should be interpreted with caution, because of potential cell injury due to the closeness of this area to the step.

Discussion

The propensity for atherosclerotic lesion development in the regions near the arterial branches and curvatures suggests that hemodynamic factors play an important role in the initiation and progression of atherosclerosis [1, 3, 7, 10, 16, 20, 28]. It has been proposed that the complex flow characteristics at arterial branches and curvatures, such as flow separation, oscillating shear stress, and large shear stress gradient, could contribute to the observed pathology of the artery wall [3]. These local complex flow patterns cause changes in EC morphology [1, 26]. In the present study using a VSF, the cells in the regions with a low shear stress environment became more rounded in plane view, whereas the cells under high shear stress became elongated and aligned with the local flow direction. The changes in cell shape and orientation in response to shear found in the present study on confluent ECs are less than those reported in a previous study on subconfluent ECs [41], suggesting that such shear-induced responses are affected by the cultured cell density. This could be due to the interactions of surrounding ECs in cultures with a high cell density. Our findings also suggest that the cells become more rounded with an increase in shear stress gradient (Table 1).

We observed some oscillation in shear stress in the region of flow reattachment, but the characteristics and the implications of this unsteadiness need elucidation by further investigations. Although it is difficult to evaluate the detailed dynamic forces on the cell surface imposed by the flow and the intercellular adhesion forces between the cells, it is possible that the rounded cells at sites of flow disturbance are subjected to shear stresses with greater spatial and temporal variations than the aligned cells, which experience steady high-shear stress.

It has been postulated that the rounded (in plane view) ECs in the vicinity of flow separation in arteries *in vivo* tend to have a higher turnover rate and that the intercellular junction around mitotic cells may become leaky to allow for a local influx of low-density lipoproteins (LDL) [31, 45]. Experimental studies have demonstrated transient increases in permeability during cell turnover [27]. The increased uptake of BrdU (Table 1) suggests that DNA synthesis rate is increased as a consequence

of flow disturbance, particularly in the regions of flow reattachment. If the accelerated EC turnover induced by variations in shear stress, as observed *in vitro* in the present investigation and by DePaola et al. [9] are also operative *in vivo*, this may contribute to the local increase in LDL permeability and focal development of atherosclerosis [4, 6], as postulated in the cell turnover–leaky junction hypothesis [45].

Reorganization of cell cytoskeletal proteins under fluid shearing forces, especially the redistribution of actin filaments, may result in altered cell-to-cell adhesion [14] and consequent regulation of intercellular permeability [46]. In our experiments, long and thick parallel stress fibers together with prominent peripheral *F*-actin filaments were shown to be present in the regions of fully developed flow. In the regions of flow reattachment with low shear stress and high shear stress gradient, the actin filaments tended to localize mainly at the cell periphery; in some cases a complete or partial loss of peripheral *F*-actin microfilaments was coupled with the formation of random and short actin bundles in cells in these regions. These results suggest that cell–substratum adhesion and cell-to-cell junctions may become more stabilized by well-organized cytoskeletal structure in the fully developed flow areas. In contrast, the redistribution of cytoskeletal proteins in regions of flow reattachment may be associated with an increase in endothelial permeability.

Atherosclerosis is a multifactorial disease involving a complex array of circulating blood cells and plasma components, their interactions with the cells and matrix proteins of the arterial wall, and the effects of flow pattern on mass transfer. We have made no attempt to identify all the events in the context of atherosclerotic lesion formation and progression; instead, the present experiments are meant to provide one step in understanding the endothelial responses in the complex flow environment existing near the arterial branches and bends.

We reported studies on the effect of disturbed flow on endothelial cell morphology and cytoskeletal organization by the use of a vertical step flow chamber. This step flow chamber can be useful in future experiments for examining local gene expression with *in situ* hybridization or other methods. From the present study we postulate that the reorganization of cytoskeletal structure coupled with the alteration of cell morphology are important responses to the local hemodynamic environment. The rounded cells observed in the regions near the arterial branches are exposed to shear stress fluctuations and thus vulnerable to injury or death. The flow reversal and reattachment in the separation region will enhance the effects of mass transfer and blood–artery wall interactions. Low flow states may cause the accumulation and aggregation of macromolecules and/or cells in the artery wall. Cell division may also be triggered in the regions of flow disturbance by fluctuating shear stress distributions and flow patterns. The enhanced cell division and the reorganization of cell cytoskeletal structure in the regions of flow disturbance may cause alteration of transendothelial permeability and allow the influx of lipoproteins or other blood component into the artery wall. These may be critical factors in the formation and development of atherosclerosis.

Acknowledgments

This work was supported by National Science Council Grant NSC 84-2331-B-001-015 and NSC 85-2331-B-001-021 from The Republic of China, and USPHS research grant HL 19454 from the National Heart, Lung, and Blood Institute of the National Institutes of Health, USA.

References

1 Buss, H., "Morphology and fluid-dynamics of endothelial cells at the site of vascular bifurcation," *Fluid Dynamics as Localizing Factor for Atherosclerosis*, Shettler, G., Nerem, R. M., and Schmid-Schönbein, H., eds., Springer-Verlag, Heidelberg, 1983, pp. 168–172.

2 Car, R. T., and Kotha, S. L., "Separation surfaces for laminar flow in branching tubes—Effects of Reynolds number and geometry," ASME JOURNAL OF BIOMECHANICAL ENGINEERING, Vol. 117, 1995, pp. 442–447.

3 Caro, C. G., Fitz-Gerald, J. M., and Schroter, R. C., "Atheroma and arterial wall shear. Observation, correlation and proposal of a shear-dependent mass transfer mechanism for atherogenesis," Proc. Roy. Soc., London, Vol. B 177, 1971, pp. 109–159.

4 Chien, S., Lin, S. J., Weinbaum, S., Lee, M. M., and Jan, K. M., "The role of arterial endothelial cell mitosis in macromolecular permeability," Adv. Exper. Med. Biol., Vol. 242, 1988, pp. 99–109.

5 Cornhill, J. F., Levesque, M. J., Herderick, E. E., Nerem, R. M., Kilman, J. W., and Vasko, J. S., "Quantitative study of the rabbit aortic endothelium using vascular casts," Atherosclerosis, Vol. 35, 1980, pp. 321–337.

6 Davies, P. F., Remuzzi, A., Gordon, E. J., Dewey, C. F., Jr., and Gimbrone, M. A., Jr., "Turbulent fluid shear stress induces vascular endothelial cell turnover in vitro," Proc. Natl. Acad. Sci., Vol. 83, 1986, pp. 2114–2117.

7 Davies, P. F., Robotewskyj, A., Griem, M. L., Dull, R. O., and Polacek, D. C., "Hemodynamic forces and vascular cell communication in arteries," Archives of Pathology & Medicine, Vol. 116, 1992, pp. 1301–1306.

8 Davies, P. F., and Tripathi, S. C., "Mechanical stress mechanisms and the cell: an endothelial paradigm," Circ. Res., Vol. 72, 1993, pp. 239–245.

9 DePaola, N., Gimbrone, M. A., Jr., Davies, P. F., and Dewey, C. F., Jr., "Vascular endothelium responds to fluid shear stress gradients," Arteriosclerosis and Thrombosis, Vol. 12, 1992, pp. 1254–1257.

10 Dewey, C. F., Jr., Gimbrone, M. A., Jr., Bussolari, S. R., White, G. E., and Davies, P. F., "Response of vascular endothelial to unsteady fluid," Fluid Dynamics as Localizing Factor for Atherosclerosis, Shettler, G., Nerem, R. M., and Schmid-Schönbein. H., eds., Springer-Verlag, Heidelberg, 1983, pp. 182–187.

11 Dewey, C. F., Jr., Bussolari, S. R., Gimbrone, M. A., Jr., and Davies, P. F., "The dynamic response of vascular endothelial cells to fluid shear stress," ASME JOURNAL OF BIOMECHANICAL ENGINEERING, Vol. 103, 1981, pp. 177–185.

12 Eskin, S. G., Ives, C. L., McIntire, L. V., and Navarro, L. T. "Response of cultured endothelial cells to steady flow," Microvasc. Res., Vol. 28, 1984, pp. 87–94.

13 Franke, R. P., Grafe, M., Schnittler, H., Seiffge, D., Mittermayer, C., and Drenckhahn, D., "Induction of human vascular endothelial stress fibers by fluid shear stress," Nature, Vol. 307, 1984, pp. 648–649.

14 Geulieb, A. I., Spector, W., Wong, M. K. K., and Lacey, C., "In vitro reendothelialization: Microfilament bundle reorganization in migrating porcine endothelial cells," Arteriosclerosis, Vol. 4, 1984, pp. 91–96.

15 Gimbrone, M. A., Jr., "Culture of vascular endothelium," Progress in Hemostasis and Thrombosis, Vol. III, Spaect, T. H., ed., Grune and Stratton, 1976, pp. 1–28.

16 Gimbrone, M. A., Jr., Kume, N., and Cybulsky, M. I., "Vascular endothelial dysfunction and the pathogenesis of atherosclerosis," Atherosclerosis Review, Weber, P. C., and Leaf, A., eds., Raven Press, Ltd., New York, 1993, pp. 1–9.

17 Helmlinger, G., Geiger, R. V., Schreck, S., and Nerem, R. M., "Effects of pulsatile flow on cultured vascular endothelial cell morphology," ASME JOURNAL OF BIOMECHANICAL ENGINEERING, Vol. 113, 1991, pp. 123–131.

18 Kim, D. W., Gotlieb, A. I., and Langille, B. L., "In vivo modulation of endothelial F-action microfilaments by experimental alterations in shear stress," Arteriosclerosis, Vol. 9, 1989, pp. 439–445.

19 Kim, D. W., Langille, B. L., Wong, M. K. K., and Gotlieb, A. I., "Patterns of endothelial microfilament distribution in the rabbit aorta in situ," Cir. Res., Vol. 64, 1989, pp. 21–31.

20 Ku, D. N., Giddens, D. P., Zarins, C. K., and Glagov, S., "Pulsatile flow and atherosclerosis in the human carotid bifurcation," Arteriosclerosis, Vol. 5, 1985, pp. 293–302.

21 Langille, B. L., and Adamson, S. L., "Relationship between blood flow direction and endothelial cell orientation at arterial branch sites in rabbits and mice," Cir. Res., Vol. 48, 1981, pp. 481–488.

22 Langille, B. L., Reidy, M. A., and Kine, R. L. "Injury and repair of endothelium at sites of flow disturbances near abdominal aortic coarctations in rabbits," Arteriosclerosis, Vol. 6, 1986, pp. 146–154.

23 Lee, D., and Chiu, J. J., "Covariant velocity based calculation procedure with non-staggered grids for computation of pulsatile flows," Numerical Heat Transfer, Vol. 21, Part B, 1992, pp. 269–286.

24 Lee, D., and Chiu, J. J., "Computation of physiological bifurcation flow using a patched grid," Comput. & Fluids, Vol. 21, 1992, pp. 519–535.

25 Lee, D., and Chiu, J. J., "A numerical simulation of intimal thickening under shear in arteries," Biorheology, Vol. 29, 1992, pp. 337–351.

26 Levesque, M. J., and Nerem, R. M., "The elongation and orientation of cultured endothelial cells in response to shear stress," ASME JOURNAL OF BIOMECHANICAL ENGINEERING, Vol. 107, 1985, pp. 341–347.

27 Lin, S.-J., Jan, K.-M., Schuessler, G., Weinbaum, S., and Chien, S., "Enhanced macromolecular permeability of aortic endothelial cells in association with mitosis," Arteriosclerosis, Vol. 73, 1988, pp. 223–232.

28 Nerem, R. M., "Vascular fluid mechanics, the arterial wall, and atherosclerosis," ASME JOURNAL OF BIOMECHANICAL ENGINEERING, Vol. 114, 1992, pp. 274–282.

29 Nerem, R. M., Harrison, D. G., Taylor, W. R., and Alexander, R. W. "Hemodynamics and vascular endothelial biology," J. Cardiovascular Pharmacology, Vol. 21, 1993, pp. 6–10.

30 Nerem, R. M., Levesque, M. J., and Cornhill, J. F., "Vascular endothelial morphology as an indicator of blood flow," ASME JOURNAL OF BIOMECHANICAL ENGINEERING, Vol. 103, 1981, pp. 172–176.

31 Nerem, R. M., and Levesque, M. J., "Fluid mechanics in atherosclerosis," Handbook of Bioengineering, Skalak, R., and Chien, S., eds., McGraw-Hill, New York, 1987, pp. 21.1–21.22.

32 Nollert, M. N., Diamond, S. L., and McIntire, L. V., "Hydrodynamic shear stress and mass transport modulation of endothelial cell metabolism," Biotechnology and Bioengineering, Vol. 38, 1991, pp. 588–602.

33 Olesen, S. P., Clapham, D. E., and Davies, P. F., "Hemodynamic shear stress activates a K current in vascular endothelial cells," Nature, Vol. 331, 1988, pp. 168–170.

34 Ookawa, K., Sato, M., and Ohshima, N., "Changes in the microstructure of cultured porcine aortic endothelial cells in the early stage after applying a fluid-imposed shear stress," ASME JOURNAL OF BIOMECHANICAL ENGINEERING, Vol. 25, 1992, pp. 1321–1328.

35 Pedley, T. J., The Fluid Mechanics of Large Blood Vessels, Cambridge U. Press, 1980.

36 Remuzzi, A., Dewey, C. F., Jr., Davies, P. F., and Gimbrone, M. A., Jr. "Orientation of endothelial cells in shear fields in vitro," Biorheology, Vol. 21, 1984, pp. 617–630.

37 Ross, R., "The pathogenesis of atherosclerosis: a perspective for the 1990s," Nature, Vol. 362, 1993, pp. 801–809.

38 Schutte, B., Reynders, M. M. J., van Assche, C. L. M. V. J., Hupperets, P. S. G. J., Bosman, F. T., and Blijham, G. H., "An improved method for the immuno-cytochemical detection of bromodeoxyuridine labeled nuclei using flow cytometry," Cytometry, Vol. 8, 1987, pp. 372–376.

39 Shyy, Y. J., Hsieh, H. J., Usami, S., and Chien, S., "Fluid shear stress induces a biphasic response of human monocyte chemotactic protein 1 gene expression in vascular endothelium," Proc. Nat. Acad. of Sci. U.S.A., Vol. 91, 1994, pp. 4678–4682.

40 Skalak, R., Ozkaya, N., and Skalak, T. C., "Biofluid mechanics," Ann. Rev. Fluid Mech., Vol. 21, 1989, pp. 167–204.

41 Truskey, G. A., Barker, K. M., Rober, K. M., Oljver, L. A., and Combs, M. P., "Characterization of a sudden expansion flow chamber to study the response of endothelium to flow recirculation," ASME JOURNAL OF BIOMECHANICAL ENGINEERING, Vol. 117, 1995, pp. 203–210.

42 Uematsu, M., Kitabatake, A., Tanouchi, J., Doi, Y., Masuyama, T., Fujii, K., Yoshida, H., Ito, K., Ishihara, M., Hori, M., Inoue, X. X., and Kamada, T., "Reduction of endothelial microfilament bundles in the low-shear region of the canine aorta: association with intimal plaque formation in hypercholesterolemia," Arteriosclerosis and Thrombosis, Vol. 11, 1991, pp. 107–115.

43 Usami, S., Chen, H. H., Zhao, Y., Chien, S., and Skalak, R., "Design and construction of a linear shear stress flow chamber," Annals of Biomed. Eng., Vol. 21, 1993, pp. 1–7.

44 Wechezak, A. R., Wight, T. N., Viggers, R. F., and Sauvage, L. R., "Endothelial adherence under shear stress is dependent upon microfilament reorganization," J. Cell Physiol., Vol. 139, 1989, pp. 136–146.

45 Weinbaum, S., Tsagai, G., Ganatos, P., Pfeffer, S., and Chien, S., "Effects of cell turnover and leaky junctions on arterial macromolecular transport," Am. J. Physiol., Vol. 248, 1985, pp. 945–960.

46 Wysolmerski, R., and Lagunoff, D., "The effect of etchchlorvynol on cultured endothelial cells: A model for the study of the mechanism of increased vascular permeability," Am. J. Pathol., Vol. 119, 1985, pp. 505–512.

APPENDIX

Blood flows in vivo and in vitro can be described with good approximation by solutions of the Navier–Stokes equations [35]. Although blood is non-Newtonian, this feature is not expected to have major effects on flow patterns and shear stress distributions in disturbed flow regions. The governing equations of incompressible, Newtonian fluid flow can be expressed in the following unified form as:

$$\frac{\partial}{\partial t}(\rho\phi) + \nabla \cdot J = R^\phi \tag{1}$$

where

$$J = \rho \vec{V} \phi - \Gamma^\phi \nabla \phi \tag{2}$$

In these equations, ϕ is a general dependent variable, Γ^ϕ is the effective diffusion coefficient of ϕ, and R^ϕ denotes the volumetric source (or sink) of ϕ. In Eq. (2), J represents the total flux of ϕ, i.e., it includes both the convective and diffusive fluxes. The governing Eq. (1) can be transformed into a curvilinear coordinate system, which leads to the general form:

$$\frac{\partial}{\partial t}(\rho\phi) + \frac{1}{\sqrt{g}}\frac{\partial}{\partial q^j}(\sqrt{g}\rho V^j\phi)$$

$$= \frac{1}{\sqrt{g}}\frac{\partial}{\partial q^j}\left(\sqrt{g}\Gamma^\phi g^{ik}\frac{\partial\phi}{\partial q^k}\right) + R^\phi \tag{3}$$

where q^j are the curvilinear coordinates, V^j are the contravariant components of velocity, g^{ik} is the contravariant metric tensor, and $\sqrt{g}$ is its Jacobian. General curvilinear coordinates are used so a grid that fits the geometry well can be used.

In this study, the convection and diffusion terms in the governing equations are discretized by using the second-order upwind and the second-order central schemes, respectively. The discretized equations after some algebraic manipulations can be written in general form as:

$$a_P\phi_P = a_E\phi_E + a_W\phi_W + a_N\phi_N + a_S\phi_S + a_T\phi_T + a_B\phi_B + b_\phi$$

$$(4)$$

where a_E, etc., are known coefficients, and E, W, N, S, T, and B (east, west, north, south, top, bottom) represent the neighboring grid node locations of the central node P.

In the present study, the variables ϕ are the velocity components u, v, w to generate the Navier–Stokes equations and $\phi = 1$ to represent the continuity equation. For the continuity equation of an incompressible fluid, Γ^ϕ and R^ϕ are set equal to zero. For the velocity components u, v, w the diffusion coefficient of momentum is the dynamic viscosity, i.e., $\Gamma^\phi = \mu$ and the source term R^ϕ is the pressure gradient.

In the algorithm employed in this study, two pressure correction equations are employed. The first correction equation is used to update the pressure itself, and the second is used to correct the predicted velocities to satisfy mass conservation. The derivation of pressure correction equations has been described in detail in the references [23–25]. In each case reported here, each grid contained approximately 34,000 nodal points was utilized and the unknown variables of p, u, v, w were computed at each point. The general computational procedure can be summarized as follows:

1 Generate a grid system for the flow field by using a grid-generation method.

2 Assign estimated values of dependent variables at every internal grid point.

3 Compute the coefficients and source terms of the momentum equations.

4 Solve the first pressure correction equation to update the pressure field.

5 With the updated pressure field, solve the momentum equations to obtain the update velocity field.

6 Solve the second pressure correction equation to correct the velocity field (so that the velocity field satisfies the continuity equation).

7 Steps (3)–(6) are repeated until a convergence criterion is satisfied. After the velocity fields are converged, the wall shear stress distributions (τ) can be calculated using the following formula:

$$\tau = \mu\sqrt{\Pi/2} \qquad (5)$$

where μ is the dynamic shear viscosity,

$$\Pi = \sum_{i=1}^{3}\sum_{j=1}^{3} \dot{\gamma}_{ij}\dot{\gamma}_{ji} \qquad (6)$$

and

$$\dot{\gamma}_{ij} = \frac{\partial u_i}{\partial x_j} + \frac{\partial u_j}{\partial x_i} \qquad (7)$$

where $u_i = u$, v, w for $i = 1, 2, 3$ respectively. $\dot{\gamma}_{ij}$ is the rate of strain tensor and Π is the second invariant of the rate of strain tensor. The shear stress τ given by Eq. (5) is, in general, the shear stress in a simple unidirectional shear flow that produces the same rate of energy dissipation as the actual flow. Since the flow over a plane boundary approaches a simple shear flow close to a solid boundary, the estimate given by Eq. (5) is a good approximation of the wall shear stress.

CHAPTER 2

PERSPECTIVES OF BIOMECHANICS

Y. C. FUNG

Department of Bioengineering, University of California, San Diego

1. Introduction

Physics studies how matters move. Chemistry studies how matters move. Engineering studies how matters move. So must biology. The science of movement of matter is mechanics. Molecular biology has molecular biomechanics. Cell biology has cell biomechanics. Tissue biology has tissue biomechanics. Organ biology has organ biomechanics. Whole individuals have animal mechanics, plant mechanics, sport mechanics, rehabilitation mechanics, etc.

Biomechanics is the study of the effects of forces in a living organism. It is a discipline of engineering science. It is a discipline that is needed in the design of any medical devices. It is a discipline needed in the understanding of the physiology of living organisms. It is a discipline the study of which requires lots of design of experiments and instruments.

Biomechanics is a basic discipline in biology, but is often neglected by biologists. It cannot be neglected by bioengineers, because forces and motion are two of the most important factors an engineer must consider to invent and design any devices to work with or work in a living organism.

Today, biomedical engineering has given us the following:

Better understanding of physiology,

Mathematical models, computational methods,

Artificial internal organs,

Artificial limbs, joints, and prostheses,

Better understanding of hearing, vision, speaking,

A number of implantable materials,

New biosensors,

New imaging techniques: CT, NMR, PET, etc.,

Clinical devices, instruments, techniques,

Remote sensors, virtual reality,

Minimally invasive surgery techniques,

And many more.

Most of the successes listed above are still limited in scope and have not achieved their full potential. Real progress will depend on further research.

It is easy to see the role played by mechanics in the items listed in the paragraphs above. Artificial hearts and heart assist devices are fluid mechanical devices. Artificial kidney is a device for mass transport. Pacemaker must be structurally sound. Design of new implantable materials depends on mechanics. Design of implantable biosensors requires fluid mechanics and mass transport analysis. Imaging and clinical devices and virtual reality, and the minimally invasive surgery techniques such as angioplasty rely on biomechanics. Other well-known contributions include sports and sports medicine, orthopedics, treatment of automobile injuries, burn injuries, and equipment for rehabilitation. Biomechanics of hearing, seeing, and speaking are extremely important subjects. Lengthening of bone to solve gait problems can be done with proper understanding of biomechanics.

To illustrate the use of biomechanics in bioengineering, it is best to consider a concrete example. In an article presented on pp. 35–56, I give you an example of the lung. In that article, it is shown how the consideration of biomechanics lead to the improvement of our understanding of the anatomy, mechanical properties, functions, and diseases of the lung. With the mastery of a quantitative approach, one can begin to talk about engineering the lung.

In the present chapter, however, I would like to dig a little at the foundation. I would like to deal with an *ad hoc* hypothesis in biomechanics which has been so universally accepted that it has almost become an axiom. The hypothesis is that when all external loads acting on an organ are removed, there is no stress in the organ. I shall show that this is generally untrue, and that removal of this hypothesis will lead us to rethink many things.

2.   The Zero-stress State of a Blood Vessel

First, let me describe a very simple experiment. In Fig. 1, an aorta is sketched. If we cut an aorta by two consecutive sections perpendicular to the longitudinal axis of the vessel,

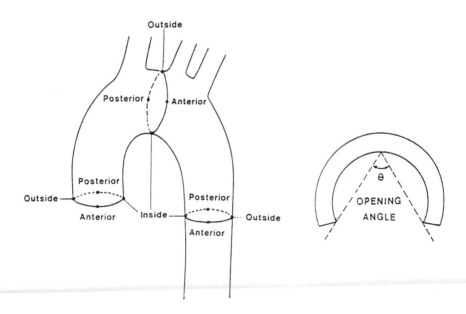

Fig. 1. Sketch of an aorta with an indication of the cutting positions. Right: Schematic cross-section of a cut vessel segment at zero-stress, defining the opening angle θ.

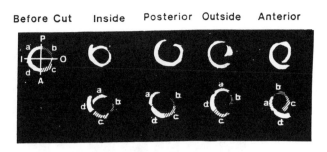

Fig. 2. The figures in the upper row show an arterial segment of a rat before cut and after cutting at four positions. The lower row shows the same vessel cut into 4 pieces and reassembled in 4 ways. It appears that one cut is sufficient to reduce an arterial segment at no-load to the zero-stress state. From Fung and Liu (2), by permission.

we obtain a ring. If we cut the ring radially, it will open up into a sector (1, 8). By using equations of static equilibrium, we know that the stress resultants and stress moments are zero in the open sector. Whatever stress remains in the vessel wall must be locally in equilibrium. If one cut the open sector further, and can show that no additional strain results, then we can say that the sector is in zero-stress state. We did a simple experiment illustrated in Fig. 2 (from Ref. 2). Five consecutive segments (rings), each 1 mm long, were cut from a rat aorta. The first four segments were then cut radially and successively at the positions indicated in Fig. 1, namely, inside, outside, anterior, posterior; designated as I, O, A, P, respectively. The fifth segment was cut in all four positions, resulting in four pieces designated a, b, c, d. The open sectors of the first four rings are shown in the upper row on the right. When the four pieces of the fifth ring were reassembled in the order of abcd, bcda, cdab, bcda, with tangents matched at successive ends, we obtain four configurations shown in the lower row of Fig. 2. They resemble the shape of the four cut segments of the first row quite well. This tells us that the arterial wall is not axisymmetric, that different parts of the circumference are different, and that one cut is almost as good as four cuts in relieving the residual stress. Hence, we may say that one cut of the ring reduces the ring into zero-strain state within the first order of infinitesimals.

Having been assured that the open sector represents zero-stress state of a blood vessel, we conclude that the zero-stress state of an artery is not a tube. It is a series of open sectors. To characterize the open sectors, we define an *opening angle* as the angle subtended by two radii drawn from the midpoint of the inner wall (endothelium) to the tips of the inner wall of the open sections (see Fig. 1).

Although the opening angle is a convenient simple measure of the zero-stress state of an artery, it is not a unique measure of the artery, because its value depends on where the cut was made. This is clearly illustrated in Fig. 2. The four segments were not identical. The blood vessel shown in Fig. 2 was not uniform circumferentially. Hence, in stating an opening angle, one must explain where the cut was made. This requirement is a limitation to the usefulness of the opening angle as a simple measure of a complex phenomenon.

The photographs in the first column of Fig. 3 show a more complete picture of the zero-stress state of a normal young rat aorta (2). The entire aorta was cut successively into may segments of approximately one diameter long. Each segment was then cut radially at

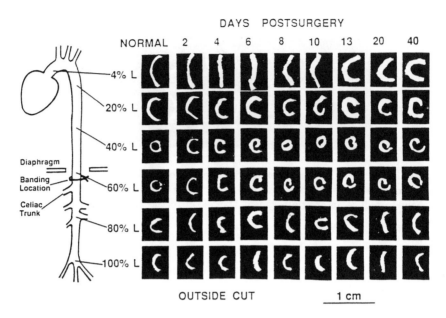

Fig. 3. Photographs of the cross-sections of a rat aorta cut along "outside" line shown in Fig. 1. The 1st column shows zero-stress state of normal aorta. Other columns show changed zero-stress states a number of days after a sudden onset of hypertension. Successive rows correlate with locations on the aorta expressed in percentage of total length of aorta, L, from the aortic valve. The aortic cross-sectional area was clamped 97% by a metal band below the diaphragm to induce the hypertension. From Liu and Fung (2), by permission.

the "outside" position indicated in Fig. 1. It was found that the opening angle varied along the rat aorta: it was about 160° in the ascending aorta, 90° in the arch, 60° in the thoracic region, 5° at the diaphragm level, 80° toward the iliac bifurcation point.

Following the common iliac artery down a leg of the rat, we found that the opening angle was in the 100° level in the iliac artery, dropped down in the popliteal artery region to 50°, then rose again to the 100° level in the tibial artery. In the medial plantar artery of the rat, the micro arterial vessel 50 μm diameter had an opening angle of the order of 100° (4).

There are similar spatial variations of opening angles in the aorta of the pig and a dog in pulmonary arteries, veins and trachea.

3.   Hypertension Causes Change of the Opening Angle of Aorta

We created hypertension in rats by constricting the abdominal aorta with a metal clip placed right above the celiac trunk (2). The clip severely constricted the aorta locally, reduced the cross-sectional area of the lumen by 97%, with only about 3% of the normal area remaining. This causes a 20% step-increase of blood pressure in the upper body, and a 55% step-decrease of blood pressure in the lower body immediately following the surgery. Later, the blood pressure increased gradually, following a course shown in Fig. 4. It is seen that in the upper body the blood pressure rose rapidly at first, then more gradually, tending to an asymptote at about 75% above normal. In the lower body, the blood pressure rose to normal value in about four days, then gradually increased further to an asymptotic value of

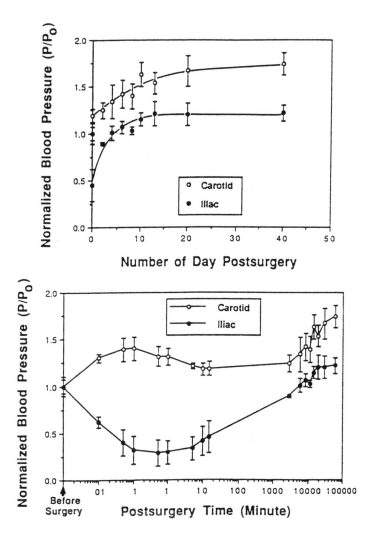

Fig. 4. The course of change of blood pressure (normalized with respect to that before surgery) after banding the aorta. From Fung and Liu (2), by permission.

25% above normal. Parallel with this change of blood pressure, the zero-stress state of the aorta changed. The changes are illustrated in Fig. 3 in which the location of any section on the aorta is indicated by the percentage distance of that section to the aortic valve measured along the aorta divided by the total length of the aorta. Successive columns of Fig. 3 show the zero-stress configurations of the rat aorta at $0, 2, 4, \ldots, 40$ days after surgery. Successive rows refer to successive locations on the aorta.

Figure 5 shows the course of change of the opening angle of the rat aorta in greater detail. Figures 3 and 5 together show that the blood vessel changed its opening angle in a few days following the blood pressure change. We found similar changes in pulmonary arteries after the onset of pulmonary hypertension by exposing rats to hypoxic gas containing 10% oxygen, 90% nitrogen, at atmospheric pressure (3).

Since opening angle changes reflect structural changes, we conclude that blood vessels remodel significantly with modest blood pressure changes.

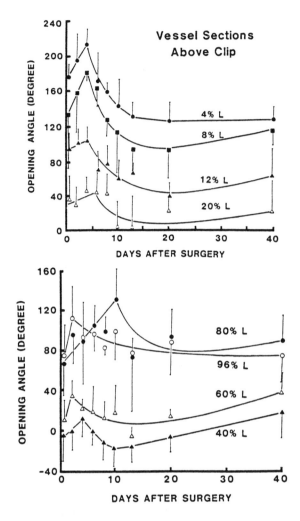

Fig. 5. The course of change of the opening angle of the rat aorta at the zero-stress state following aortic banding to change blood pressure. L is the length of aorta. %L indicates location of section from aortic valve. (a) Locations above the constriction. (b) Locations below the constriction. From Fung and Liu (2), by permission.

4.   What Does the Change of Opening Angle Mean?

The open sector configuration of an artery at zero-stress looks like a curved beam and mechanically can be analyzed as a curved beam. (See Figs. 1 and 6.) A beam can change its curvature only if one side of the beam lengthens while the other side of the beam shortens. If the opening angle increases due to tissue remodeling, then the endothelial side of the blood vessel wall must have an increase in circumferential strain, while the adventitial side of the blood vessel wall must have a decrease of circumferential strain, see Fig. 6 (5). Since these increases and decreases are not due to external loads, but are due directly to growth and resorption of the tissue in remodeling, we can conclude without much ado that the change of opening angle of blood vessels due to change of blood pressure is due to *nonuniform* remodeling throughout the thickness of the vessel wall.

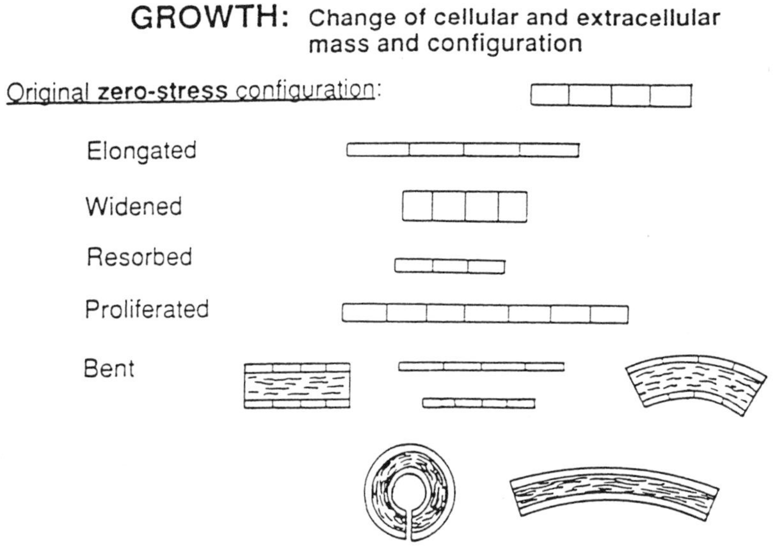

Fig. 6. Illustration that the remodeling of a blood vessel is best described by change of its zero-stress state. From Fung (6, p. 528), by permission.

From the point of view of studying tissue remodeling, the zero-stress state is significant because it reveals the configuration of the vessel in the most basic way, without being complicated by elastic deformation. If cellular or extracellular growth or resorption occurs in the blood vessel due to any physical, chemical and biological stimuli, they will be revealed by the change of zero-stress state.

5. What is the Residual Stress Doing There?

The state of a body on which all the external load are removed is called a *no-load state*. The internal stress existing in a body at the no-load state is called the *residual stress*. For a blood vessel, if there is no longitudinal tension and no transmural pressure, then it is at a no-load state. From Fig. 6, we see that a blood vessel at no-load state can be obtained by bending the vessel wall in its zero-stress state into a tube and then welding the edges into a seamless tube. The residual stresses can be calculated according to this point of view.

The *in vivo* state of a blood vessel can be obtained from the no-load state by stretching it longitudinally and then put the blood pressure on the inner wall and external pressure on the outer wall. Follow through on this thought, one can easily show that the residual circumferential stress is compressive at the inner wall of the blood vessel, and tensile at the outer wall of the blood vessel. On introducing the blood pressure, and working out the mechanics problem, one will see that the residual stresses will make the stress distribution much more uniform in the vessel wall at the *in vivo* state. Therefore, we found that the state of stress in a blood vessel *in vivo* is very much affected by the residual stress. Accordingly, it is clear that we must know the zero-stress state of all organs in our body in order to calculate the stress in our body *in vivo*. Thus, the very simple experiment illustrated in Fig. 1 is indeed fundamental and far-reaching.

6.   Tissue Remodeling Revealed by Change of Zero-stress State

The stress and strain in our body change normally or pathologically. In principle, the reason for these changes is very simple. By molecular mechanism, living cells can make proteins to enlarge themselves or build up intercellular matrix. Hence, new materials are made from a molecular pool according to the laws of molecular biomechanics. These proteins can also be metabolized and carried away by blood and lymph flow. Hence, the mass of the tissue can vary with time. It follows that the structure of the tissue can change with time. This changes the zero-stress state of the tissue. The stress-strain laws of the tissue will change, and the stress distribution will change (6, 7).

The changing stress then feed back onto the cells, causing them to react to the changing stress field by producing new materials, or move, or proliferate by cell division, or be resorbed. This logical sequence of events has been demonstrated. In the following, a brief sketch is given to indicate the current status of our knowledge.

7.   Tissue Remodeling of Arteries Under Stress

Figure 7, from Fung and Liu (3), shows the history of tissue remodeling of rat pulmonary artery when the pulmonary blood pressure was raised from the normal value at sea level to a higher value obtained by breathing a gas with 10% O_2 and 90% N_2 at normal atmospheric pressure. The left hand side figure is a sketch of the pulmonary artery of the left lung. The first row shows the morphological changes of the cross-section of the arterial wall in the arch region. It is seen that significant changes occurred already in 2 hours. The intima changed first, followed by the media in which the vascular smooth muscle reside. The adventitia changed slower. The succeeding rows show the changes in the vessel walls of smaller arteries.

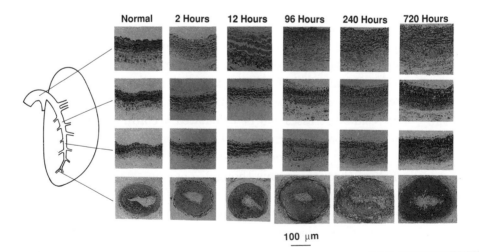

100 μm

Fig. 7. Remodeling of rat pulmonary arteries when the animal is subjected pulmonary hypoxic hypertension by breathing hypoxic gas to the length of time shown in the figure. Photographs of histological slides from 4 regions of main pulmonary artery of a normal rat and hypertensive rats with different periods of hypoxia. Specimens were fixed at no-load condition. From (3), by permission.

Figure 8, from Liu and Fung (9), shows how the stress-strain relationship of rat thoracic aorta is changed by tissue remodeling during the development of diabetes following an injection of streptozocin. The general trend remains the same, but the elastic constants at any given strains are changed.

Figure 9, from Fung and Liu (10), shows our current method of determining *in vivo* the incremental elastic constants of the arterial wall when it is regarded as a two-layered

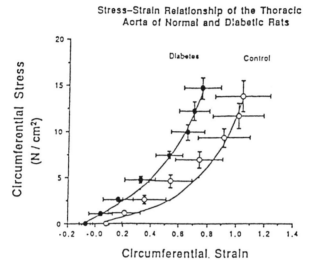

Fig. 8. Change of the stress-strain relationship of rat thoracic aorta during the development of diabetes 20 days after an injection of AZT. From Liu and Fung (9), by permission.

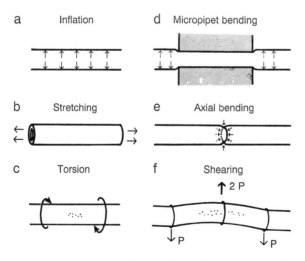

Fig. 9. Six types of *in vivo* experiments that can be used to determine the eight incremental elastic moduli of the two layers of a blood vessel at a homeostatic state *in vivo*. Counting an experiment of type *a* with variable longitudinal stretch, and another one of type *b* with variable internal pressure as two additional experiments, we have a total of eight experiments for the eight unknowns. From Fung and Liu (10), by permission.

material: an intima-media inner layer, and an adventitia outer layer. (a) refers to the measurement of the vessel diameter when the blood pressure is changed while the length of the vessel is held fixed. (b) refers to changing longitudinal tension due to longitudinal stretch. By varying the length in (a) and the blood pressure in (b), we cover a combined variation of the two variables, length and pressure. (d) refers to bending of the vessel wall with a pair of micropipets of rectangular mouth pushing or pulling diametrically while measuring the blood pressure and the micropipet force and displacement. (e) refers to axial bending with a silk thread, measuring force in the thread and deflection of the vessel. These four experiments are sufficient to obtain the incremental. Young's modulus in the circumferential and longitudinal directions.

Figure 10, from Fung (5), shows the length-tension relationship of the vascular smooth muscle in the pig coronary arterioles deduced by the present author (5) from the experimental data of Kuo, Chilian and Davis (14–16). This is the homeostatic length-tension relationship of the vascular smooth muscle. The relationship is strongly influenced by the shear stress imposed on the blood vessel wall by the flowing blood, τ. The upper panel shows the relationship in vessels without flow. The lower panel shows the effect of flow-induced shear,

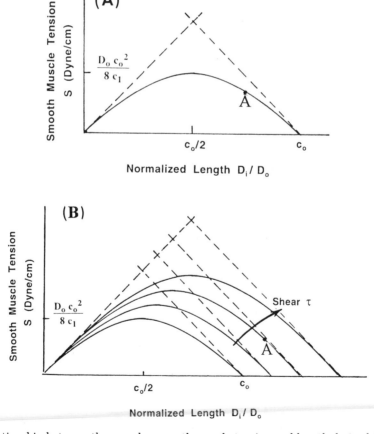

Fig. 10. (A) Relationship between the vascular smooth muscle tension and length derived in Ref. 5 from the experimental results given in Refs (12–16). (B) The influence of shear stress, τ, acting on the endothelium. From Fung (5), by permission.

τ. Note the similarity of these curves to the length-tension curve of the tetanized skeletal muscle, and to the curve of muscle length vs the peak tension in isometric twitch of heart muscle. The major distinction of the vascular smooth muscle is that it normally works at lengths corresponding to the right leg of an arch shaped curve, whereas the heart and skeletal muscles normally work at lengths corresponding to the left leg of the arch shaped curve. Knowing the curves of Fig. 6, we can now understand the autoregulation of the blood flow; and hyperemia, the increased blood flow in a vessel upon release from compression; and the Bayliss phenomena. Autoregulation is the tendency for blood flow to remain constant in the face of changes in arterial pressure to the organ. It is seen in virtually all organs of the body. It is most pronounced in the brain and kidney. (See Johnson (17) for detailed discussion and literature.) Each organ has a steady state flow-and-pressure relationship. Suppose in an autoregulated region, say at a point A in Fig. 10, you made a sudden increase of arterial blood pressure. By the elasticity of the artery, the arterial diameter will be increased in response to the sudden increase of pressure, thereby, the resistance to blood flow is transiently reduced and the flow increases. Then the tension in the vascular smooth muscle is increased to a new value above that of the point A in Fig. 10. The curve Fig. 10 shows that at equilibrium, such an upward movement of tension will have a muscle length that is shorter than that of point A. The smooth muscle shortens to reach the new equilibrium state. The vessel diameter is reduced, the flow resistance increases, and the flow falls back toward the normal value. This is autoregulation, which used to be considered mysterious. With Fig. 10, we see it as a revelation of the basic feature of the smooth muscle property, and the fact that the normal working condition of the artery corresponds to a point lying on the right leg of the arch-like curve of Fig. 10.

Reactive hyperemia is the period of elevated flow that follows a period of circulatory arrest, e.g. by inflating a cuff on an arm as we use a manometer to measure our blood pressure. On releasing the cuff pressure, the blood flow burst forth far above the normal value. (See Johnson, (17) for details.) The explanation based on Fig. 10 is that under compression the tension in the smooth muscle is reduced: the condition is represented by a point lower than the normal condition of point A. Hence, the smooth muscle lengthened and the vessel circumference enlarged while the vessel was compressed. Upon release of the compression and the return of the blood flow, the enlarged blood vessel causes a large flow. Later, the muscle length adjustments take place to return the flow to the normal condition. Reactive hyperemia is often used to detect if a vessel has an arteriosclerotic plaque, because the plaque abolishes reactive hyperemia.

8. Theory and Experiment. Design and Explanation

The example above illustrates how a simple experiment can lead to a broad and deep theoretical investigation with fundamental implications. These theoretical investigations lead to new experiments; new experiments lead to new theory. The spiral leads to greater understanding of nature. In the present example, the results of several rounds of theory and experiment has led to the concept of tissue engineering, because if tissues remodel under stress, then the stress can be used as a tool to control or "engineer" the tissue. This concept has led to significant medical advances.

Thus, you can see how theory and experiment couple together, how design and explanation couple together. Historically, I thought of doing the experiment shown in Fig. 1 in 1982 because new papers kept appearing which showed that there is a strong stress concentration in heart and blood vessels due to internal pressure. These stress concentrations were demonstrated by the theory of linear elasticity. When people improved the theory with the nonlinear stress-strain law, and taking the nonlinear finite strains-deformation gradient relationship into account, the stress concentrations did not decrease. Not only did they not decrease, they became much worse. Thus, these studies suggested an intrinsic risk factor in our bodies. Before accepting that conclusion, I thought the fault might lie in the assumption that the heart and blood vessels are stress free when the pressure loading is removed. To test this suspicion, I cut up the specimen to see if there were residual strains. There were! The evidence presented in Ref. (1) shows them. Paul Patitucci, a graduate student at that time, made lots of contributions. Almost simultaneously, without us knowing it, Vaishnav and Vossoughi made similar investigation and obtained the same results (12, 13). Following this, other important papers followed. I have listed most of them in the list of References at the end of this chapter (11–44). New investigations have turned to cellular and molecular details to explain the macroscopic observations. The better we know how our body works, the better we feel.

9. Conclusion

What impact does this kind of biomechanics have on the classical theory of continuum mechanics? The impact lies in calling for vigorous renewal. Biomechanics differs from other branches of applied mechanics by (1) nonconservation of mass, (2) nonconservation of structure, (3) nonlinearity of the equations describing the mechanical properties, (4) the stress-strain relationship changes with the changing state of stress or disease, (5) the stress-growth laws define the outcome of the external or internal loading, and (6) by probing more deeply into the basic reasons of the features named above, one is lead to the study of biomechanics of the cells and biomolecules. One can expect a vigorous new development of applied mechanics to answer this calling!

The point of this chapter is to show that a very simple experiment may open up a large gate for research. All depends on penetrating thoughts.

For an account of a more complete story of applying biomechanics to the study of one organ, the lung, see the article on pp. 35–56.

Acknowledgments

This work was supported in part by grant HL43026 from the National Heart, Lung, and Blood Institute.

References

1. Fung, Y. C., What principle governs the stress distribution in living organs? *Biomechanics in China, Japan, and USA*, Science Press, Beijing (1983), pp. 1–12.

2. Fung, Y. C. and Liu, S. Q., Change of residual strains in arteries due to hypertrophy caused by aortic constriction, *Circulation Res.* **65**, 1340–1349 (1989).

3. Fung, Y. C. and Liu, S. Q., Changes of zero-stress state of rat pulmonary arteries in hypoxic hypertension, *J. Appl. Physiol.* **70**(6), 2455–2470 (1991).

4. Fung, Y. C. and Liu, S. Q., Strain distribution in small blood vessels with zero-stress state taken into consideration, *Am. J. Physiol.* **262**, H544–H552 (1992).

5. Fung, Y. C., *Biodynamics: Circulation*, Springer-Verlag, New York (1984), 2nd edition, changed title to *Biomechanics: Circulation*, Springer-Verlag, New York (1996).

6. Fung, Y. C., *Biomechanics: Motion, Flow, Stress and Growth*, Springer-Verlag, New York (1990).

7. Fung, Y. C., *Biomechanics: Mechanical Properties of Living Tissues*, Springer-Verlag, New York (1981), 2nd edition (1993).

8. Liu, S. Q. and Fung, Y. C., Relationship between hypertension, hypertrophy, and opening angle of zero-stress state of arteries following aortic constriction, *J. Biomech. Eng.* **111**, 325–335 (1989).

9. Liu, S. Q. and Fung, Y. C., Influence of STZ-induced diabetes on zero-stress states of rat pulmonary and systemic arteries, *Diabetes* **41**, 136–146 (1992).

10. Fung, Y. C. and Liu, S. Q., Determination of the mechanical properties of the different layers of blood vessels *in vivo*, *Proc. Natl. Acad. Sci. USA* **92**, 2169–2173 (1995).

11. Xie, J. P., Liu, S. Q., Yang, R. F. and Fung, Y. C., The zero-stress state of rat veins and vena cava, *J. Biomech. Eng.* **113**, 36–41 (1991).

12. Vaishnav, R. N. and Vossoughi, J., Estimation of residual strains in aortic segments, In: *Biomedical Engineering, II. Recent Developments* (ed. C.W. Hall), Pergamon Press, New York (1983) pp. 330–333.

13. Vaishnav, R. and Vossoughi, J., Residual stress and strain in aortic segments, *J. Biomech.* **20**, 235–239 (1987).

14. Kuo, L., Davis, M. J. and Chilial, W. M., Myogenic activity in isolated subepicardial and subendocardial coronary arterioles, *Am. J. Physiol.* **255**, H1558–H1562 (1988).

15. Kuo, L., Davis, M. J. and Chilial, W. M., Endothelium-dependent, flow-induced dilation of isolated coronary arteries, *Am. J. Physiol.* **259**, H1063–H1070 (1990).

16. Kuo, L., Chilian, W. M. and Davis, M. J., Interaction of pressure- and flow-induced responses in porcine coronary resistance vessels, *Am. J. Physiol.* **261**, H1706–H1715 (1991).

17. Johnson, P. C., Peripheral circulation, Wiley, New York (1978).

18. Xie, J. P., Zhou, J. and Fung, Y. C., The pattern of coronary arteriolar bifurcations and the uniform shear hypothesis, *Ann. Biomed. Eng.* **23**, 13–20 (1995).

19. Han, H. C. and Fung, Y. C., Longitudinal strain of canine and porcine aortas, *J. Biomech.* **28**, 637–641 (1995).

20. Debes, J. C. and Fung, Y. C., Biaxial mechanics of excised canine pulmonary arteries, *Am. J. Physiol.* **269**, H433–H442 (1995).

21. Fung, Y. C., Stress, strain, growth, and remodeling of living organisms, *Z. Angew. Math. Phys.* (special issue) **46**, S469–S482 (1995).

22. Han, H. C. and Fung, Y. C., Direct measurement of transverse residual strains in aorta, *Am. J. Physiol.* **270**, H750–H759 (1996).

23. Fung, Y. C., New trends in biomechanics. Keynote Lecture, *Proc. 11th Conf. Engineering Mechanics.* (eds, Y. K. Lin and T. C. Su), American Society of Civil Engineers (1996), pp. 1–15.

24. Liu, S. Q. and Fung, Y. C., Indicial functions of arterial remodeling in response to locally altered blood pressure, *Am. J. Physiol.* **270**, H1323–H1333 (1996).

34

25. Gregersen, H., Kassab, G., Pallencaoe, E., Lee, C., Chien, S., Skalak, R. and Fung, Y. C., Morphometry and strain distribution in guinea pig duodenum with reference to the zero-stress state, *Am. J. Physiol.* **273**, G865–G874 (1997).

26. Zhou, J. and Fung, Y. C., The degree of nonlinearity and anisotropy of blood vessel elasticity, *Proc. Natl. Acad. Sci. USA* **94**, 14255–14260 (1997).

27. Liu, S. Q. and Fung, Y. C., Changes in the organization of the smooth muscle cells in rat vein grafts, *Ann. Biomed. Eng.* **26**, 86–95 (1998).

28. Skalak, R. and Fox, C. F., eds, *Tissue Engineering*, Alan R. Liss, Inc. New York (1988).

29. Woo, S. L.-Y. and Seguchi, Y., eds, *Tissue Engineering — 1989*, ASME Publications No. BED-Vol. 14, *Am. Soc. Mech. Eng.*, New York (1989).

30. Bell, E., ed, *Tissue Engineering: Current Perspectives*, Birkhäser, Boston (1993).

31. Abé, H., Hayashi, K. and Sato, M., eds, *Data Book on Mechanical Properties of Living Cells, Tissues, and Organs*, Springer, Tokyo (1996).

32. Berry, J., Rachev, A., Moore, J. E. and Meister, J. J., Analysis of the effect of a noncircular two-layer stress-free state on arterial wall stresses, *Proc. 14th Ann. Int. Conf. IEEE Eng. Med. Bio. Sec.* (1992), pp. 65–66.

33. Rachev, A., Theoretical study of the effect of stress-dependent remodeling on arterial geometry under hypertension condition, *13th Southern Biomed. Eng. Conf. U. District. Columbia*, Washington, D.C. (1994).

34. Adachi, T., Tanaka, M. and Tomita, Y., Uniform stress state in bone structures with residual stress, *J. Biomech. Eng.* **120**, 342–348 (1998).

35. Han, H.-C., Shao, L., Huan, M., Hou, L.-S., Huang, Y.-T. and Kuang, Z.-B., Postsurgical changes of the opening angle of canine autogenous vein graft, *J. Biomech. Eng.* **120**, 211–216 (1998).

36. Niklason, L. E., Gao, J., Abbott, W. M., Hirshi, K. K., Houser, S., Marini, R. and Langer, R., Functional arteries grown *in vitro*, *Science* **284**, 489–493 (1999).

37. Rachev, A., Stergiopulos, N. and Meister, J.-J., A model for geometric and mechanical adaptation of arteries to sustained hypertension, *J. Biomech. Eng.* **120**, 9–17 (1998).

38. Vossoughi, J., Hedjazi, Z., Boriss, F. S., II. Intimal residual stress and strain in large arteries, *BED-24, Bioeng. Conf. ASME* (1993), pp. 4394–4437.

39. Liu, S. Q., Influence of tensile strain on smooth muscle cell orientation in rat blood vessels, *J. Biomech. Eng.* **120**, 313–320 (1998).

40. Liu, S. Q. and Fung, Y. C., Changes in the organization of the smooth muscle cells in rat vein grafts, *Ann. Biomed. Eng.* **26**, 86–95 (1998).

41. Price, R. J. and Skalak, T. C., Distribution of cellular proliferation in skeletal muscle transverse arterioles during maturation, *Microcirculation* **5**, 39–48 (1998).

42. Taber, L. A., A model for aortic growth based on fluid shear and fiber stresses, *J. Biomech. Eng.* **120**, 348–354 (1998).

43. Liu, S. Q., Biomechanical aspects of vascular tissue engineering, *Crit. Rev. Biomed. Eng.* **27**, 75–148 (1999).

44. Langille, B. L., Arterial remodeling: Relation to hemodynamics, *Can. J. Physiol. Pharm.* **74**, 834–841 (1996).

17

Pressure, Flow, Stress, and Remodeling in the Pulmonary Vasculature

Yuan-Cheng B. Fung, Ph.D.

Professor Emeritus, Bioengineering and Applied Mechanics, University of California-San Diego, La Jolla, California

My Personal Experience

Personal experience has nothing to do with science but it has a lot to do with one's choice of topics and the approach one takes. To explain my approach I shall say a few words of my experience. I entered college in 1937 when Japanese militarists started the last big push to conquer China. I took my college entrance examination in Shanghai at the time Japanese troops landed in Shanghai. I chose to study airplane design because that seemed to be needed most by China to fight for its survival.

In wartime (1937–45) Chongqing, China had virtually no air defense. Our classes were usually held at the crack of dawn. Regularly, by 10:00 A.M. the Japanese air raid would arrive, and students and teachers would stay in the caves on the banks of Jialing River. One thing I saw most in those years around the caves was the clouds in the foggy sky of Chongqing. It was natural that my first publication was a small book on soaring and gliding in clouds.[7]

I entered the California Institute of Technology in 1946, obtained my Ph.D. in aeronautics and mathematics in 1948, and stayed on as a faculty member until 1966. My specialty was the mathematical theory of elasticity and nonstationary aerodynamics. The combined field is called *aero-*

From: Wagner WW, Jr, Weir EK (eds): The Pulmonary Circulation and Gas Exchange. © 1994, Futura Publishing Co Inc, Armonk, NY, by permission.

elasticity. It deals with the phenomenon of *flutter,* which is a dynamic instability of airplanes, supersonic aircraft, spacecraft, and birds when their flight speed exceeds their respective critical speed for flutter. It deals also with phenomena that occur when these flying objects encounter wind shear, gusts, clouds, clear air turbulences, or thunderstorms. It applies equally well to wind blowing on stationary structures. Indeed, the first paycheck I earned was for checking the safety of the design of the cantilever roof of the stadium of the University of Washington in Seattle against wind. I presented a systematic survey of the field in my book *An Introduction to the Theory of Aeroelasticity,* published in 1955.[8] My other papers were concerned with structures, vibrations, elastic waves, stochastic processes, protection of structures against nuclear bombs (the base-hardening problem). My endeavor to teach better led me to publish a book *Foundations of Solid Mechanics,* in 1965,[9] and *A First Course in Continuum Mechanics* in 1969, 1977, and 1993.[13] I was lucky to be recognized by my colleagues in engineering who elected me a fellow of the American Institute of Aeronautics and Astronautics in 1969, a fellow of the American Society of Mechanical Engineering in 1978, and a member of the National Academy of Engineering in 1979. Later I was even luckier and was elected a senior member of the Institute of Medicine of the National Academy of Sciences in 1990.

I began my self-study of physiology in 1957 because my mother had glaucoma, and out of concern and gratitude I periodically translated newly published articles on glaucoma into Chinese, which I sent to her and her surgeon in China. My sabbatical leave in Göttingen and Brussels in 1957–58 provided an excellent chance for me to read physiology papers. On returning to Caltech I joined Sidney Sobin, Wally Frasher, and Ben Zweifach in their studies of microcirculation. We wrote a

few papers together.[10,11,35,51] In 1966 I resigned my professorship of aeronautics from Caltech and went to the University of California, San Diego, to devote myself full time to physiology and bioengineering.

How do you seduce a person comfortably established in one field to leave it and enter another field in which he is completely unknown? I suppose that there had to be a feeling that there was something in the new field for me to do. My first chance came when Ben Zweifach told me that capillary blood vessels are rigid. Looking at capillaries' ultrastructure, I could not believe this from my solid mechanics background. That led me to think about the contribution of surrounding gellike tissues that support the capillaries, and publish my first two papers in biology.[10,51] The same thought made me want to study the capillaries in the lung as a counterexample, because the pulmonary capillaries have virtually no surrounding tissue in the direction perpendicular to the interalveolar septa; hence, the distensibility of pulmonary capillaries must be very different from the systemic capillaries of the mesentery. The second chance came when Alan Burton and his students, Rand and Prothero, published their experiments on the deformation of red blood cells. I was a professional thin-walled shells man; I worked out a theory of red cell deformation and compared the theoretical results with experimental data to infer the bending and stretching rigidity of red cell membrane. This led to my third and fourth papers in biology.[11,46] Subsequently I made rubber models of red cells and studied the hemodynamics of red cell flow in capillaries.[59] Again I began to think of using the lung as a counterexample because Sid Sobin had shown me how different the pulmonary capillaries looked in comparison with the systemic capillaries. Then Wally Frasher introduced me to the determination of the mechanical properties of blood vessels.[58] I improved the experimental method and formulated a new theory to describe the mechanical properties of soft tissues.[14] So as soon as I spied the fringes of the new field, I knew how rich it was and was convinced that there would be plenty of things there for me to do.

Was something wrong with the old field that it repelled me? Not too much. At that time aerospace engineering still held sway. The government and military were pouring money into the field. The available financial, human, technological, and computational resources were almost infinite compared with individual initiatives. To me that suggested it was time to clear out.

Further, I did not need much imagination to see what contribution engineering can offer medicine. New instruments, devices, and prostheses are needed; new insight to physiology and pathology would be welcome. I was convinced that the science of biomechanics must be developed. The concerns of those in biomechanics—force, motion, flow, stress, strength, and remodeling—pervade the living world, yet the literature of biology has largely ignored these words. Hence I resolved to dedicate myself to the development of biomechanics.

Cooperative Research with Sid Sobin and Mike Yen

Having Sidney Sobin and Michael Yen as close friends and collaborators is an extraordinary good luck for me. Sid was a prosperous cardiologist who accepted an NIH career professorship in 1965 to devote himself full time to physiology. Mike was my former graduate student and later my colleague at the University of California, San Diego. Our views on physiological research are similar. We would like to separate the work of searching for basic principles from that of solving boundary-value problems. We would like to identify the basic principles with as few ad hoc hypotheses as possible and base the boundary-value problems on the geometry and material properties of the real structure, i.e., geometry based on morphometry, material properties based on rigorous biorheological investigations. Then, for each clearly formulated boundary-value problem, we would like to solve the mathematical problem accurately so as to eliminate any inadvertent introduction of approximations, which are equivalent to additional ad hoc hypotheses. We thought this kind of approach would lead to greater understanding. Physicists, mathematicians, and engineering mechanicists call this kind of approach a *rational* approach. We would like to see if it works in physiology.

We chose to use the rational approach to study pulmonary circulation because in 1965 Sid had already made excellent progress in using the polymer casting method to study the morphology of pulmonary microvasculature. We saw that the pulmonary capillary blood vessels form a dense two-dimensional network in each alveolar wall. (See Figure 1, left panel, which is a photograph of the capillary network in a cat pulmonary interalveolar septum.) The capillary blood vessels occupy about 80% of the area in this plan view. Between the vessels are solid bits of tissue, which we call "posts."[35] The structure of pulmonary capillaries is so different from the systemic capillaries that we expect the hemodynamics to be different. We call the unique network a *capillary sheet*. We presented a theoretical study of pulmonary circulation at an ACEMB conference in November 1967[35] and at a FASEB meeting in April 1968.[36] Our first full theoretical paper was published in the *Journal of Applied Physiology* in 1969.[37] To complete the research proposed by this theory, we set up a program of study consisting of the following steps:

346 • *THE PULMONARY CIRCULATION AND GAS EXCHANGE*

1. Measure the morphology of the vascular tree to understand the geometric structure.

2. Obtain basic data on the properties of matter involved in the system, including all of the gases, fluids, and tissues. Some mass transport data such as the diffusion constants, permeability, and solubility existed already. But rheological data on the apparent viscosity of the blood in small blood vessels and capillary network and on the mechanical properties of the pulmonary blood vessels of all generations were lacking and had to be measured.

3. Identify the applicable basic principles for the analysis. We decided to use as few ad hoc hypotheses as possible. We wanted to take the attitude of the reductionist to the extreme and allow only the following principles:
   • the conservation of mass,
   • the conservation of momentum,
   • the conservation of energy,
   and nothing else. For example, we know that waterfall phenomenon exists, as several authors before us have shown, but we are not going to make it a hypothesis. It should fall out as a result or conclusion of some specific boundary-value problem.

4. Formulate problems of pulmonary blood flow according to specific boundary conditions and appropriate geometric and material properties data. By "boundary conditions" is meant the external conditions applied to the lung, e.g., the pressure or flow at arterial inlet or venous outlet, the pleural pressure, any constraints imposed on the lung (e.g., a deformation due to a tumor, a movement of the diaphragm or chest wall), an external load applied on the lung (e.g., gravitation, inertial force due to acceleration or sports), and the mutual influence of blood vessels and airway. Each problem must have a special set of boundary conditions.

5. The basic equations and the boundary conditions together form a mathematical problem. In physics and engineering, these are called boundary-value problems. The next step is to solve the boundary-value problems of physiology.

6. Perform experiments and compare experimental results with theoretical predictions. If agreement is obtained, one gains confidence in the mathematical solution. If agreement is not obtained, one must reexamine the theory and experiment to identify the reasons for the disagreement. Inaccuracy of the material properties data, oversimplification of the mathematical analysis, neglect of extraneous factors in the experiment, mismatch of theory and experiment (e.g., dissimilar boundary conditions, wrong boundary values, etc.) are the usual culprits.

7. Finding genuine disagreement between theory and experiment provides opportunities for advancing scientific knowledge. Rectification often calls for major improvements in experimental control or accuracy and changes in theoretical concept or its mathematical description. One cannot rest until agreement is obtained.

8. The validated theory is then used to solve new problems and predict new events, which leads to new experiments, new comparisons between experiment and theory, new opportunities, and new understanding.

This program is straightforward, but it needed lots of work and documentation. It took us 12 years before we could close the first round of comparison between theory and experiment on the pressure-flow relationship of the whole lung[5-50,68-74,89-92,98-100]. But we had lots of fun on the way and found many pretty pebbles right and left. Furthermore, because a master plan exists, we know the value of every link in the chain, and the plan gives us greater patience in working out the details. A few highlights are given below.

Anatomy and Morphometry

Much of the anatomy of the lung was well understood long before we began our work in mid-1960s. Data were incomplete, however, and a great deal of work was needed to fill the gaps. In mid-1960s, the data base was as follows: the branching pattern of human airway and arteries in the lung were known from the work of Drs. Weibel, Horsfield, Cumming, and their associates.[6,57,79] A lot of human clinical data existed but the majority of experimental physiological data was obtained from the dog, whose pulmonary arterial tree branching pattern is unknown. Elasticity data was known for only a few arterial segments of the rabbit and dog. The elasticity of the pulmonary capillary blood vessels was being measured by Sobin on the cat. We estimated that the measurement of the distensibility of pulmonary vessels of all generations would be a major effort. Because Sobin's work was progressing well, we chose the cat as our experimental animal.

The cat alveolar wall appears to be similar to that of man. The cat alveoli are almost twice as large as the dog alveoli and are closer to the human alveoli in size. Weibel had already idealized the capillary blood vessel network as made up of short circular cylindrical tubes arranged in a hexagonal pattern.[79] We looked at his picture and wondered how we could compute the pressure-flow relationship in such a network of tubes. Being schooled in fluid mechanics, we remembered the hypotheses under which Poiseuille's formula was derived. They are (1) the flow is laminar (i.e., not turbulent, no separation or secondary flow), (2) the tube is circular cylindrical in shape and is infinitely long, or, if it is of finite length, the velocity distribution at the entry and exit sections are parabolic over the radius, exactly as in

the long tube, and (3) the pressure distribution in any section perpendicular to the longitudinal axis is uniform. If these conditions are not met, Poiseuille's formula is not valid. That is why books on hydraulics are full of empirical formulas that are modifications of Poiseuille's formula: to take into account the finite length of the tube, the curvature of the tube axis, the existence of converging or diverging branches, and any special entry and exit conditions. There is no question that the individual tubes of the pulmonary capillary network do not satisfy the premise under which Poiseuille formula was derived. Modification is necessary, and trying to modify is in no sense disrespectful to Jean Poiseuille. Yet, strange it may seem, the resistance I encountered from the medical circle to our new theory came in large measure from our modification of Poiseuille formula.[54,68]

In deriving the pressure-flow relationship in pulmonary capillaries, we took advantage of the fact that the Reynold's number of capillary flow is very small (<0.001), so that the Navier–Stokes equation can be replaced by the Stokes equation. Even for the Stokes equation, we cannot handle the mathematical analysis of flow in the Weibel hexagonal-tube network.[79] We can manage the mathematics, however, to analyze a similar but different model of flow between parallel plates obstructed by round posts arranged in a periodic hexagonal pattern by means of the doubly periodic Weierstrass elliptic functions. In giving a name to this model we called it a *sheet-flow model*. J. S. Lee, Mike Yen, and I had little difficulty publishing the fluid mechanical analysis and its experimental verification in fluid mechanics and engineering journals.[12,15,19,44,45,60,76,83,84] Having obtained the basic solution, Sobin and I introduced further modifications to allow the sheet to be of finite size and to be distensible under positive transmural pressure and collapsible under negative transmural pressure.[38,39,72] In the meantime, we pursued morphometry of the alveolar sheet and compared our data with the data of other authors.[38,39,74] We believe that morphologically our model is as close to nature as Weibel's and is in some respects much better. In the first place, in the plane of the interalveolar septum the blood vessels of the real lung and our model have no corners, whereas Weibel's has. In the second place, with increased blood pressure the blood vessels of the real lung and of our model remain smooth, whereas the vessel walls of Weibel's model would have structural instability and buckle at the inner corners of tube junctions. With regard to the deformation of the capillary bed, when the blood pressure is varied, we can derive a theoretical relationship between the sheet thickness, transmural pressure, and tissue stress in the sheet model on the basis of the known geometrical and material properties.[38] This theoretical result

fits our experimental results very well. On the other hand, if we used the tube network model, we would obtain some conclusions on the deformation pattern that contradict experimental results.

Knowing the capillary network in each interalveolar septum alone is useless to hemodynamics unless we know how the capillary networks are supplied and drained with blood. In other words, we would have to know the geometry of the arterioles that supply blood to the capillaries and the geometry of the venules that drain them. From photomicrographs such as those shown in Figure 1, we know that the capillary sheet of one interalveolar septum is connected to the neighboring sheets. In a sense the huge number of sheets of interalveolar septa form one big sheet. The arterial entry and venous drainage of each septum are rarely seen in alveolar micrographs, suggesting that they are very sparse relative to the capillaries.

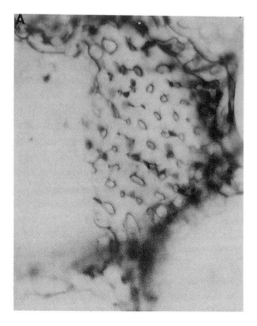
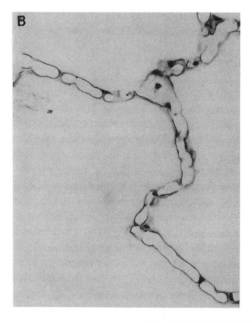

Figure 1. Two views of pulmonary alveolar wall (interalveolar septum) in a cat's lung. Left: Looking at the septum en face. Right: Cross section of a sheet. (Reproduced with permission from reference 37. Copyright 1969 American Physiological Society.)

Later we learned[92] that, on average, each arteriole supplies 24.5 alveoli and each venule drains 17.8 alveoli. This explains the experimental difficulty. To see the arterioles and venules along their length requires thick sections and low magnification. To identify the relationship of these vessels to the capillaries requires thin sections and high magnification.

A compromise method of identifying the capillary-arteriole-venule relationship is shown in Figure 2, which is a montage of micrographs representing a cross section of a cat lung.[69] On these micrographs every blood vessel larger than the capillaries was examined individually in the histological slide under an optical microscope to determine whether it is an artery (or arteriole) or a vein or (venule).[69] If a vessel was found to be arterial, then it was marked with a white dot. If it was venular, then it was marked with a bull's eye. The result shows a remarkable segregation of territories occupied by the arteries and veins. An artery is more likely to be surrounded by other arteries. A vein is more likely to be surrounded by other veins. An overlay of colored cellophane was used to cover areas in which there are only arteries. Looking at the colored cellophane against the background, we see a two-colored map. A geographic map of two colors can only represent islands and ocean. Hence our morphometric question becomes very simple: which are the islands? Arterial areas or venous areas? Which is the ocean? The montage gives the answer immediately: The arterial areas are the islands, the venous area is the ocean.[69]

How can we interpret this topological result in three-dimensional geometry? For this purpose, I can offer an analog. Imagine a tree, like a pine or a bonsai. The tree trunk branches and branches again and again. Each terminal branch occupies some volume of space. Suppose you find a specimen which is so shaped that when you pass a plane to cut the tree, the termi-

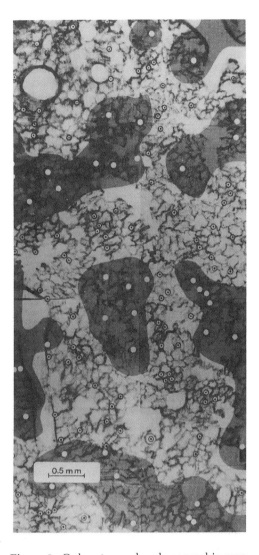

Figure 2. Only a two-colored geographic map can represent islands in an ocean. This is a lung map, representing a cross section of a cat lung, in which every arterial vessel is marked by a white dot and every venous vessel is marked by a bull's eye. The domain of the pulmonary arteries are darker and the individual alveoli are seen. The arterial domains are like islands in a lung cross section. Such a map has many uses, e.g., it shows the relationship between microcirculation and respiration, it helps to locate the sites of "sluicing" or "waterfalls" in zone 2 condition, and explains the recruitment after capillary collapse in zone 2 condition. (Reproduced with permission from Sobin, Fung, Lindal, Tremer, and Clark, *Microvas Res* 19: 224, 1980.)

nal branch volumes are reasonably separated; that specimen's geometry is an analog of the geometry of the pulmonary artery. On the other hand, a garden variety analog of the pulmonary venous tree is hard to find. Perhaps a multitrunk banyan tree whose top covers a large area is a possible analog. In any case, the top branches of the venous tree must fill all the space left by the arterial tree within the envelope of the total volume.

Having clarified the topologial relation between the pulmonary arterioles, venules, and capillaries, we can easily do some stereological measurements to obtain some needed morphometric data. Thus we obtain[69] that, for the cat lung, the average "diameter" of the arteriolar islands is 0.918 ± 0.156 (SD) mm, the average "width" of the venous zone is 1.158 ± 0.410 (SD) mm; the shortest path length between an arteriolar inlet into the capillary sheet to a venular outlet is 0.556 ± 0.285 (SD) mm. Furthermore, by counting the number of arteries, veins, and alveoli in several maps like that of Figure 2, Zhuong, Yen, Fung, and Sobin[92] obtained the ratios of arterioles, venules, and alveoli as mentioned earlier.

Turning to the relationship between pulmonary circulation and respiration, we need to know how the interalveolar septa are put together to form the alveoli and alveolar ducts and how the arterioles and venules are placed relative to the alveolar ducts. Hence I studied the models of the alveoli and alveolar ducts,[26] and the distribution of arterioles and venules relative to the alveolar ducts.[28,30]

Many people have made alveolar casts of wax, lead, vinyl, etc., to observe the geometry and structure of the alveoli and alveolar ducts. Miller[63] has reviewed earlier observations of historical importance, including his own contributions. More recently Hanson and Ampaya[55,56] presented a remarkably comprehensive set of data on the geometry of human alveoli and alveolar ducts. Oldmixon and

Hoppin[64] made sophisticated measurements. Mercer, Laco, and Crapo[62] used advanced computational techniques to make serial reconstruction of electron microscopic images to obtain data on alveolar geometry, surface area, etc. Ciurea and Gil,[5] Silage and Gil[67] made similar measurements of alveolar ducts. Budiansky et al.[3] have made theoretical calculations of alveolar elasticity on the basis of a pentagonal dodecahedron model. Wilson and Bachofen[81] have analyzed lung elasticity on the assumption that the structural elements of collagen and elastin are concentrated in the alveolar mouths, that the alveolar mouths lining the ducts can be represented by simple spirals, and that the alveolar walls contribute virtually nothing to lung tissue elasticity. But all this does not provide a clear three-dimensional mathematical description of the airway at the ducts-alveoli level.

I attempted to construct a mathematical model[26] with the following assumptions: (1) Alveolar ducts are formed by removing certain walls of the alveoli so that every alveolus can be ventilated to the atmosphere. (2) Before removal of walls to form ducts, all alveoli are equal and together fill the whole space of the lung. (3) Ducts are formed in a way that is efficient for gas exchange and structurally sound.

To begin constructing the duct model, we use a mathematical fact. There are only five known regular polyhedrons: the tetrahedron, the cube, the regular octahedron, the pentagonal dodecahedron, and a 20-faced icosahedron, of which each face is a triangle.[53,80] Of these, the pentagonal dodecahedron and the icosahedron are not space filling, and the tetrahedron and octahedron are space filling only if they are used in combination. Two nonregular polyhedra are known to be space filling: the rhombic dodecahedron and the 14-sided tetrakeidecahedron (14-hedron). The choice of space-filling polyhedron is thus limited to four. But because the interalveolar septa of the reconstructed alveoli given

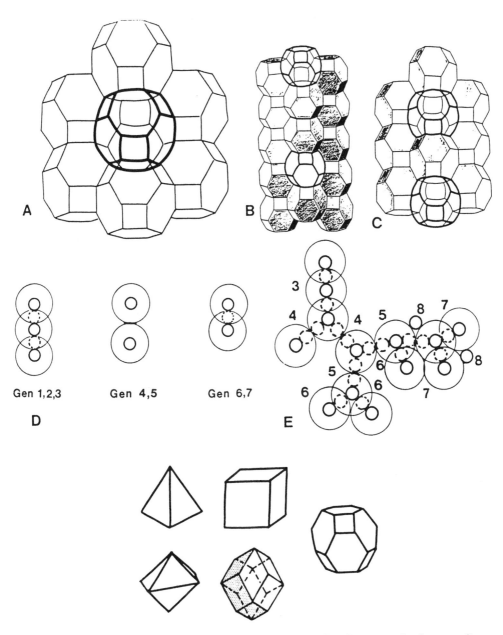

Figure 3. My model of pulmonary alveoli and alveolar ducts. At the bottom, five polyhedrons are shown. The three on the right-hand side are space filling. In (A), an order-2 14-hedron serves as my basic model. Fourteen 14-hedrons representing 14 alveoli are ventilated to a central duct, which is an empty 14-hedron. In (B), (C), the basic units are joined to form higher order structures. Fig. (D) shows 7 generations of ducts consisting of structures shown in (B), (A.A), and (C). Fig. (E) shows eight generations of alveolar ducts. (Reproduced with permission from reference 26. Copyright 1988 American Physiological Society.)

in references[5,55,56,62–64,67] do not appear to be all triangles or all squares or all parallelopipeds, we have to reject the 4-, 6-, and rhombic 12-hedrons as suitable models. Hence I choose the 14-hedron as the basic alveolar model,[26] Figure 3. Fourteen 14-hedrons surrounding a central 14-hedron form a unit called a second order 14-hedron. A second order 14-hedron with the walls of the central 14-hedron removed forms a basic unit of alveolar duct. The lung parenchyma is composed of second order and first order 14-hedrons. Successive second order 14-hedrons can be joined together to form longer alveolar ducts. My ducts model[26] is illustrated in Figure 3. The quantitative attributes derived from this model have been compared in detail with the data given by Hanson and Ampaya,[55,56] and I consider the validation successful.[26]

With this model and the data on pulmonary arterioles and venules, I deduced

Figure 4: Conceptual illustration of how circulation and respiration systems are joined at the microvascular level. Arterioles are indicated by white. Venules are indicated by black. Alveoli are indicated by small circles. The basic units of duct, the order-2 14-hedra, are indicated by large circles. Alveoli in venular region are not shown. A similar drawing can be found in references 28 and 30.

the picture of the connection between the arterioles, venules, and capillaries, i.e., between the macrocirculation and microcirculation and between circulation and respiration, as illustrated in Figure 4.

In summary, I found the anatomical and morphometric study most rewarding. This study yielded a new model of the pulmonary capillary blood vessel network, laid the foundation for the sheet-flow theory, led to the basic concept of directional compliance of interalveolar septa: high in the direction perpendicular to the septal plane, low in the plane of the septa. We recognized for the first time the island-in-the-ocean type of arterial region distribution in an organ. We identified a mathematical model of the pulmonary alveolar ducts and alveoli. These results are new and fundamental, and we would not have studied them if they were not part of the information needed in the rational approach to pulmonary circulation. Yet they are basic to any boundary-value problems of pulmonary circulation. Hence, the rational approach is usually an economical one.

Mechanical Properties of the Materials

A great deal was known about the properties of the gases and fluids in the lung and the mass transport characteristics across membranes. Least known in the 1960s were the mechanical properties of the pulmonary blood vessels. As I have explained earlier, due to the experimental work of Baez[1] on the capillaries in the mesentery, the capillary blood vessels were generally believed to be very rigid. Fung et al.[51] explained this rigidity in the mesentery on the basis of the tunnel-in-gel concept. Bouskla and Wiederhielm[2] showed that the capillaries in a bat's wing, in which the gel tissue surrounding the vessel is thin (about equal to the capillary diameter), are distensible. Now, in the pulmonary alveolar septa, capillaries are separated from the alveolar gas by a layer of tissue whose thickness is only on the order of 1μm (a small fraction of the capillary diameter), so the distensibility of the capillaries can be expected to be large. To document this expectation, Sobin and I worked out a theory of alveolar elasticity[38] and then set out to measure the change in the thickness of the pulmonary capillary sheet with the transmural pressure (i.e., the difference of blood pressure and alveolar gas pressure). We built a spherically rotatable microscopic stage to do it. The capillaries were perfused with a catalized liquid polymer. The static pressure was controlled while the flow was stopped before and during the solidification of the polymer. After solidification, histological slides were prepared and the thickness of the capillary vascular space was measured. We obtained the results shown in

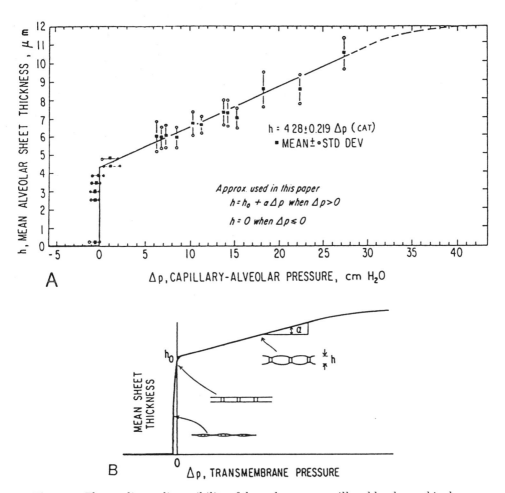

Figure 5: The nonlinear distensibility of the pulmonary capillary blood vessel is shown in the figure. The abscissa is the transmural pressure, Δp. the ordinate is the vascular sheet thickness, h. The distensibility curve starts at h = 0 for negative Δp, jumps up to h_O at $\Delta p = 0$, then follows a straight line at positive Δp until $\Delta p = 3$kPa (about 30 cm H_2O), and swings toward an upper limit at greater Δp. Theoretical deflection shapes are sketched. (Reproduced with permission from references 38 and 39. Copyright 1972 American Heart Association.)

354 • *THE PULMONARY CIRCULATION AND GAS EXCHANGE*

Figure 5, which shows that when the transmural pressure is positive, the sheet is distensible. When the transmural pressure is negative, the sheet is collapsed.

Then Yen and I and our students[88,89] made morphometry of the branching patterns of the pulmonary blood vessels. We found that the arteries and veins are tree-like, whereas the capillaries are not. The fractal concept is applicable to arteries and veins, but not to capillaries. We then measured the pressure-diameter relationship of the small arteries and veins (with diameter less than 100 µm) by the silicone polymer cast method.[82,85] We measured also the pressure–diameter relationship of the larger vessels by x-ray angiography of isolated perfused lung.[82,85] Altogether we obtained the distensibility of the blood vessel of every generation.

The most important result we obtained in these studies, besides the nonlinear elastic behavior of the capillaries shown in Figure 5, was the observation that the pulmonary veins remain patent (not collapsed) when the difference of the local blood pressure minus the airway pressure turns negative in the physiological range.[43,82] This pressure difference is the net outward pressure acting on the blood vessel wall: when it is negative the vessel wall is compressed. For the vena cava or a vein without external tethering, the vessel wall becomes unstable when the critical "buckling" pressure is reached, and the vein will be collapsed if the critical pressure is exceeded. The critical "buckling" pressure of systemic veins is less than 1 cm of water. But the critical buckling pressure of the pulmonary vein is larger than 10 cm of water. This is because pulmonary veins are embedded in lung tissue, i.e., tethered by the parenchyma, which is composed of the alveolar walls—the interconnected interalveolar septa. In an inflated lung the interalveolar septa are stretched (in tension), and they give elastic support to the blood vessel wall. Figure 6 illustrates our experimental

results.[82] It refers to cat lung with pleural pressure equal to the values indicated in the figure (1 cm H_2O is about 100 Pa). The airway pressure was zero (atmospheric). The blood pressure in pulmonary veins was varied, and the diameters of the veins were measured by x-ray angiography. The veins were perfused with a radio opaque material osmium tetraoxide, and the flow was stopped to make sure that the venous pressure was uniform everywhere. The horizontal axis of Figure 6 shows the value of blood pressure minus airway pressure. The vertical axis shows the ratio of the measured vessel diameter divided by its value when the pressure difference is 10 cm H_2O at which the vessel cross section is circular. Thus, it is seen that the pressure–diameter curve is a straight line whose slope does not change as the blood and airway difference changes from positive to negative values.

Reference 43 presents photographic evidence of the patency of pulmonary veins under negative transmural pressure. Here we also provided the data from direct measurement of the diameter, length, and number of branches of the first three orders of the smallest pulmonary veins subjected to a wide range of negative transmural pressure.

This patency of pulmonary veins is of great importance to the understanding of pulmonary blood flow. In standing humans, part of the lung is above the level of the heart. The blood pressure in the left atrium is not far from the atmospheric value. For a pulmonary vein at a sufficient height above the left atrium, the hydrostatic pressure due to gravitational acceleration would decrease the blood pressure in the pulmonary vein to a value below the airway pressure. This is said to be in "zone 2."[65] If we do not know that the pulmonary veins can remain patent in this condition, we would worry about whether the veins in zone 2 would be collapsed. Now we know that they would not. We do know that the pulmonary capillaries will

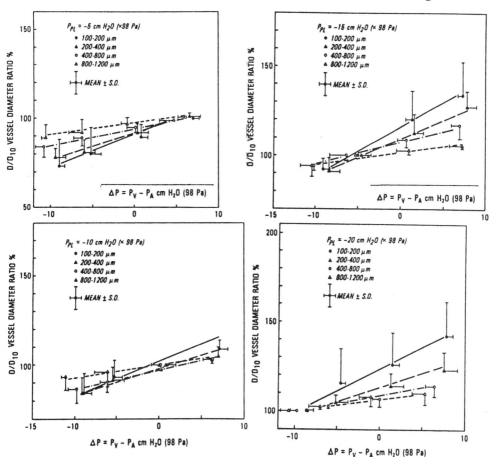

Figure 6. If waterfall or sluicing occurs in pulmonary circulation, the theoretical sites of action must be located at the ends of the capillaries where blood enters the venules. I reached this conclusion from the facts shown in this figure. Here the percentage change of diameter of pulmonary veins of the cat is plotted against the blood pressure, p_V. The airway pressure, p_A is zero. The pleural pressure, p_{PL}, is listed in the figures. Note that the curves remain straight and they go through the point $p_V - p_A = 0$. The pulmonary veins do not collapse when the transmural pressure $p_V - p_A$ turns negative, but the capillaries (see Figure 5). Blood pressure decreases downstream. Hence the first change for the blood vessels to collapse is at the ends of the capillaries. (Reproduced with permission from reference 82.)

collapse when the alveolar gas pressure exceeds the capillary blood pressure, such as at heights where the blood pressure is less than the airway gas pressure. For the blood flow analysis it is very important to know the exact sites where such collapse occurs. With the knowledge that the pulmonary veins do not collapse, our search for the collapsing sites is narrowed down to finding out where in the capillaries the

collapse will occur. The answer is pretty straightforward: due to viscous dissipation, blood pressure decreases downstream. Therefore the exit section where the capillary enters a venule is the place of the lowest blood pressure in the capillaries. That must be the site of first collapse in zone 2 condition.[38,39,41,43,48,50,91]

The sluicing of blood at a site of collapse of blood vessels leads to the "waterfall phenomenon" mentioned earlier. Observations and explanations of this phenomenon have been made by Permutt, Bromberger, Barnes, Bane, Riley, Zieler, and others.[65] The concept of waterfall is brilliant. A number of authors attributed the cause of waterfall to the action of vascular smooth muscle. Holding onto the idea that capillaries are rigid, they searched the sites of these waterfalls in arteries and veins. I considered the smooth muscle action as an extra hypothesis and proceeded to find out if it is possible to explain this phenomenon without the extra hypothesis. I recognize as an experimental fact that the capillaries will collapse under negative transmural pressure, whereas the pulmonary veins will not. A theoretical investigation on the stability of the capillaries[50] and the flow through the sluicing gates[48] led to the following conclusions for blood vessels in zone 2:

1. There is no waterfall in capillaries in the arterial "islands" of Figure 2.
2. All possible sites of waterfalls are located in the white "venous ocean" area of Figure 2.
3. A pulmonary capillary connected to a venule can be in one of the following three forms:
   a. completely collapsed (no flow),
   b. completely open: blood pressure in capillary remains larger than the alveolar gas pressure; there is no sluicing; while a rapid local pressure drop occurs after blood gets into a venule, or
   c. has a waterfall or a sluicing gate, and the pressure-flow relationship obeys the sluicing flow rule.
4. The number of capillaries in each of the three categories, a, b, c, named above depends on the size of the blood flow and the preconditioning (subjecting the lung to cyclic large flow) of the lung.

With these issues pinned down, a mathematical theory of zone 2 flow is then possible, leading to several simple, definitive formulas for the pressure-flow relationship of pulmonary blood flow.[29,30,48,86]

Boundary-Value Problems

Once the geometry and material properties are known[14–16,29,31,82–85,93,94,98] we formulated many large and small problems by specifying various kinds of boundary conditions. We solved these problems and compared the theoretical results with experiments. The basic laws are the conservation of mass, momentum, and energy. Ad hoc hypotheses are avoided in deriving the basic equations.

The freedom of formulating small problems and testing the results along the way is the advantage of the rational approach we adopted. For example, our ultimate goal is to validate a theory leading to a computing program that can explain or test clinical conditions of patients. The system is the patient, the boundary conditions are the environment. This problem is too big to tackle all at once. Rushing ahead without proper preparation would not be productive. A more modest problem is determining the relationship between the pressures (arterial, venous, airway, pleural) and total flow in an isolated lung with nerves cut and environmental conditions maintained constant. The specified pressure and flows are the boundary values of the isolated lung. But the system is still very big. To further reduce the scope of the problem, we might ask: "When we keep on reducing the static pressure of the blood relative to the airway pressure, which blood vessel in the lung will be the first to collapse?" The system would be even smaller if the question were "For a pulmonary vein of order n in an isolated lung subjected to specified values of airway and pleural pressures, what is the critical value of the static pressure of the blood that will cause the collapse of that vein?" The answer to the last question is dependent on the length of the vein, the conditions at the ends (details of bifurcation junction), and the interaction of the vein with the lung parenchyma attached to its outer wall. Further simplification can be obtained if one asked the same question of collapse of the capillary blood vessels in an interalveolar septum, because in that case no other tissue tethers the capillary sheet.

A big problem can often be broken down to a number of smaller problems. The lung is so big and complex that an infinite number of small problems can be formulated. I always turn a dozen or two such problems over in my mind, and every day I select one as my first priority. I train my students to do the same. I teach undergraduate courses in mechanics and biomechanics and ask students to formulate problems and solve them. Some students hate it, some love it. Many prefer to follow examples in the textbook to be assured of better grades. I try to get them to experience the pleasure of formulating problems for research or application.

Some little problems lead to nice little experiments. Some lead to nice little mathematics. My usual experience with theoretical problems is that at first the problems appears terribly difficult to solve. After tackling it for some time, I suddenly understand, and it becomes very simple, often too trivial to tell anybody. But I got lots of pleasure out of the experience anyway.

As I said at the outset, our objective is to obtain a validated theory that will enable us to compute the pressure, flow, stress, and strain in the whole lung, or anywhere in the lung, under specified boundary conditions. Our mathematical theory can be summarized quite succinctly, as I have done in reference 30, or, in greater detail as in references 23 and 29.

A brief summary may suffer a loss of clarity or precision. Hence I refer the readers to references 23, 29, and 30.

Dynamic Remodeling of Pulmonary Blood Vessels

The stress-strain relationship of blood vessels mentioned above is an instantaneous relationship. Living tissues respond to stresses acting in them not only by an instantaneous deformation, but also by a slower process of remodeling their structures. I would like to show a picture of how fast the pulmonary blood vessels remodel themselves when the blood pressure is changed.

Liu and I[94,95] induced pulmonary hypertension in the rat by hypoxia. We used a commercial hypoxic chamber with a noncirculating gas mixture of 10% O_2, 90% N_2 at atmospheric pressure, 20°C, and 50% humidity. When a rat is placed in the chamber, its pulmonary blood pressure increases within minutes like a step function of time. Figure 7 shows a series of photographs of histological slides from four regions of the pulmonary artery of a normal rat and five hypertensive rats of the same initial age and body weight when put into the hypoxic chamber and kept there for different periods of hypoxia. The specimens were dissected and excised with the aid of a stereomicroscope, transferred into an aerated Krebs solution at room temperature (20 °C), fixed with 2.5% glutaraldehyde in phosphate buffer with pH 7.4 and osmotic pressure 310 mOsmol for 12 hours, postfixed with 2% Os O_4 for 2 hours, washed with distilled water, dehydrated, embedded in Medcast resin, cut into 1-μm sections, and stained with Toluidine Blue O. Marked swelling and thickening of intima were seen in the early period of exposure to hypoxia. Blebs appeared in intima in 2 hours and disappeared after 48 hours of exposure to hyp-

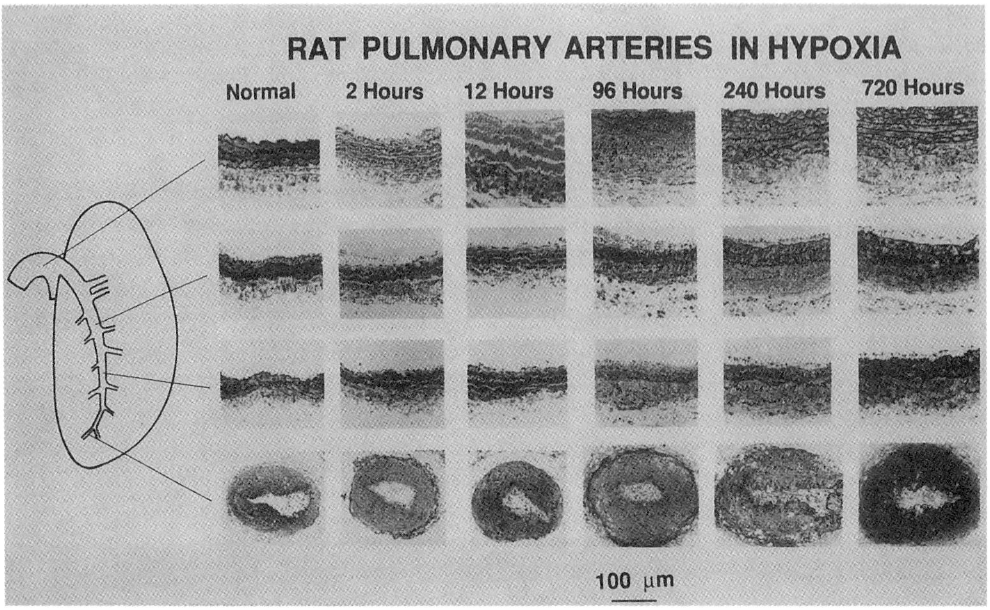

Figure 7. Photographs of histological slides from four regions of the main pulmonary artery of a normal rat and hypertensive rats with different periods of hypoxia. (Reproduced with permission from reference 34. Copyright 1991 American Physiological Society.)

oxia. The thickness of the medial layer increased slightly after 12 hours of exposure to hypoxia, more rapidly with a variable rate up to 240 hours, and then remained relatively stable in most regions of the vessel. The adventitia was thinner than the media in the normal group, changed little in the first 48 hours of exposure to hypoxia, but increased its thickness from 48 to 240 hours. It exceeded intimal-medial thickness by 96 hours. From 240 to 720 hours, the course of change of the thickness of both the endothelial-medial and adventitial layers ran parallel to each other. Thus the different layers of the vessel wall remodel at different rates.

Hypertension causes circumferential tensile stresses of variable magnitude in the blood vessel wall. Hence, one factor that can be correlated with the remodeling is the increased tensile stress. Other factors may be more directly responsible for tissue growth and remodeling and may be changed because of stress changes, but our concern is with the stress.

To study the stress-growth relationship, we made use of another new observation. This is the finding that the zero-stress state of a blood vessel is not a tube but consists of segments of open-sectors whose opening angles vary along the length of the vessel. By cutting a vessel into short segments of rings and cutting the rings radially, they will open up into sectors. For the pulmonary artery, the opening angle of segments near the right ventricle can be as large as 360° or more. The variation of the opening angle along the pulmonary artery of a normal rat is shown in Figure 8. We can show that the opening angle correlates linearly with the thickness to radius ratio of the vessel wall if the vessel is straight. The opening angle is larger if the vessel is curved. When hypoxic hypertension is produced in the

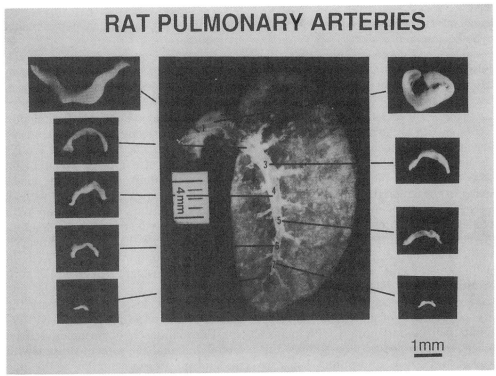

Figure 8. Photographs of the main pulmonary artery of a rat, and the zero-stress configuration of the artery at the locations indicated, with the endothelium facing downward. (Reproduced with permission from reference 34. Copyright 1991 American Physiological Society.)

rat, the opening angle at any given location changes with the length of time the animal suffered hypertension. Generally, at first the opening angle increases, then it decreases to values smaller than those of the control. This reflects the process of remodeling in the different layers of the blood vessel.[24,29,33,34,92–97]

Similar opening angle measurements and histological observations have been

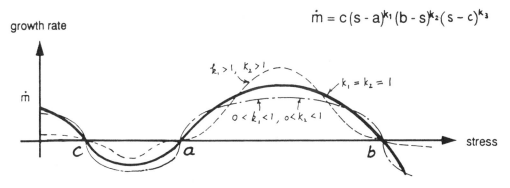

Figure 9. The author's proposed stress-growth law. (Reproduced with permission from reference 29. Copyright 1991 Springer-Verlag.)

made on rat aorta in hypertension and hypotension, and diabetes.[33,95] These experiences lead me to speculate about the existence of a stress-growth law. I would like to present a possible form of such a law.[29] Referring to Figure 9, let the solid curve represent a relationship between the rate of growth of the mass of a material in the blood vessel (such as the smooth muscle, endothelial cells, collagen, elastin, or other substances) and the stress or strain acting in the material. The symbol $\dot{m}$ represents the rate of material growth. The symbol s may represent a component of the stress or strain tensor, or a stress or strain invariant. Let a represent a homeostatic value of s, at which the tissue can maintain a steady state. $\dot{m}$ is positive when s exceeds a. $\dot{m}$ is negative when s is less than a. The rate of growth $\dot{m}$, however, cannot increase indefinitely with increasing s. In orthopedics it is known that excessive s causes resorption. Hence we assume that another homeostatic stress state b exists beyond which resorption occurs. Similarly, the negative rate of $\dot{m}$ when $s < a$ cannot be unbounded. Suppose that resorption stops when $s = c$, where $c < a$; then c is another homeostatic stress or strain. If the rate of growth $\dot{m}$ is a continuous function of s, and this function has zeros at a, b, c, and if the trend of change of $\dot{m}$ at a, b, c, is as discussed above, I propose the following simple relation.[29]

$$\dot{m} = C\,(s\text{-}a)^{k_1}\,(b\text{-}s)^{k_2}\,(s\text{-}c)^{k_3} \qquad (1)$$

in which a, b, c, C, k_1, k_2, k_3 are positive constants. This formula implies $\dot{m}$ to be positive when $s = 0$, as in cell culture in petrie dishes. The exponents k_1, k_2, k_3 determine how fast the growth rate changes when s deviates from the homeostatic state. If $k_1 > 1$, the slope of the growth curve is zero at $s = a$, then small deviation has little influence. The slope of the growth curve is infinite at a if $k_1 < 1$. $k_1 = 1$ signals a finite slope. I am using this theoretical proposal as an experimental hypothesis.

In biomechanics of the bone, a relation between stress and growth is known as Wolff's law, which is over 100 years old.[29] Kummer[100] has posed Wolff's law in a manner similar to Eq. 1. Any formulation of growth-stress law for internal organs should refer to the literature on Wolff's law. There is also a huge literature on physical education, sports, sports medicine, and the science and art of surgery and rehabilitation, all concerned with stress and growth. thus a large stock of knowledge exists that has not, however, been distilled into mathematical form. In the new "tissue engineering" on the horizon, which aims to create tissue substitute with living cells to be used in surgery, the growth-stress law will play a decisive role.

The relationship of the topics of tissue remodeling and the growth-stress law to the study of pressure and flow in the lung is not far-fetched. Pressure and flow depend on the geometry and mechanical properties of the blood vessels, which are controlled by remodeling.

Conclusion

I am a physiologist who entered the door by self-study, without formal training, because I found the subject more interesting than others and the people in the field nice. I made a plan to take a rational approach to studying pulmonary circulation, a plan that begins with morphometry, centers on the determination of the mechanical properties of materials involved, then follows a process of formulating boundary-value problems based on well accepted basic principles, solving the problems and testing the results experimentally in order to validate the analysis and gain confidence in the results, until it becomes possible to put all the blocks together to obtain the pressure-flow relationship of the whole organ. Then we compare the theoretical predictions with experimental results of the whole organ to validate the theory and seek to improve it. Once satisfactory agreement is obtained, then we can go to the next level of understanding and practical applications. Any success I have had I owe to my collaborators, Sid Sobin, Mike Yen, Shu Qian Liu, Herta Tremer, Feng Yuan Zhuang, and many others. In this chapter I present a few figures that represent some of the more interesting things picked up along the way. The explanations presented here are abridged and incomplete. For a more complete explanation, please see my books.[23,29,31] I would hope that I and my collaborators are remembered by the road we took and the vistas we opened. The odds and ends we picked up from the roadside are treasures for us. They gave us pleasure and enthusiasm; they illustrate the beauty of the road and the fertility of the land. But they are incidental. Only the road can give us direction.

References

1. Baez, S., H. Lamport, and A. Baez. Pressure effects in living microscopic vessels. In: *Flow Properties of Blood and Other Biological Systems,* edited by Copley and Stainsky. London: Pergamon Press, 1960, p. 122–136.
2. Bouskela, E., and Wiederhielm, C. A. Microvascular myogenic reaction in the wing of the intact unanesthetized bat. *Am J. Physiol* 237: H59–H65, 1979.
3. Budiansky, B., and E. Kimmel. Elastic moduli of lungs. *J. Appl. Mech.* 109: 351–358, 1987.
4. Chuong, C. J., and Y. C. Fung. Residual stress in arteries. In: *Frontiers in Biomechanics,* (edited by G. W. Schmid-Schobein, S. L.-Y. Woo, and B. W. Zweifach. New York: Springer, 1986, p. 117–129.
5. Ciurea, D., and J. Gil. Morphometric study of human alveolar ducts based on serial sections. *J. Appl. Physiol.* 67: 2512–2521, 1989.
6. Cumming, G., R. Henderson, K. Horsfield, and S. S. Singhal. The functional morphology of the pulmonary circulation. In: *The Pulmonary Circulation and Interstitial Space.* edited by A. P. Fishman and H. H. Hecht. 1969, p. 327–340.
7. Fung, Y. C. *Gliding and Soaring in Clouds.* Chongking: Chinese Gliding Society, [In Chinese.] 1944.
8. Fung, Y. C. *An Introduction to the Theory of Aeroelasticity.* New York: John Wiley and Sons, 1955. Revised, Dover Publications, New York, 1969, and 1993.
9. Fung, Y. C. *Foundations of Solid Mechanics.* Englewood Cliffs, NJ: Prentice-Hall, 1965.
10. Fung, Y. C. Microscopic blood vessels in the mesentery. In: *Biomechanics,* edited by Y. C. Fung, New York: American Society of Mechanical Engineers, 25: 1966, p. 151–166.
11. Fung, Y. C. Theoretical considerations of the elasticity of red cells and small blood vessels. *Fed Proc* 25: 1761–1772, 1966.
12. Fung, Y. C. Blood flow in the capillary bed. *J. Biomech.* 2: 353–373, 1969.
13. Fung, Y. C. *A First Course in Continuum Mechanics.* 3rd ed. Englewood Cliffs, NJ: Prentice-Hall, 1993.
14. Fung, Y. C.: Stress-strain-history relations of soft tissues in simple elongation. In: *Biomechanics: Its Foundations and Objectives,* edited by Y. C. Fung, Englewood Cliffs, NJ: Prentice-Hall, 1971, Chap. 7, 181–208.
15. Fung Y. C. Theoretical pulmonary microvascular impedance. *Ann. Biomed. Eng.* 1: 221–245. 1972.
16. Fung, Y. C.: Biorheology of soft tissues. *Biorheology* 10: 139–155, 1973.
17. Fung, Y. C. Stochastic flow in capillary blood vessels. *Microvasc. Res.* 5: 34–49, 1973.
18. Fung, Y. C. Fluid in the interstitial space of the pulmonary alveolar sheet. *Microvasc. Res.* 7: 89–113, 1974.
19. Fung, Y. C. A theory of elasticity of the lung. *J. Appl Mech.* 41E: 8–14, March 1974.
20. Fung, Y. C. Does the surface tension make the lung inherently unstable? *Circ. Res.* 37: 497–502, 1975.
21. Fung, Y. C. 1975 Eugene M. Landis Award Lecture: Microcirculation as seen by a red cell. *Microvasc. Res.* 10: 246–264, 1975.
22. Fung, Y. C. Stress, deformation, and atelectasis of the lung. *Circ. Res.* 37: 481–496, 1975.

23. Fung, Y. C. *Biodynamics: Circulation.* New York: Springer, 1984.

24. Fung Y. C. What principle governs the stress distribution in living organs? In: *Biomechanics in China, Japan, and USA: Proceedings of an International Conference held in Wuhan, China, in May, 1983.* (edited by Y. C. Fung, E. Fukada, J. J. Wang), Beijing: Science, 1984, p. 1–13.

25. Fung, Y. C. Microrheology and constitutive equation of soft tissue. *Biorheology* 25: 261–270, 1988.

26. Fung, Y. C. A model of the lung structure and its validation. *J. Appl. Physiol.* 64: 2132–2141, 1988.

27. Fung, Y. C. Connecting incremental shear modulus and Poisson's ratio of lung tissue with morphology and rheology of microstructure. *Biorheology* 26: 279–289, 1989.

28. Fung, Y. C. Connection of micro- and macromechanics of the lung. In: *Microvascular Mechanics: Hemodynamics of Systemic and Pulmonary Microcirculation,* (edited by J. S. Lee and T. C. Skalak). New York: Springer, 1989, Chap. 13, p. 191–217.

29. Fung, Y. C. *Biomechanics: Motion, Flow, Stress, and Growth.* New York: Springer, 1990.

30. Fung, Y. C. Dynamics of blood flow and pressure-flow relationship. In: *The Lung: Scientific Foundations,* edited by R. G. Crystal and J. B. West. New York: Raven, 1991, Chap. 5.2.2.,p. 1121–1134.

31. Fung, Y. C. *Biomechanics: Mechanical Properties of Living Tissues.* New York, Springer Verlag, 1st ed. 1981 2nd ed. 1993.

32. Fung, Y. C., K. Fronek, and P. Patitucci. Pseudoelasticity of arteries and the choice of its mathematical expression. *Am. J. Physiol.* 237: H620–H631, 1979.

33. Fung, Y. C. and S. Q. Liu, Change of residual strains in arteries due to hypertrophy caused by aortic constriction. *Cir. Res.* 65: 1340–1349, 1989.

34. Fung, Y. C. and S. Q. Liu, Changes of zero-stress state of rat pulmonary arteries in hypoxic hypotension. *J. Appl. Physiol.* 70: 2455–2470, 1991.

35. Fung, Y. C., and S. S. Sobin. A sheet-flow concept of the pulmonary alveolar microcirculation: mophometry and theoretical results (Abstract). In: *Proceedings of the Conference on Annual Engineering in Medicine and Biology,* Vol. 9. Boston, MA, 1967, p. 97.

36. Fung, Y. C., and S. S. Sobin. Sheet-flow concept of pulmonary alveolar microcirculation: theory (Abstract). In: *Federation of the American Society for Experimental Biology,* 1968, p. 578.

37. Fung, Y. C., and S. S. Sobin. Theory of sheet flow in the lung alveoli. *J. Appl. Physiol.* 26: 472–488, 1969.

38. Fung, Y.C., and S. S. Sobin. Elasticity of the pulmonary alveolar sheet. *Circ. Res.* 30: 451–469, 1972.

39. Fung, Y. C., and S. S. Sobin. Pulmonary alveolar blood flow. *Circ. Res.,* 30: 470–490, 1972. Including an appendix by Y. C. Fung, R. T. Yen, and E. Mead, Model experiment on the stability of a collapsed Starling resistor in flow at low Reynolds number, p. 487–490.

40. Fung, Y. C., and S. S. Sobin. Mechanics of pulmonary circulation. In: *Cardiovascular Flow Dynamics and Measurements,* edited by N. H. C. Hwang and N. A. Norman, Baltimore: University Park Press, 1977, Chap. 17, p. 665–730.

41. Fung, Y. C., and S. S. Sobin. Pulmonary alveolar blood flow. In: *Bioengineering Aspects of Lung Biology,* edited by J. B. West, New York: Marcel Dekker, 1977, Chap. 4, p. 267–358.

42. Fung, Y. C., and S. S. Sobin. The retained elasticity of elastin under fixation agents. *J. Biomech. Eng.* 103: 121–122, 1981.

43. Fung, Y. C., S. S. Sobin, H. Tremer, R. T. Yen, and H. H. Ho. Patency and compliance of pulmonary veins when airway pressure exceeds blood pressure. *J. Appl. Physiol. 54 (Respirat. Environ. Exer. Physiol).* 1538–1549, 1983.

44. Fung, Y. C., and H. T. Tang, Solute distribution in the flow in a channel bounded by porous layers (a model of the lung). *J. Appl. Mech.* 41: 531–535 (*Trans. ASME,* Vol. 97, Ser. E), 1975.

45. Fung, Y. C., H. T. Tang. Longitudinal dispersion of tracer particles in the blood flowing in a pulmonary alveolar sheet. *J. Appl. Mech.* 42: 536–540 (*Tans. ASME,* Vol. 97, Ser. E), 1975.

46. Fung, Y. C., and P. Tong. Theory of the sphering of red blood cells. *Biophys. J.* 8: 175–198, 1968.

47. Fung, Y. C., P. Tong, and P. Patitucci. Stress and strain in the lung. American Society of Civil Engineers. *J. Eng. Mech. Div.* 104(EM1): 201–223, 1978.

48. Fung, Y. C., and R. T. Yen. A new theory of pulomonary blood flow in zone 2 condition. *J. Appl. Physiol.* 60: 1638–1650, 1986.

49. Fung, Y. C., R. T. Yen, Z. L. Tao, and S. O. Liu. A hypothesis on the mechanism of

trauma of lung tissue subjected to impact load. *J. Biomech. Eng.* 110: 50–56, 1988.

50. Fung, Y. C., and F. Y. Zhuang. An analysis of the sluicing gate in pulmonary blood flow. *J. Biomech. Eng.* 108: 175–182, 1986.

51. Fung, Y. C., B. W. Zweifach, and M. Intaglietta. Elastic environment of the capillary bed. *Circ. Res.* 19: 441–461, 1966.

52. Fung, Y. C., and S. Q. Liu. Strain distribution in small blood vessels with zero-stress state taken into consideration. *Am. J. Physiol., Heart and Circ.* 262: H544–H552, 1992.

53. Gasson, P. C. *Geometry of Spatial Forms.* New York: Wiley, 1983.

54. Gunteroth, W. G., D. L. Luchtel, and I Kawabari. Functional implications of the pulmonary microcirculation. *Chest,* 101: 1131–1134, 1992.

55. Hansen, J. E., and E. P. Ampaya. Human air space shapes, sizes, areas, and volumes. *J. Appl. Physiol.* 38: 990–995, 1975.

56. Hansen, J. E., E. P. Ampaya, G. H. Bryant, and J. J. Navin. The branching pattern of airways and air spaces of a single human terminal bronchiole. *J. Appl. Physiol.* 38: 983–989, 1975.

57. Horsfield, K., and G. Cumming. Morphology of the bronchial tree in man. *J. Appl. Physiol.* 24: 373–383, 1968.

58. Lee, J. S., W. G. Frasher, and Y. C. Fung. Comparison of the elasticity of an artery in vivo and in excision. *J. Appl. Physiol.* 25: 799–801, 1968.

59. Lee, J. S., and Y. C. Fung. Modeling experiments of a single red blood cell, moving in a capillary blood vessel. *Microvasc. Res.* 1: 221–243, 1969.

60. Lee, J. S., and Y. C. Fung, Stokes' flow around a circular cylindrical post, confined between two parallel plates. *J. Fluid Mech.* 37: 657–670, 1969.

61. Matsuda, M., Y. C. Fung, and S. S. Sobin. Collagen and elastin fibers in human pulmonary alveolar mouths and ducts. *J. Appl. Physiol.* 63: 1185–1194, 1987.

62. Mercer, R. R., and J. D. Crapo. Three-dimensional reconstruction of the rat acinus. *J. Appl. Physiol.* 63: 785–794, 1987.

63. Miller, W. S. *The Lung.* 2nd ed. Springfield, Il: Charles C. Thomas, 1947.

64. Oldmixon, E. H., J. P Butler, and F. G. Hoppin, Jr. Dihedral angles between alveolar septa. *J. Appl. Physiol.* 64: 299–307, 1988.

65. Permutt, S., B. Bromberger-Barnea, and H. N. Bane. Alveolar pressure, pulmonary venous pressure, and the vascular waterfall. *Med. Thorac.* 19: 239–260, 1962.

66. Rosenquist, T. H., S. Bernick, S. S. Sobin, Y. C. Fung. The structure of the pulmonary interalveolar microvascular sheet. *Microvasc. Res.* 5: 199–212, 1973.

67. Silage, D. A., and J. Gil. Morphometric measurement of local curvature of the alveolar ducts in lung mechanics. *J. Appl. Physiol.* 65: 1592–1597, 1988.

68. Sobin, S. S., and Y. C. Fung. Response to Guntheroth et al.'s challenge to the Sobin-Fung approach to the study of pulmonary microcirculation. *Chest,* 101: 1135–1143, 1992.

69. Sobin, S. S., Y. C. Fung, R. G. Lindal, H. M. Tremer, and L. Cleark. Topology of pulmonary arterioles, capillaries, and venules in the cat. *Microvasc. Res.* 19: 217–233, 1980.

70. Sobin, S. S., Y. C. Fung, and H. M. Tremer. The effect of incomplete fixation of elastin on the appearance of pulmonary alveoli. *J. Biomech. Eng.* 104: 68–71, 1982.

71. Sobin, S. S., Y. C. Fung, and H. M. Tremer. Collagen and elastin fibers in human pulmonary alveolar walls. *J. Appl. Physiol.* 64: 1659–1675, 1988.

72. Sobin, S. S., Y. C. Fung, H. M. Tremer, and T. H. Rosenquist. Elasticity of the pulmonary alveolar microvascular sheet in the cat. *Circ. Res.* 30: 440–450, 1972.

73. Sobin, S. S., R. G. Lindal, Y. C. Fung, H. M. Tremer. Elasticity of the smallest noncapillary pulmonary blood vessels in the cat. *Microvasc. Res.* 15: 57–68, 1978.

74. Sobin, S. S., H. M. Tremer, and Y. C. Fung. Morphometic basis of the sheet-flow concept of the pulmonary alveolar microcirculation in the cat. *Cir. Res.* 26: 397–414, 1970.

75. Tanaka, T. T. and Y. C. Fung, Elastic and inelastic properties of the canine aorta and their variation along the aortic tree. *J. Biomech.* 7: 357–370, 1974.

76. Tang, H. T., and Y. C. Fung. Fluid movement in a channel with permeable walls covered by porous media: a model of lung alveolar sheet. *J. Appl. Mech.* 42: 45–50 (*Trans. ASME,* Vol. 97, Ser. E), 1975.

77. Vawter, D. L., Y. C. Fung, and J. B. West. Elasticity of excised dog lung parenchyma. *J. Appl. Physiol.* 45: 261–269, 1978.

78. Wall, R. J., S. S. Sobin, M. Karspeck, R. G. Lindal, H. M. Tremer, and Y. C. Fung. Computer derived image compositing. *J. Appl. Physiol. (Respirat. Environ. Exer. Physiol.)* 51(1): 84–89, 1981.

79. Weibel, E. R. *Morphometry of the Human Lung.* New York: Academic, 1963.

364 • *THE PULMONARY CIRCULATION AND GAS EXCHANGE*

80. Weyl, H. *Symmetry.* Princeton, NJ: Princeton University Press, 1952.
81. Wilson, T. A., and H. Bachofen. A model for mechanical structure of the alveolar duct. *J. Appl. Physiol.* 52: 1064–1070, 1982.
82. Yen, R. T. and L. Foppiano. Elasticity of small pulmonary veins in the cat. *J. Biomech. Eng.* 103: 38–42, 1981.
83. Yen, R. T., and Y. C. Fung. Model experiments on apparent blood viscosity and hematocrit in pulmonary alveoli. *J. Appl. Physiol.* 35: 510–517, 1973.
84. Yen, R. T., and Y. C. Fung. Effect of velocity distribution on red cell distribution in capillary blood vessels. *Am. J. Physiol.* 235: H251–H257, 1978.
85. Yen, R. T., Y. C. Fung, and N. Bingham. Elasticity of small pulmonary arteries in the cat. *J. Biomech. Eng.* 102: 170–177, 1980.
86. Yen, R. T., Y. C. Fung, H. H. Ho, and G. Butterman. Speed of stress wave propagation in lung. *J. Appl. Physiol.* 61: 701–705, 1986.
87. Yen, R. T., Y. C. Fung, and S. O. Liu. Trauma of lung due to impact load. *J. Biomech.* 21: 745–753, 1988.
88. Yen, R. T., F. Y. Zhuang, Y. C. Fung, H. H. Ho, H. Tremer, and S. S. Sobin. Morphometry of cat's pulmonary arterial tree. *J. Biomech. Eng.* 106: 131–136, 1984.
89. Yen, R. T., F. Y. Zhuang, Y. C. Fung, H. H. Ho, H. Tremer, and S. S. Sobin. Morphometry of cat pulmonary venous tree. *J. Appl. Physiol.* 55: *Respirat. Environ. Exer. Physiol.* 236–242, 1983.
90. Zeng, Y. J., D. Yager, and Y. C. Fung. Measurement of the mechanical properties of the human lung tissue. *J. Biomech. Eng.* 109: 169–174, 1987.
91. Zhuang, F. Y., Y. C. Fung, and R. T. Yen. Analysis of blood flow in cat's lung with detailed anatomical and elasticity data. *J. Appl. Physiol.* 55: *(Respirat. Environ. Exer. Physiol.)* 1341–1348, 1983.
92. Zhuang, F. Y., M. R. T. Yen, Y. C. Fung, and S. S. Sobin. How many pulmonary alveoli are supplied by a single arteriole and drained by a single venule? *Microvasc. Res.* 29: 18–31, 1985.
93. Debes, J. C., and Fung, Y. C. Effect of temperature on the biaxial mechanics of excised parenchyma of the dog. *J. Appl. Physiol.* 73: 1171–1180, 1992.
94. Fung, Y. C., and Liu, S. Q. Elementary mechanics of the endothelium of blood vessels. *J. Biomechanical Engineering.* 115: 1–12, 1993.
95. Liu, S. Q., and Fung, Y. C. Influence of STZ-induced diabetes on zero-stress states of rat pulmonary and systemic arteries. *Diabetes.* 41: 136–146, 1992.
96. Liu, S. Q., and Fung, Y. C. Changes in the rheological properties of blood vessel tissue remodeling in the course of development of diabetes. *Biorheology.* 29: 443–457, 1992.
97. Liu, S. Q., and Fung, Y. C. Changes in the structure and mechanical properties of pulmonary arteries of rats exposed to cigarette smoke. *Am. Rev. Respir. Dis.* 148: 768–777, 1993.
98. Yager, D., Feldman, H., and Fung, Y. C. Microscopic vs. macroscopic deformation of the pulmonary alveolar duct. *J. Appl. Physiol.* 72(4): 1348–1354, 1992.
99. Yen, R. T., and Sobin S. S. Pulmonary blood flow in the cat: correlation between theory and experiment. In *Frontiers in Biomechanics* (ed. by G. W. Schmid-Schoenbein, S. L.-Y. Woo, and B. W. Zweifach, Springer-Verlag, New York, pp. 365–376, 1986.
100. Kummer, B. K. F. Biomechanics of bone: Mechanical properties, functional structure, functional adaptation. In *Biomechanics: Its Foundations and Objectives,* (ed. by Y. C. Fung, N. Perrone, M. Anliker), Englewood Cliffs, NJ, Prentice-Hall, pp. 237–269, 1972.

CHAPTER 3

THE IMPLANTABLE GLUCOSE SENSOR: AN EXAMPLE OF BIOENGINEERING DESIGN

DAVID A. GOUGH

Department of Bioengineering, University of California, San Diego

1. Introduction

The development of an implantable glucose sensor for use in diabetes has long been an important bioengineering objective. In this chapter, I explain how such a sensor would be used and review the work of my student collaborators and myself over the past decade. Our efforts have led to the development of a prototype glucose sensor and telemetry unit that has been implanted in animals and currently holds the record for long-term operation. A miniaturized version of the sensor system is being prepared for use in humans. The developmental process, from the conception of the sensor to its eventual application in humans, is an example of the application of design principles in bioengineering.

1.1. *What is the problem in diabetes?*

The pancreas is a finger-sized organ that lies just below the stomach. The pancreas has two major functions: 98% of its mass is composed of *acinar cells* that produce enzymes which are secreted into the intestinal tract to aid in digestion, and the remaining 2% of its mass is organized in small, dispersed cell clumps called *islets* that contain pancreatic *beta cells* which synthesize and secrete the hormone *insulin*. Insulin is often described as the master hormone of the body because of its many roles. Most importantly, insulin regulates the metabolic breakdown of the sugar *glucose*, the main fuel of the body, and its conversion into the energy needed by all cells to carry out their respective functions (e.g. thinking, muscle contraction, growth, etc.). The beta cells detect blood glucose concentration and secrete just the right amount of insulin, which allows glucose to enter muscle cells and the metabolic production of energy to proceed.

In the most severe type of diabetes, the beta cells of the pancreas are selectively obliterated by a misdirected *autoimmune reaction*. Although the causes of this reaction are not clear, the result is that the pancreas eventually looses the ability to synthesize and secrete insulin. Without insulin, glucose cannot readily enter most cells and builds up in the bloodstream, leading eventually to a paradoxical state of cellular starvation in the face of an overabundance of glucose in the blood. If this situation, known as *hyperglycemia*, is allowed to progress unchecked, it can lead to serious problems. Hyperglycemia can easily be reversed by an injection of insulin, but if too much is given, blood glucose can fall too rapidly leading to life-threatening *hypoglycemia*, or low blood glucose. A person with diabetes must constantly adjust his/her glucose consumption, activity levels and insulin injections to try and keep blood glucose within the normal range. This delicate process is carried out automatically by the healthy pancreas.

Although hypoglycemia is immediately dangerous and is thought to occur frequently in many people with diabetes, it is usually not a cause of death as it can be perceived by most people and rapidly reversed by glucose consumption. More serious are the long-term problems that result from hyperglycemia, including retina and nerve damage, kidney failure, large blood vessel disease and reduction of longevity. Diabetes is a leading cause of loss of vision, kidney failure and, amputation, and has a strong correlation with cardiovascular diseases. Diabetes is a major health problem: it affects 2 to 4% of the population and accounts for over 15% of the total national health care expenditures, in addition to an inestimable loss of human resources (1).

1.2.   *How could an implantable glucose sensor help?*

At present, a person with diabetes must determine his/her blood glucose concentration by collecting a drop of blood by "fingersticking" many times a day and measuring the glucose concentration with a home blood glucose detector. The amount and timing of insulin injections, eating, and activity schedules are then adjusted to maintain blood glucose as close to normal as possible (around 90 mg/dl). The problem is that blood glucose can rise or fall to dangerous levels within 15 minutes in the worst case, but most people are (understandably) unwilling to perform more than 2 to 4 fingersticks and 1 or 2 insulin injections per day to avoid these excursions. (You can imagine the unfortunate difficulties of a parent dealing with a two-year old diabetic!) It is not realistically possible to monitor blood glucose and adjust insulin administration frequently enough with present technology to achieve close blood glucose control that approaches normal (1).

A miniature glucose sensor that can be completely implanted in the body along with a small implanted radio telemetry system could continuously record blood glucose and display the value on an external receiver similar to a pager. The device could be programmed to warn of hypo- or hyperglycemia and indicate the actual minute-by-minute blood glucose value. This would allow users to adjust their insulin injections much more effectively if they wished to continue with that form of therapy, or the sensor could eventually be coupled to an implanted insulin pump to deliver the correct amount of insulin automatically, to make a *mechanical artificial beta cell* (2). This affords the possibility of achieving close control of blood glucose and avoiding the long-term complications of the disease. The implantable insulin pump has already been developed by others and is awaiting the advent of a reliable blood glucose sensor.

1.3.   *Alternative approaches*

There are other experimental approaches to treatment in which bioengineering design may be important. Functioning beta cells can be isolated from animals, grown in culture and entrapped in an implantable *membrane device* intended to provide isolation of the functioning beta cells from the body's immune system. The membrane has to be designed so that glucose and insulin are freely permeable, but antibodies and *cytokines* (molecules that regulate cell growth) cannot interfere with beta cell function. Pancreatic *transplants* are also being carried out on a limited basis, but the rejection process is still not easily managed and there is a substantial shortage of transplantable tissue. *Xenotransplantation*, or transplants from animal donors, may eventually address the tissue availability problem, but there is much to be done before cross-species transplanting becomes possible. Studies into diabetes *prevention* involve identifying parts of the genome related to diabetes and alerting carriers. This may lead to approaches for avoiding the disease and, eventually, to new strategies for intervention, but prevention is of little value for those who already have the disease. These and other exploratory approaches, while potentially exciting in the long-term, are unlikely to be available in the near future (3). There are plenty of challenges for bioengineers in these areas.

2.   The Sensor from Concept to Prototype

Of the several hundred physical principles that have been tried as the basis of a continuous glucose sensor, only three or four presently show much promise (4). The main problem is achieving selectivity for glucose over the many other molecules found in blood that have similar size, concentration and molecular properties.

2.1.   *The sensing prinicple*

The most promising methods have employed an *enzyme*, or biological catalyst, known as glucose oxidase, which catalyzes the following reaction:

$$\text{glucose} + O_2 + H_2O \rightarrow \text{gluconic acid} + H_2O_2\,.$$

In our sensor, the enzyme can be *immobilized*, or bound to a membrane, and coupled to an electrochemical sensor that detects oxygen consumption (5). Glucose and oxygen from the body diffuse into the membrane, the reaction occurs consuming some of the oxygen, and the remaining oxygen is detected by the oxygen sensor in the form of an electron flux or current. This is a glucose-modulated, oxygen-dependent current. A second oxygen sensor without the enzyme membrane is needed to determine the ambient or background oxygen concentration, producing an oxygen reference current. The two currents are then subtracted to give a glucose-dependent difference current, which under certain conditions, indicates glucose concentration. The enzyme is used because there are few other means of obtaining a potentially specific glucose measurement. The most promising alternative sensor approaches also involve the enzyme in some way.

2.2.   *Design challenges and performance requirements*

Although this approach was potentially feasible, there were a number of substantial challenges that had to be addressed (5). Originally, the biggest obstacle was *understanding* the sensing mechanism — the enzyme reaction, oxygen sensor and other aspects — in sufficient detail to permit engineering design. This required some basic scientific studies described below. Second, it was widely known that most enzymes are unstable at body temperatures and lose their catalytic activity over the period of hours to days. Reasons for this loss of activity would have to be determined and methods would have to be found for prolonging activity for months to years for use of the enzyme to be feasible. Third, the electrochemical oxygen sensors that were then available were widely known to be unstable and require frequent recalibration. The reasons for instability would have to be determined and improved sensors would have to be developed. Fourth, we discovered that there was an *oxygen deficit* in the body due to the relatively high glucose concentration (5.0 mM and usually higher) and low oxygen concentration (at most 0.2 mM). Unless this concentration disparity could somehow be reversed, the equimolar enzyme reaction would be limited by oxygen rather than glucose and impede glucose sensing. This previously unrecognized problem had confounded previous attempts by others to make a sensor by this approach. Fifth, the sensor would have to be *biocompatible*, or acceptable to the body as an implant. This would require minimizing or avoiding the universal response of the body to an implanted foreign object. Sixth, the sensor would have to respond to glucose rapidly enough that feedback control action by adjustment of insulin administration would be feasible.

Once ways were found to meet these challenges, several specific engineering design requirements could be addressed. The sensor would have to provide a specific response to glucose over a range of about 2 to 20 mM, independent of oxygen variations. The sensor must not need frequent recalibration, a process that is likely to require the aid of a professional. The implanted sensor would have to be capable of following a maximal 0.2 mM/min rise or fall in blood glucose concentration. The sensor, its associated implantable telemetry and battery should be small enough to be inserted with a minimal surgical procedure, and potentially expendable elements such as the enzyme may have to be replaced every 6–12 months with a simple nonsurgical procedure. The implant would have to be capable of transmitting glucose values to the external receiver every 1–2 minutes and the implanted battery should last for at least a year. The overall device must be acceptable for use in small children, who are most difficult to treat and who stand to gain the most from controlled blood glucose over a lifetime.

2.3.   *The engineering design approach*

The idea of developing an implantable glucose sensor was first mentioned in the late 1960's and many investigators have been involved since. There are over 10,000 papers and many books that mention some aspect of the implantable glucose sensor (4). However, the vast majority of these studies emphasize a "trial-and-error" approach in which sensors are built a given way, implanted in animals or humans, the response tested, deficiencies noted, sensor properties adjusted, and the cycle repeated. With a device as complex as an implantable sensor, this approach is very slow, labor intensive, costly, and inefficient.

The use of the classic engineering design approach has been a deciding factor in our ability to make more rapid progress (5). We emphasized understanding the physics of the sensing process in detail. This led to the development of *a priori* mathematical models that were solvable for the glucose signal under arbitrary conditions (e.g. 6). The parameters required for these models were determined independently by physical experiments. This allows the sensor design to be optimized by computation and modeling, with the need to actually build and test only the best candidate sensors. The sensor signal is predictable and improvements can be rapidly made. This approach is analogous to modern engineering methods for design of complex devices such as spacecraft, bridges, computers, automobiles, etc.

We also believe that sensors should be thoroughly understood and tested *in vitro*, that is, on the bench under simulated physiologic conditions, before *in vivo* testing as an implant can be justified. The biological environment is much more complicated and implant experiments are more costly and difficult to interpret. Experiments with animals are essential, but must be minimized and very carefully conceived. Experiments with animals must be thoroughly mastered before experiments with humans are justified.

Finally, a broad interdisciplinary approach is essential. The development has been based on principles of mathematical modeling, chemical engineering, materials engineering, electrical engineering, dynamic systems analysis, enzymology, organic chemistry, electrochemistry, physiology, tissue engineering, and manufacturing. Research groups that have taken narrow disciplinary approaches limited to only a few of these areas have been ineffective.

3.   Benchtop Studies (5, 6)

3.1.   *Membrane synthesis and characterization*

It was first necessary to synthesize enzyme-containing membranes and characterize their permeability to glucose and oxygen. Synthesis of the most basic membrane is simple, but measuring permeability can be complicated in this case because of *concentration boundary layer effects*, in which the stagnant layer of fluid adjacent to the membrane contributes to the mass transfer resistance and masks membrane permeability. This is a concern for small diffusable molecules such as oxygen and glucose and relatively permeable membranes. A novel rotated disc electrode system and electrochemical methods were therefore developed (obviously not for implantation!) for membrane characterization, with which boundary layers could be precisely controlled and accounted for in the membrane permeability measurements. Characterization was done with and without the enzyme reaction operating, allowing the effect of the reaction to be determined.

3.2.   *Immobilized enzyme stability*

A new technique was then devised for using the rotated disc electrode for nondestructive estimation of the kinetic constants and intrinsic catalytic activity of the immobilized enzyme. With this technique, the rate of enzyme inactivation could then be estimated. Repeated kinetic determinations were made on membranes stored at 37°C under quasi-physiologic conditions, giving the *in vitro* rate constant of inactivation. The main mechanism of inactivation was found to be consistent with peroxide complexation to certain forms of the enzyme. It was discovered that the lifetime of glucose oxidase can be extended substantially by coimmobilization another enzyme catalase, which breaks down peroxide. A second, slower form of inactivation occurred in the absence of peroxide. The rate constants of inactivation were used to simulate the change in signal resulting from enzyme inactivation during long-term operation. These studies confirmed that the signal can remain constant and insensitive to enzyme inactivation while enzyme activity remains high, but these conclusions needed to be verified *in vivo*.

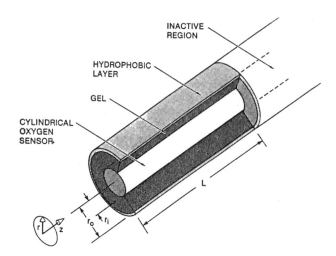

Fig. 1. The Two-Dimensional Glucose Sensor (6). Glucose and oxygen enter from the exposed end, but only oxygen can enter through the curved surface. The reference oxygen sensor without the enzyme can be similar, or it can be incorporated into the glucose sensor cylinder.

3.3. *Accounting for oxygen effects* (5)

All previous sensor designs required that both oxygen and glucose diffuse into the membrane from the same direction. Diffusion to and within the membrane has been mainly perpendicular to the plane of the membrane for both molecules. This posed a problem because it is difficult to synthesize membranes that eliminate the oxygen deficit by having high oxygen permeability and low glucose permeability, while still allowing excess enzyme loading to promote a long lifetime. A new design shown schematically in Fig. 1 is a small tube with the glucose sensor at one end. The core of the cylinder at the exposed end is the oxygen sensor, surrounded by a concentric layer of immobilized enzyme. On the outside is a tube of silicone rubber, which is impermeable to solvated molecules such as glucose but highly permeable to oxygen. Both glucose and oxygen can diffuse into the gel layer through the exposed annular end parallel to the axis of the oxygen sensor, but only oxygen has radial access to the gel through the silicone rubber covered surface. Thus, there is a two-dimensional supply of oxygen to the enzyme region (radial and axial) and only a one-dimensional supply of glucose (axial). *In vitro* experimental results have shown that glucose can be determined with this sensor over a clinically useful range, even for very low physiologic oxygen concentrations. The two-dimensional sensor design is a major advance: it solves the oxygen deficit problem in a simple way, a problem that had been considered by some to be intractable.

3.4. *The oxygen sensor*

An essential component of the sensor is a stable oxygen sensor. Previous electrochemical oxygen sensors used in medical applications required frequent recalibration and were never intended for continuous, long-term operation. A new type of oxygen sensor was developed based on adaptation of classical electrochemical principles. The signal of this sensor has been documented stable to within ±10% during continuous *in vitro* operation for periods of months without recalibration, and for periods of several years in unpublished studies. Stability of this oxygen sensor is more than adequate.

3.5. *Modeling studies*

We have developed quantitative models of the steady state and transient responses to glucose and oxygen, for both 1-D and 2-D sensors. These models were based on nonlinear ordinary and partial differential equations with variable coefficients, cast in classical dimensionless variables, and solved for the sensor signal in terms of sensor parameter values. We have also developed a description of the dynamic delay and maximal dynamic error present in all continuously operated sensors. We related the dynamic delay to our previous rotated disc methods for sensor characterization. The models have been useful to help understand the sensing mechanism and describe the response of specific sensors.

4. Implant Studies in Animals

At that point, there was little more that could be discovered from *in vitro* studies and *in vivo* experiments were needed.

4.1. *Telemetry and implantable instrumentation* (7)

To test the sensor in animals, it was necessary to make an implantable battery-operated telemetry-instrumentation system. The system, shown with sensors and a wire antenna in Fig. 2, continuously recorded the sensor currents and potentials, temperature, and other system parameters (12 in all), linked short segments of the signals into a 1-second long train, produced a radio-frequency oscillation proportional to each signal component, and transmitted the train in a burst every 1, 10 or 100 seconds, as set by an external magnetic switch. An external receiver decoded and displayed the information. This prototype system

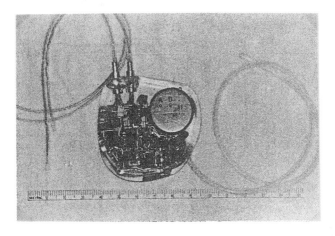

Fig. 2. The Implantable Sensor with Telemetry (7). The sensor leads project from the top and the antenna lead from the bottom, right.

was based on inexpensive commercial integrated circuits, a 3-month wristwatch battery, and epoxy encapsulation, but a much smaller, long-lived system can be developed later by industry. This system allowed recording of all parameters from the implanted sensor that could possibly be recorded of a nonimplanted sensor using external instrumentation, and was an important research tool.

4.2.   *Chronic central venous implantation in the dog* (8)

There are several possible sites in the body where the sensor could be implanted for testing, each with advantages and disadvantages. The first question is whether the sensor should be implanted in the tissues or exposed to blood. A sensor implanted in the bloodstream would report blood glucose directly and results could be validated by comparison to blood glucose assayed by an established method. However, implantation in contact with blood may pose a possibility of blood clotting. The sensor could alternatively be implanted in a tissue site, which would not cause clotting, but tissue remodeling around the sensor may occur, leading to delays in glucose transport from the local capillary network.

We chose to implant sensors in the superior vena cava, the large vein that brings blood from the upper body and has the lowest likelihood of incurring a clot. Sensors were implanted in six dogs for periods of 7 to 108 days. The sensors were introduced into the jugular vein of the anesthetized canine subject and advanced into the vena cava. The sensor tips were located in the center of the blood vessel, typically several centimeters above the entrance of the right atrium of the heart. The telemetry unit was implanted in a subcutaneous pocket which was formed nearby. An intravenous catheter attached to a subcutaneous access port was implanted simultaneously for sampling and infusion. No systemic anticoagulation was used.

Some results are shown in Fig. 3. The panels show the response to respectively one and two glucose excursions on days 87 and 108 after implantation. Blood glucose excursions

Long-term Central Venous Glucose Sensor
with Telemetry in Dogs

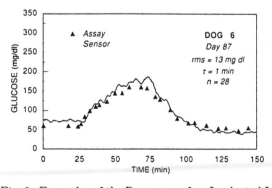 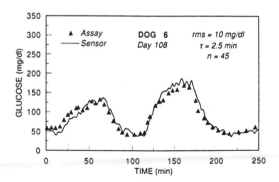

Fig. 3. Examples of the Responses of an Implanted Intravenous Sensor to Glucose Challenges (8). The solid line is the sensor signal and triangles are conventional venous blood glucose assays. (Note: 90 mg/dl glucose = 5.0 mM).

were produced by infusions of sterile glucose solutions through an intravenous catheter in a foreleg vein. There was no detectable electrochemical interference, enzyme inactivation, oxygen sensor instability, or biocompatibility problems, and, importantly, there was no need for recalibration during the entire implant period. (Note that the phase-correlated root mean square difference between the signal and blood glucose values on day 108 was 10 mg/dl.) At the time of this writing, this remains the published world record for long-term implanted glucose sensor operation. The conclusion is that the sensor is accurate and stable as a central venous implant in dogs for several months. When we began this journey, we and others saw many obstacles and did not expect to obtain favorable results.

5. From Prototype to the Patient

There are many important considerations in going from a promising prototype in animals to a safe and reliable device for use in humans. For example: What would the ideal configuration be? How would the electronics be integrated? How and where would the device be implanted? How would it be recalibrated, if necessary? Would there be replaceable components? What safety features should be included? How would the sensor be eventually integrated with the pump and controller? Some of these questions involve final design and manufacturing, and must take into account device reliability, materials compatibility, integration of components, extensive testing, and eventually, costs. At this stage, the plan must take into consideration the FDA (Food and Drug Administration) testing and approval process. Although the university is the best place to develop a prototype, the questions at this level must be addressed by a collaboration between university and industry.

6. Conclusions

How long will this take? It is difficult to say exactly, but with the right teamwork approach and resources, it may be possible within a few years. Bioengineers using the powerful tools of engineering design will continue to play a key role in this effort.

Acknowledgments

A large number of students — postdocs, graduates, undergraduates and high school students — who I have worked with over the years deserve tremendous credit. This project has also benefited substantially from the intellectual environment at UCSD. The work was supported by grants from the National Institutes of Health, the Juvenile Diabetes Foundation International, and the American Diabetes Association.

References

1. Diabetes in America, 2nd edition, National Diabetes Data Group, NIH Pub. No. 95–1468 (1995).
2. Albisser, A. M. S., Leibel, T. G., Ewart, Z., Davidovac, C. K., Botz and Zingg, W., An artificial pancreas, *Diabetes* **23**, 389–396 (1974).
3. Kahn, C. R. and Weir, G. C., eds, Joslin's Diabetes Mellitus, 13th edition, Lea and Febiger, Philadelphia, (1995).

4. Turner, A. P. F., Karube, I., Wilson, G. S., eds, Biosensors: Fundamentals and applications, Oxford University Press, London, (1987).

5. Gough, D. A., Leypoldt, J. K. and Armour, J. C., Progress toward a potentially implantable, enzyme-based glucose sensor, *Diabetes Care* **50**, 190–198 (1982).

6. Gough, D. A., Lucisano, J. Y. and Tse, P. H. S., Two-dimensional enzyme electrode sensor for glucose, *Anal. Chem.* **57**, 2351–2357 (1985) (appended).

7. McKean, B. D. and Gough, D. A., A telemetry-instrumentation system for chronically implanted glucose and oxygen sensors, *IEEE Trans. Biomed. Eng.* **35**, 526–532 (1988).

8. Armour, J. C., Lucisano, J. Y., McKean, B. D. and Gough, D. A., Application of chronic intravascular blood glucose sensor in dogs, *Diabetes* **39**, 1519–1526 (1990).

Two-Dimensional Enzyme Electrode Sensor for Glucose

David A. Gough,* Joseph Y. Lucisano, and Pius H. S. Tse

Department of Applied Mechanics and Engineering Sciences, Bioengineering Group, University of California, San Diego, La Jolla, California 92093

The enzyme electrode-type sensor holds promise as a tool for continuous monitoring of glucose concentration in physiologic systems. Previous designs based on parallel diffusion of glucose and oxygen into the enzyme-containing membrane may, however, have certain disadvantages for in vivo application. A novel sensor configuration is described in which oxygen diffuses into the membrane from two directions while glucose diffuses from only one. This results in sensitivity to glucose concentration over a wide range, even at very low oxygen concentrations.

An electrochemical sensor based on the "enzyme electrode" principle was proposed years ago as a means of monitoring glucose concentration (1, 2). A recent version of this sensor (3) operates by glucose and oxygen diffusing from the sample medium into a membrane that contains the immobilized enzymes glucose oxidase and catalase, where the following chemical reaction takes place:

$$\text{glucose} + {}^1\!/_2\text{O}_2 \rightarrow \text{gluconic acid} \qquad (1)$$

Excess oxygen that is not consumed in this reaction is detected

2352 • ANALYTICAL CHEMISTRY, VOL. 57, NO. 12, OCTOBER 1985

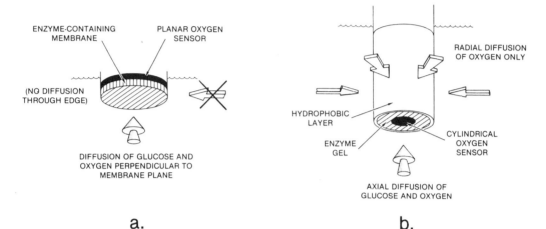

Figure 1. (a) The previous sensor design. Glucose and oxygen both diffuse in the same direction, mainly perpendicular to the plane of the electrode surface. Analogous diffusion regimes have been used for spherical sensors. (b) The two-dimensional design. Glucose and oxygen diffuse axially into the enzyme gel but only oxygen can diffuse radially into the gel through the hydrophobic membrane.

by an electrochemical oxygen sensor to which the membrane containing the immobilized enzymes is attached, giving rise to a glucose-modulated, oxygen-dependent current i_{gmo}. This current is subtracted from the current of a similar oxygen sensor without enzymes i_o, that indicates the background oxygen concentration, and a glucose-dependent difference current i_g results (3). Glucose sensors based on this principle have been made in several configurations and have been employed in various glucose monitoring situations (3–9).

One particularly advantageous application for this type of sensor that has not yet been successfully exploited is implantation in the body to monitor glucose concentration in diabetes. An acute or chronic implanted sensor would continuously indicate the glucose concentration of blood or interstitial fluid, providing information that may enable better control of blood glucose by more appropriate diet, exercise, or antidiabetes medication, thereby possibly ameliorating some of the manifestations of the disease (10).

This sensor has not been widely implemented in vivo for several reasons. One reason, which has been largely overlooked, is that on a molar basis the concentration of unbound oxygen in the body is typically much lower than the concentration of glucose, resulting in a stoichiometric limitation of the enzyme reaction by oxygen (3). Physiologic oxygen concentrations range from approximately 0.15 mM (or 100 mmHg partial pressure) in arterial blood to as low as 0.01 mM (7.0 mmHg) in venous blood and certain peripheral tissues. The blood glucose concentration of interest may be 15.0 mM (270.0 mg/dL) or higher, but the range of greatest clinical importance is below 7.0 mM (125.0 mg/dL). Glucose concentrations in tissues are thought to be proportionally lower than concentrations in blood. The site of implantation is still to be determined but will depend on other factors. For a given location and blood glucose concentration, the disparity between glucose and oxygen concentrations can be 3 orders of magnitude. This "oxygen deficit", if not overcome, would cause the sensor signal to be strongly dependent on oxygen and relatively insensitive to glucose. Some other obstacles to implantation (11) are the catalytic lifetime of the immobilized enzymes, the stability of the oxygen sensors, and the physiologic response to implantation of a foreign object.

We describe here a novel enzyme electrode design based on geometrical considerations that eliminates the oxygen deficit and may significantly reduce the magnitude of certain other problems related to implantation. Some advantages

resulting from this configuration are demonstrated in vitro using simple sensors that incorporate the basic features of the new design. With further development, this design may become the basis of a practical implantable sensor.

THE NEW DESIGN

All previous designs have required that both oxygen and glucose diffuse into the membrane from the same direction. As shown in Figure 1a, diffusion to and within the membrane has been mainly perpendicular to the plane of the membrane, for *both* substrates. This is the case for planar as well as spherical sensors. Such a diffusion regime ideally allows the enzymatic and electrochemical reactions to occur uniformly in the plane of the membrane and across the electrode surface, with substrate concentration gradients developing perpendicular to this plane.

The new design is shown schematically in Figure 1b. The sensor is a small cylinder made of concentric layers of materials open to the sample medium at one end. The core is a short segment of platinum wire, inactive at the exposed end, which serves as a cylindrical oxygen sensor. When this platinum electrode is properly polarized, oxygen is electrochemically consumed on its curved surface, resulting in a steady oxygen flux from the sample medium that is both radial toward the curved electrode surface and axial from the region in front of the sensor. Adjacent to the electrode is a concentric layer of hydrophilic gel containing the immobilized enzymes. Surrounding this gel layer is a hydrophobic, oxygen-permeable layer or tube, made of a material such as silicone rubber that is impermeable to glucose, which serves as a selectively permeable membrane. Both glucose and oxygen can therefore diffuse into the gel layer through the exposed annular end along a direction parallel to the axis of the cylinder, but only oxygen has radial access to the gel through the membrane-covered surface. This gives rise to a two-dimensional supply of oxygen to the enzyme region (radial and axial) and only a one-dimensional supply of glucose (axial). For substrate concentrations specified in the external medium, the concentrations in the gel are determined by (1) the permeability of the gel, membrane, and adjacent medium to each substrate, (2) the enzyme activity, and (3) the aspect ratio, or ratio of the cross-sectional area of the annular end of the gel layer to the area of the cylindrical face covered by the hydrophobic membrane. These factors ultimately determine the sensor response.

Table I. Nondimensional Boundary Conditions for Glucose and Oxygen

glucose	oxygen	ρ	ζ
$\partial \bar{c}_g / \partial \rho = 0$	$\bar{c}_o = 0$	$\rho = 0$	$0 \leq \zeta \leq 1$
$\partial \bar{c}_g / \partial \rho = 0$	$\partial \bar{c}_o / \partial p = \mathrm{Bi}_{o,r}(1 - \bar{c}_o)$	$\rho = 1$	$0 \leq \zeta \leq 1$
$\partial \bar{c}_g / \partial \zeta = \mathrm{Bi}_g(1/\bar{c}_o{}^* - \bar{c}_g)$	$\partial \bar{c}_o / \partial \zeta = \mathrm{Bi}_{o,z}(1 - \bar{c}_o)$	$0 < \rho \leq 1$	$\zeta = 0$
$\partial \bar{c}_g / \partial \zeta = 0$	$\partial \bar{c}_o / \partial \zeta = 0$	$0 \leq \rho < 1$	$\zeta = 1$

THEORETICAL SECTION

We consider a glucose sensor as shown in Figure 2, having an internal cylindrical oxygen sensor of radius r_i and length L that is surrounded by a gel of inner radius r_i and outer radius r_o. The oxygen sensor is electrochemically active only on its curved surface. The gel contains immobilized glucose oxidase and excess catalase. A hydrophobic layer of specified oxygen permeability is in contact with the outer cylindrical surface of the gel. The analysis here is based on diffusion and reaction of glucose and oxygen within the gel. The chemical reaction given in eq 1 can be generalized as follows

$$\nu \text{ glucose} + O_2 \rightarrow \text{products} \tag{2}$$

where ν is a stoichiometry coefficient that has a value of 2 in this case. The dimensionless steady state governing equations are

$$\frac{\partial^2 \bar{c}_g}{\partial \rho^2} + \frac{1}{\rho + r_i/R} \frac{\partial \bar{c}_g}{\partial \rho} + \epsilon^2 \frac{\partial^2 \bar{c}_g}{\partial \zeta^2} - \phi^2 V(\bar{c}_g, \bar{c}_o) = 0 \tag{3}$$

$$\frac{\partial^2 \bar{c}_o}{\partial \rho^2} + \frac{1}{\rho + r_i/R} \frac{\partial \bar{c}_o}{\partial \rho} + \epsilon^2 \frac{\partial^2 \bar{c}_o}{\partial \zeta^2} - \phi^2 V(\bar{c}_g, \bar{c}_o) = 0 \tag{4}$$

where

$$\rho = (r - r_i)/R \qquad R = r_o - r_i \qquad \zeta = z/L$$
$$\bar{c}_o = c_o/\alpha_o c_{oB} \qquad \bar{c}_g = D_g c_g / \nu \alpha_o D_o c_{oB} \qquad \epsilon = R/L \tag{5}$$

Here, c_g, c_o, D_g, and D_o are, respectively, the concentrations and diffusion coefficients of glucose and oxygen within the gel. The diffusion coefficients in the gel in the radial and axial directions are assumed to be the same for a given solute. The partition coefficient and bulk concentration of oxygen are, respectively, α_o and c_{oB}. The parameter ϕ^2, the square of the Thiele modulus (12) is given by

$$\phi^2 = \sigma^2 \frac{\kappa_o}{1 + \kappa_o + \kappa_g \bar{c}_o{}^*} \tag{6}$$

in which σ^2, the square of the relative catalytic activity, is

$$\sigma^2 = \frac{R^2 V_{\max}}{D_o K_o} \tag{7}$$

The effective substrate concentration ratio within the gel is

$$\bar{c}_o{}^* = \frac{\nu \alpha_o D_o c_{oB}}{\alpha_g D_g c_{gB}} \tag{8}$$

and the dimensionless kinetic constants are

$$\kappa_o = \frac{K_o}{\alpha_o c_{oB}} \qquad \kappa_g = \frac{D_g K_g}{\nu \alpha_o D_o c_{oB}} \tag{9}$$

The parameters c_{gB} and α_g are, respectively, the bulk concentration and partition coefficient of glucose. The kinetic parameters $V_{\max}$, K_g, and K_o are, respectively, the maximal reaction velocity per unit volume and the Michaelis constants for glucose and oxygen determined at infinite concentrations of the other substrate. The dimensionless reaction rate is given by

$$V(\bar{c}_g, \bar{c}_o) = \frac{v(c_g, c_o)}{v(\alpha_g c_{gB}, \alpha_o c_{oB})} \tag{10}$$

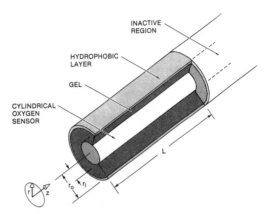

INACTIVE REGION

HYDROPHOBIC LAYER

GEL

CYLINDRICAL OXYGEN SENSOR

Figure 2. A coordinate system for modeling the two-dimensional sensor.

where the reaction rate expression applicable in this case is derived from

$$v(c_g, c_o) = \frac{V_{\max} c_g c_o}{c_g c_o + K_o c_g + K_g c_o} \tag{11}$$

This is the general ping-pong rate expression for the soluble enzyme (13).

The geometric effective substrate concentration ratio $\bar{c}_o{}'$, or the relative concentration of the two substrates within the two-dimensional reaction zone, is given by

$$\bar{c}_o{}' = \frac{\nu c_{oB}}{c_{gB}} \left(\frac{2LP_{o,r} + RP_{o,z}}{RP_g} \right) \tag{12}$$

This is the ratio of the bulk substrate concentrations c_{oB}/c_{gB}, multiplied by the stoichiometry coefficient ν, times the ratio of the products of the permeability P_i for the respective substrate i and the area normal to which diffusion of that substrate occurs. There are two permeability terms for oxygen and one for glucose defined as

$$P_{o,r} = \alpha_o D_o/R \qquad P_{o,z} = \alpha_o D_o/L \qquad P_g = \alpha_g D_g/L \tag{13}$$

Rearranging these terms, eq 12 can be written as

$$\bar{c}_o{}' = \bar{c}_o{}^* \left(\frac{2}{\epsilon^2} + 1 \right) \tag{14}$$

where $\bar{c}_o{}^*$ is given in eq 8. This latter parameter is similar to the relative substrate concentration within the enzyme region for the one-dimensional sensor design where both substrates diffuse from the same direction (3, 14). The term in parentheses in eq 14 is a geometry parameter that depends only on the aspect ratio ϵ.

The nondimensional boundary conditions for the two species are given in Table I where

$$\mathrm{Bi}_{o,r} = \frac{h_{o,r}R}{\alpha_o D_o} \qquad \mathrm{Bi}_{o,z} = \frac{h_{o,z}L}{\alpha_o D_o} \qquad \mathrm{Bi}_g = \frac{h_g L}{\alpha_g D_g} \tag{15}$$

The mass transfer coefficients h_g and $h_{o,z}$ correspond to glucose and oxygen, respectively, in the axial direction whereas $h_{o,r}$ is the mass transfer coefficient for oxygen in the radial di-

2354 • ANALYTICAL CHEMISTRY, VOL. 57, NO. 12, OCTOBER 1985

rection. Only the mass transfer parameter for oxygen in the radial direction is necessarily distinguished from its counterpart in the axial direction.

These equations with the boundary conditions given in Table I were solved by using an adaptation of a two-dimensional orthogonal collocation technique (*15*). After concentration profiles were obtained, the oxygen flux was calculated at different collocation points in the z direction at the electrode surface $r = r_i$. The fluxes at these points were then subjected to a spline fitting routine and multiplied by the electron equivalent n and the Faraday constant F and integrated over the active electrode area to give the current.

Although a large number of terms are required in this analysis, the signal expressed in dimensionless form as the ratio of the glucose-dependent difference current to the oxygen-dependent current is a function of only seven dimensionless parameters, namely

$$i_g/i_o = f(\sigma, \kappa_g, \kappa_o, \bar{c}_o', Bi_g, Bi_{o,z}, Bi_{o,r}) \qquad (17)$$

This represents a substantial simplification, since, to a first approximation, only these parameters need be included in experimental studies or numerical simulations. Equation 17 does not imply however that these parameters can always be varied independently, since in some cases, they are mutually coupled as a result of necessarily being composed of common dimensionalized parameter elements. This provides an interesting comparison to the analysis of the one-dimensional sensor, which has been described by five analogous independent dimensionless parameters (*14*).

EXPERIMENTAL SECTION

Sensor Fabrication and Testing. The electrodes were short segments of 0.01 in. diameter platinum wire, insulated at the inactive end by flame-sealing in borosilicate glass. The exposed flat end was rendered electrochemically inactive by the application of a thin glass coating. A segment of silicone rubber tubing, 0.07 in. i.d. × 0.11 in. o.d. (Dow Corning Corp.), was fitted over the electrode and cemented to the glass insulation at the closed end. The enzymes were immobilized in a gel formed in the cylindrical cavity between the electrode and the tube. The gel contained 20 g % denatured bovine achilles tendon collagen, 6 g % glucose oxidase from *A. niger* (Sigma Chemical Co., type VII), and sufficient catalase from the same biological source (Behring Diagnostics) to ensure excess catalytic activity (activity ratio > 10). These components were dissolved in 0.1 M phosphate buffer, pH 7.3, and cross-linked with glutaraldehyde (25% solution). Sensors of different aspect ratios were made by varying the active length while holding the diameter constant. Smaller sensors having comparable aspect ratios were also fabricated. A separate counter electrode and high-impedance reference electrode were used. The cathodes were polarized at −350 mV vs. Ag/AgCl using a PAR Model 173 potentiostat. Currents were recorded digitally and displayed on a Tektronix 4051 computer. Although a slightly different version of the sensor may be required for implantation, other features can be readily incorporated later and are not important to this discussion.

The sensors were placed in a sealed, thermostated (37 °C) vessel containing 0.1 M phosphate buffer, pH 7.3, and equilibrated with the gas mixture. The oxygen-dependent current was determined before the addition of glucose and used as a basis for calculation and normalization of the glucose-dependent difference current. Small background currents produced in the absence of both glucose and oxygen were also taken into account.

Values for the Parameters. One of the difficulties in ascertaining if experimental observations correspond closely to the predicted response is determining the values of the parameters to be used in the simulation. The values used here are listed in Table II. We have previously described methods based on a rotated disk electrode apparatus for determining the mass transfer and catalytic properties of gel membranes of the same composition, applying a one-dimensional diffusional analysis (*3, 14*). Results from those studies were used as a basis for the calculation of the dimensionless parameters of the present geometry. Reasonable

Table II. Values of Parameters Used in the Calculations

$K_g = 0.125$ M (*13*)		$K_o = 0.001$ M (*13*)
$Bi_g = 60.0$ (*3*)	$Bi_{o,r} = 5.0 - 38.0$	$Bi_{o,z} = 10.0$ (*3*)
$D_g = 4.0 \times 10^{-6}$ cm^2/s		$D_o = 1.0 \times 10^{-5}$ cm^2/s
	$\sigma = 2.0 - 10.0$	

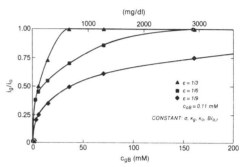

Figure 3. Normalized glucose-dependent different current plotted as a function of bulk glucose concentration for three sensors of different length.

estimates were made for values of the mass transfer coefficients. The actual values of σ or the relative catalytic activity obtained here were approximately 200, based on formulation and on previous results for the yield of immobilized catalytic activity. These values were not used in simulations however, because the computational method is not effective for σ greater than approximately 10. It will be shown that this value is sufficiently high for acceptable simulation and that there is little effect of variation in the value of σ above 10.

RESULTS AND DISCUSSION

Experimental results from sensors having different aspect ratios are compiled in Figure 3. The glucose-dependent difference current normalized by the oxygen-dependent current is plotted as a function of bulk glucose concentration for sensors of three different values of aspect ratio or ϵ, corresponding to sensors of different lengths with all other characteristics held constant. Thus, the parameters $\sigma, \kappa_g, \kappa_o$, and $Bi_{o,r}$ were the same for these sensors. The bulk oxygen concentration was fixed at 0.1 mM, while the glucose concentration was varied over a broad range. In each case, the signal rises nonlinearly and approaches unity with increasing glucose concentration. For the sensor of aspect ratio 1/3, the normalized difference current is mildly nonlinear with glucose concentration over the entire range of response. The other two sensors, however, show a relatively large variation in the normalized difference current at low glucose concentrations and a small variation at higher concentrations. Although the range of glucose concentration that can be detected is far greater here than would be required in clinical applications, these results illustrate the general behavior of two-dimensional sensors and the extent to which the range of sensitivity can be controlled by specifying the aspect ratio. The sensors respond to glucose in spite of substantial oxygen deficits (in some cases, more than 3 orders of magnitude). Control of the range of sensitivity to a comparable extent is virtually impossible in a one-dimensional design (*3, 14*).

Effect of Oxygen Concentration. The dependence of the response to glucose on oxygen concentration is shown in Figure 4. The glucose-dependent difference current normalized as above is plotted as a function of the ratio of bulk glucose and oxygen concentrations. The experiments were carried out with a sensor of aspect ratio 1/3 by recording the response to increasing glucose concentrations at each oxygen concentration. The value of the abscissa indicates the oxygen deficit. The symbols correspond to different oxygen concentrations

71

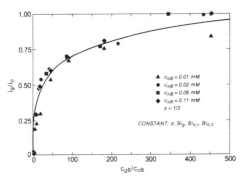

Figure 4. Normalized glucose-dependent difference current plotted as a function of the ratio of the bulk glucose to bulk oxygen concentration.

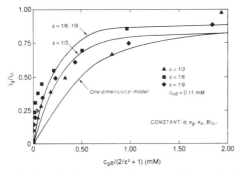

Figure 5. Normalized glucose-dependent current plotted as a function of the bulk glucose concentration divided by the geometry parameter for sensors of three different lengths. The lines are calculations based on the model.

over the range of 0.01–0.11 mM and the solid line is a fit of the experimental results. The parameters σ, Bi_g, $Bi_{o,r}$, and $Bi_{o,z}$ were held constant. Plotting the results in this fashion, suggested by previous studies of the one-dimensional sensor design (14), greatly simplifies the interpretation, since here the response to glucose can be visualized independent of oxygen concentration. This approach is also readily amenable to automatic computation in the following way. A reference oxygen electrode is used in practice to indicate i_o, from which c_{oB} is determined. The glucose electrode produces i_{gmo}, which is subtracted from i_o to give i_g. A given current ratio i_g/i_o corresponds to a ratio of concentrations as shown in the figure from which c_{gB} can be determined, irrespective of oxygen concentration. These results indicate that this particular sensor can overcome an oxygen deficit of greater than 500, allowing detection to at least 5.0 mM glucose concentration in the worst case of lowest oxygen concentration and up to 50.0 mM at the highest oxygen concentration used here. An aspect ratio of 1/3 would be suitable for most implant applications. Thus, potentially useful sensitivity to glucose concentration can be obtained. The sensor response can be displayed in a way that is functionally independent of oxygen concentration.

Dependence on Aspect Ratio, ϵ. The experimental results of Figure 3 are replotted in Figure 5, showing the normalized glucose-dependent difference current as a function of bulk glucose concentration divided by the geometry parameter. The symbols correspond to experimental results from sensors of different aspect ratios as before, with the bulk oxygen concentration and the parameters σ, κ_g, κ_o, and $Bi_{o,r}$ held constant. The two upper lines are theoretical calculations based on the model for the indicated aspect ratios. Calculated results from the one-dimensional model (14) with the same

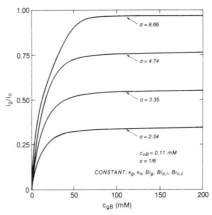

Figure 6. Calculated effect of variable σ on the sensor response.

values of all corresponding parameters are given by the lower line. The theoretical results for the sensors of aspect ratios 1/6, 1/9, and smaller are coincident and form a calculated upper bound for the experimental results, whereas the line based on the one-dimensional model can be taken as a lower bound over the range shown here. As the aspect ratio increases, the two-dimensional model reduces computationally to the one-dimensional model, although there is no physical basis for such reduction. After an initial rise, the two upper curves approach unity only very slowly compared to the curve for the one-dimensional model and are eventually crossed by the latter. The experimental results approximate the two upper lines reasonably well, considering the broad absolute range of glucose concentration involved.

The relative concentration of oxygen in the two-dimensional design is enhanced over that of the one-dimensional design by the geometry parameter $2/\epsilon^2 + 1$. This effect can be substantial. For example, for a value of ϵ of one-third, the relative oxygen concentration is increased by approximately 20-fold, and for a value of ϵ of one-sixth, the oxygen concentration is enhanced over 70-fold. Each doubling of the length at constant radius enhances the relative oxygen concentration by a factor of 4, proportionally extending the range of sensitivity to glucose. This indicates that changes in the aspect ratio (or gel length) can be used in a simple way to substantially extend the range to glucose.

The results of Figure 5 suggest that experimental observations can be reasonably approximated in certain cases by judicious use of the less complicated one-dimensional model, scaled by the appropriate geometry parameter. The two-dimensional model is nevertheless essential in other cases and required for a full understanding of the system.

Effect of Relative Catalytic Activity, σ. The calculated effect of variations in relative catalytic activity is shown in Figure 6. The normalized difference current is plotted as a function of bulk glucose concentration for various values of relative catalytic activity. The aspect ratio, the bulk oxygen concentration, and the parameters κ_g, κ_o, Bi_g, $Bi_{o,r}$, and $Bi_{o,z}$ are held constant. These calculations correspond to experimental variations in enzyme loading over a relatively low range of loading. The signal rises relatively rapidly at low glucose concentrations, but only very gradually as concentration is increased further. In contrast to the response of the one-dimensional sensor (14), the response at low values of σ may not approximate unity with increasing glucose concentration. Qualitative experimental observations have confirmed this prediction. Over the range of σ shown here, the signal at high glucose concentrations is approximately proportional to the value of σ, converging to a limiting response with increasing

2356 • ANALYTICAL CHEMISTRY, VOL. 57, NO. 12, OCTOBER 1985

values of σ. The experimental results shown in Figure 5, where the value of σ is approximately 200, fit the theoretical calculations reasonably well, even though these calculations are based on a value of 10 for σ. This suggests by analogy to the one-dimensional sensor that there is little effect of varying σ above a characteristic value. This is as expected for a system that is largely diffusion-limited and may be of significant advantage in forestalling the effects of enzyme inactivation during operation.

Other Parameters. The fact that certain of the dimensionless parameters are mutually coupled imposes an additional degree of complexity over the one-dimensional model. For example, the coupled parameters κ_g and $\bar{c}_o'$ both contain the parameter c_{oB}. In our previous analysis of the one-dimensional sensor (14), it was possible to express the analogous one-dimensional parameters independently by employing a simplified kinetic expression in which

$$K_g \approx K_g + c_g \tag{18}$$

This approximation, useful in the one-dimensional system where the bulk glucose concentration is always much less than the value of K_g (0.125 M, ref 13), led to a simplified rate expression and the definition of a dimensionless parameter containing the ratio K_g/K_o, but independent of concentration. In the present system, where the value of c_{gB} can be comparable to that of K_g, this approximation could lead to some degree of error, although probably small. The observation that certain experimental results can be approximated by a scaled one-dimensional model indicates that the assumption given by eq 18 is reasonable in some cases. Nevertheless, the full rate expression was retained for generality. The relationship between the mass transfer parameters and $\bar{c}_o'$ is another example of coupling that is not found in the one-dimensional model. The mass transfer parameters can be varied for example, by changing the gel dimensions or altering the thickness or composition of the hydrophobic layer. This may also have an effect on the aspect ratio and $\bar{c}_o'$.

One aspect not addressed here is the consequence of sensor size. Dimensions used in the model are given as ratios and experiments were conducted with sensors having a limited range of sizes. The close correspondence between experimental results and calculations suggests that the effects of size on the general behavior of sensors may be of relatively minor importance for sensors of practical dimensions at steady state.

Concentration Contours. The sensing mechanism can be visualized in terms of the spatial distributions of glucose and oxygen within the gel layer. Figure 7 shows the glucose and oxygen concentration in a plane through the gel extending radially from the curved electrode surface and running the length of the cylinder. In each case, the concentration increases in the upward direction and the boundaries of the figure base are the oxygen electrode in the front, the hydrophobic membrane in the back, the end of the gel in contact with the sample medium at the right facing away, and the opposite sealed end of the gel at the left. The effects of external concentration boundary layers are not shown. The steep glucose concentration contour is a result of glucose diffusing from the sample medium at the right into the gel and consumption by the enzymatic reaction. Glucose is almost totally depleted within a short distance from the right end of the gel for the conditions used. Since there is no flux of glucose through the hydrophobic membrane and no electrochemical reaction of glucose at the electrode surface, no substantial glucose concentration gradient develops in the radial direction. The more complex oxygen concentration contour is a result of the following: (1) oxygen diffusion into the gel, both from behind through the hydrophobic membrane, and directly from the sample medium at the right; (2) elec-

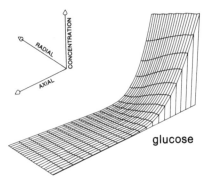

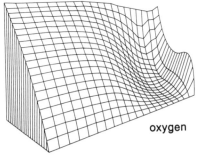

Figure 7. Concentration contours in the enzyme gel. The substrate concentration is plotted in the vertical axis over an axial–radial plane through the enzyme gel. The active end of the gel contacting the sample medium is the right border of each figure, the insulated end is at the left, the oxygen electrode surface is the front border, and the hydrophobic membrane surface is at the back. Values used in the calculation are as follows: σ, 9.70; c_{gB}, 0.50 mM; c_{oB}, 0.11 mM; others, as in Table II.

trochemical oxygen consumption at the electrode surface in the front; and (3) enzymatic oxygen consumption, occurring mainly near the right end of the gel. The enzymatic reaction causes a sharp depression in the oxygen contour near the right border corresponding to the region of the steep glucose concentration gradient. The concentration fluctuations along the top and right edges of the oxygen contour correspond to oxygen gradients that extend slightly into respectively the hydrophobic membrane and the sample medium. For certain other combinations of external substrate concentration the contours vary more gradually and are therefore less useful for illustrative purposes.

Catalytic Lifetime. Although not explored in this study, the two-dimensional design offers the potential for significantly extending the effective catalytic lifetime. It has been shown that a major cause of inactivation of these enzymes in the immobilized form is the intermediate hydrogen peroxide (16, 17). Inactivation takes place in regions of the gel where the chemical reaction occurs and significant quantities of hydrogen peroxide are produced. As the enzyme near the entrance becomes inactive, the limiting substrate subsequently must diffuse farther axially before encountering active enzyme. Compared to the one-dimensional sensor, the two-dimensional design provides substantially greater diffusional length for the limiting substrate glucose, which may result in an extension of the catalytic lifetime. The calculations in Figure 6 suggest that the steady state characteristics may not change significantly during this inactivation process until the value of σ drops to below approximately 10. The transient response may, however, be altered. The result may depend on the effectiveness of coimmobilized catalase in removing hydrogen

ANALYTICAL CHEMISTRY, VOL. 57, NO. 12, OCTOBER 1985 • 2357

peroxide, since radial diffusion of hydrogen peroxide from the gel region would be prevented by the hydrophobic membrane. If the intermediate were allowed to accumulate, inactivation may actually occur more rapidly in this design. These aspects are being studied quantitatively.

Transient Response. A full study of the transient response remains to be carried out. Qualitative observations indicate that the response time to step changes in bulk glucose concentration is several minutes and comparable to that of the one-dimensional design when the glucose concentrations involved are both relatively low. When either or both of the glucose concentrations are relatively high, however, the response time can be as long as 40 min. The response time may also be comparably long at low glucose concentrations when the enzyme loading is low. A reduction in sensor size may be employed to reduce the response time if this is determined to be problematic in specific physiologic monitoring situations. Systematic studies of these effects are under way.

CONCLUSIONS

The two-dimensional sensor design has interesting features. With the additional degree of freedom afforded by this design, sensors can be made that overcome the oxygen deficit to the extent required for implant applications. The model presented here is useful to explain the sensing mechanism and provides a basis for interpretation of the response and incorporation of further improvements.

NOMENCLATURE

Bi = mass transfer Biot number, dimensionless transfer parameter

c = concentration

$\bar{c}$ = dimensionless concentration

$\bar{c}_o{}'$ = geometric effective substrate concentration ratio, two-dimensional case

$\bar{c}_o{}^*$ = effective substrate concentration ratio, one-dimensional case

D = diffusion coefficient

F = Faraday constant

h = mass transfer coefficient

i = current

K = Michaelis constant

L = axial length

n = electron equivalent

P = permeability

r = radial distance

V = dimensionless reaction rate

v = reaction rate

V_{max} = maximum velocity of reaction per unit volume

z = axial distance

Greek Letters

α = partition coefficient

ϵ = aspect ratio

ν = stoichiometry coefficient

κ = dimensionless Michaelis constant

ρ = dimensionless radial distance

σ = dimensionless relative catalytic activity

ϕ = Thiele modulus

ζ = dimensionless axial distance

Subscripts

B = value in the bulk solution

g = glucose

gm = glucose-modulated

o = oxygen

r = radial

z = axial

Registry No. D-Glucose, 50-99-7; glucose oxidase, 9001-37-0; catalase, 9001-05-2.

LITERATURE CITED

(1) Clark, L. C., Jr.; Lyons, C. *Ann. N. Y. Acad. Sci.* **1962**, *102*, 29–45.
(2) Updike, S. J.; Hicks, G. P. *Nature (London)* **1967**, *214*, 986–988.
(3) Gough, D. A.; Leypoldt, J. K.; Armour, J. C. *Diabetes Care* **1982**, *5*, 190–198.
(4) Layne, E. C.; Schultz, R. D.; Thomas, L. J.; Slama, G.; Slayer, D. F.; Bessman, S. P. *Diabetes* **1976**, *25*, 81–89.
(5) Threvenot, D. R.; Sternberg, R.; Coulet, P. R.; Laurent, J.; Gautheron, D. C. *Anal. Chem.* **1979**, *51*, 96–100.
(6) Lobel, E.; Rishpon, J. *Anal. Chem.* **1981**, *53*, 51–53.
(7) Clark, L. C., Jr.; Duggan, C. A. *Diabetes Care* **1982**, *5*, 174–179.
(8) Fischer, U.; Abel, P. *Trans.—Am. Soc. Artif. Intern. Organs* **1982**, *28*, 245–248.
(9) Shichiri, M.; Yamasaki, Y.; Kawamori, R.; Abe, H. *Lancet* **1982**, *2*, 1129–1131.
(10) Cahill, G. F.; Etzwiler, D. D.; Freinkel, N. *New Engl. J. Med.* **1976**, *294*, 1004–1005.
(11) Threvenot, D. R. *Diabetes Care* **1982**, *5*, 184–189.
(12) Aris, R. "The Mathematical Theory of Diffusion and Reaction in Permeable Catalysts"; Clarendon Press: Oxford, 1975.
(13) Weibel, M. K.; Bright, H. J. *J. Biol. Chem.* **1971**, *246*, 2734–2744.
(14) Leypoldt, J. K.; Gough, D. A. *Anal. Chem.* **1984**, *56*, 2896–2904.
(15) Villadsen, J.; Michelson, M. L. "Solution of Differential Equation Models by Polynomial Approximation"; Prentice-Hall: Englewood Cliffs, NJ, 1978.
(16) Greenfield, P. F.; Kittrell, P. F.; Laurence, R. L. *Anal. Biochem.* **1975**, *65*, 109–124.
(17) Tse, P. H. S. Ph.D. Dissertation, University of California, San Diego, 1984.

RECEIVED for review January 14, 1985. Accepted June 5, 1985. This work was suppoted by a grant from the National Institutes of Health.

CHAPTER 4

DESIGN AND DEVELOPMENT OF ARTIFICIAL BLOOD

MARCOS INTAGLIETTA

Department of Bioengineering, University of California, San Diego

1. Introduction

Blood is a key necessity in surgery and treatment of injury. Its availability is a critical factor in insuring survival in the presence of severe blood losses. Its use in the general public, under conditions of optimal medical care delivery is one unit (0.5 liters) per 20 person-year. However, statistics from the World Health Organization indicate that at present it is used at the rate of one unit per 100 person-year on a world basis. The gap between need and availability is further aggravated by the fact that blood *per se* has a number of inherent risks (Dodd, 1992), ranging from about 3% (three adverse outcome per 100 units transfused) for minor reactions, to a probability of 0.001% of undergoing a fatal hemolytic reaction. Superposed to these risks is the possibility of transmission of infectious diseases such as hepatitis B (0.002%) and hepatitis non-A non-B (0.03%). The current risk of becoming infected with the human immunodeficiency virus (HIV) is about 1 chance in 400,000 under optimal screening conditions. These are risks for the transfusion of one unit of blood. It should be noted that many surgical interventions and trauma victims require multiple transfusions, with a proportional increase in risk of adverse outcomes.

The HIV epidemic caused many fatalities due to blood transfusion before stringent testing was introduced in 1984. The potential contamination of the blood supply led the United States military to develop blood substitutes starting in 1985 with the objective of finding a product for battlefield conditions that was free of contamination, could be used immediately and that did not need special storage. The end of the Cold War, changes in the nature military engagements, the slow rate of progress, and the perception that the blood supply was again safe and abundant, significantly lowered the interest of the U.S. military in the development of artificial blood. After a decade of work, about a dozen blood substitutes had been developed, and many of these are presently undergoing clinical trials.

There are several products that are about to enter clinical use. They restore circulating volume, i.e. they are plasma expanders, and are able to deliver oxygen to the tissue, thus restoring the two main functions of blood. However, the majority of these product has shown unexpected side effects, most commonly hypertension. This unwanted feature appears is the result of the philosophy of development, which emphasized systemic, whole organ studies in order to assess efficacy and biocompatibility, without quantitative analysis of transport phenomena inherent to blood function, or data from the microcirculation, the microscopic system of vessels that permeates the body where all the exchange processes take place.

Realistic studies on oxygen exchange in the microcirculation become possible in 1990's through technology developed at UCSD that allowed to measure optically how oxygen is

Fig. 1. Hamster skin fold preparation. This experimental model allows to study the microcirculation of skeletal muscle and subcutaneous connective tissue, at the microscope, for several days, and in the absence of anesthesia. The effect of blood substitutes can be evaluated by measuring blood flow in the microscopic vessels, the reaction of arterioles and venules, oxygen delivery from the microvessels, and oxygenation distribution in the microvessels and the surrounding tissue.

distributed and exchanged from blood vessels whose dimensions are 10 times thinner than human hair. This technology was used in specialized animal model called the chamber window preparation of the hamster, which allows to study the microcirculation *in vivo*, without anesthesia, and for periods of up to one week (Fig. 1). Advances in pharmacology established the fundamental role played by blood flowing over the cells that lined the inside surface of the microvessels (the endothelium) in producing substances that control blood flow and exchange through the mechanism of autoregulation. Data obtained from the transport phenomena and pharmacological experiments in the microcirculation obtained at UCSD is presently used in bioengineering models of the circulation and the tissues exposed to the artificial blood substitutes. This mathematical analysis have provided a new gauge with which to formulate artificial blood, and how to use the products available and those to be developed, to optimize efficacy and cost and eliminate side effects. Therefore, a new generation of artificial blood products, based on rational design established in microscopic models of the circulation is now ready for clinical trials. This new formulations also have a significant price advantage over blood and conventionally developed "artificial blood".

2. Modern History of Blood Transfusion and Blood Substitutes

The need for blood transfusion for replacing lost circulating volume in severe blood loses was recognized since the discovery of the circulation. The earliest successful volume replacement fluid, or blood substitute, was physiologic saline which was introduced around 1875. Blood transfusion remained the major objective in restoring tissue oxygenation, and the use of human blood become possible when substances responsible for incompatibility reactions and hemolysis of blood were discovered in 1900, which led to blood typing. The finding that sodium citrate and glucose prevented coagulation of blood was the final step in the introduction and relatively safe use of blood transfusions. World War II caused the development of the blood banking system, and the American Red Cross begun establishing blood banks in 1947.

The addition of potassium and calcium ions significantly improves the effectiveness of saline in maintaining the viability of tissue, and lactate is added to this composite solution, which is gradually converted into sodium bicarbonate, which prevents alkalosis. This constitutes "lactate-Ringers" which is the most widely used solution for blood volume restitution. This fluid must be given in volumes that are as much as 3 times the blood loss to be treated, since the dilution of plasma proteins lowers the plasma osmotic pressure causing an imbalance of the fluid exchange favoring microvascular fluid loss and edema. The advantage of crystalloid solutions is that large volumes can be given over short periods, and excess volumes are rapidly cleared from the circulation by diuresis which is beneficial in the treatment of trauma.

Blood plasma (the fluid in which red blood cells are suspended) and serum (blood plasma with the coagulation factors and platelets removed) were recognized earlier on to be superior to crystalloid solutions in maintaining tissue viability, and this effect was correctly attributed to the osmotic pressure due to albumin whose molecular weight is 69 kDa. In 1947, it was shown that bovine plasma could be used if globulins were removed. Dextrans, high molecular weight polysaccharides consisting of linked glucose molecules with molecular weights 40–70 kDa, have proven to be safe as a colloidal substitute for albumin and were used for reestablishing blood volume with a noncellular nonoxygen carrying blood substitute. This compound was regarded as the colloid of choice for volume replacement and shock treatment until recently in Europe. It is now gradually replaced worldwide by hydroxyethyl starch, a chemically modified starch with molecular weight ranging from 40 to 450 kDa.

Perfluorocarbons were introduced in the 1960's. These materials are fluorinated polymers, similar to Teflon which bind reversibly with oxygen. They are water insoluble and are dispersed in water as an emulsion. They dissolve oxygen as a linear function of pO_2 and in physiological conditions they carry about 2–3 times the oxygen than is carried by water in solution. By comparison, the arterial blood of a person breathing normal air carries about 20 times the oxygen present in the same volume of water (or plasma). Therefore, for these products to deliver sufficient oxygen the lungs must be exposed to high oxygen atmospheres.

Hemoglobin, the protein present in red blood cells that reversibly binds oxygen is the oxygen carrier of choice because of its high oxygen carrying capacity. The introduction of stroma-free hemoglobin, i.e. pure hemoglobin void of the cellular debris remaining from the breakdown of red blood cells from which it is extracted, eliminated many of the toxic effects noted with the use of these solutions. Pure hemoglobin, however, cannot be used to carry oxygen in the circulation because it dissociates into dimers which cause renal toxicity, a problem corrected by chemical modifications that bind the dimers.

3. Nonoxygen Carrying and Oxygen Carrying Blood Volume Replacement

In general, it is possible to survive losses of the red blood cell mass of the order of 70%. However, our ability to compensate for comparatively smaller losses of blood volume is limited. A 30% deficit in blood volume leads to irreversible shock if not rapidly corrected.

As a consequence, maintenance of normovolemia is the objective of most forms of blood substitution or replacement, leading to the dilution of the original blood constituents or hemodilution. This process causes systemic and microvascular effects that underlie all forms of blood replacement and therefore is a reference point for the understanding of the behavior of blood substitutes.

The fluids for volume restitution are crystalloid solutions, colloidal solutions and oxygen carrying solutions. The end point of any change of the properties of blood is whether tissue is adequately oxygenated, a phenomenon that takes place in the microvasculature. Therefore hemodilution must be analyzed not only in terms of systemic effects but also in terms of how these, coupled with the altered composition of blood, influence the exchange of oxygen in the microcirculation. Data for this kind of analysis was not available when plasma expanders and hemoglobin based blood substitutes were developed in the 1980's.

An important characteristics of blood is its viscosity which is determined by the concentration of red blood cells or hematocrit (% of red blood cells by volume vs the volume of whole blood). Normal blood has a viscosity of about 4 centipoise (*cp*) while plasma is approximately 1 *cp*. Blood viscosity is approximately proportional to the hematocrit squared, therefore small changes in red blood cell concentration have important effects on blood viscosity at normal hematocrits. Colloidal solutions at about 10% concentration have viscosities similar to that of plasma (7% plasma proteins concentration), therefore hemodilution with most plasma expanders and hemoglobin solutions significantly reduces blood viscosity.

The lowered blood viscosity due to hemodilution allows the heart to pump blood at higher volumetric rate without increased expenditure of energy, since the power (or metabolic) requirement of the heart, when blood pressure is maintained, is equal to the product of volumetric output (cardiac output) and blood viscosity. Physiologically, lowered blood viscosity increases central venous pressure and the return of venous blood to the heart, which improves cardiac performance and increases cardiac output. This causes increased blood flow velocity and shear rate at the vessel walls.

The decrease in hematocrit lowers the intrinsic oxygen carrying capacity of blood, but this effect is compensated by the increased blood flow velocity due to the lowered blood viscosity, which increases the rate at which the fewer oxygen carrying red blood cells are delivered to the microcirculation. As a consequence, both systemic and capillary oxygen carrying capacities remain approximately normal up arterial hematocrit of about 25% (normal is 45%). This explains why the organism can easily sustain blood losses that lower the number of red blood cells to half of its original value.

During hemodilution with a nonoxygen carrying colloid the organism compensates for the lost oxygen carrying capacity and there is no advantage in restituting blood oxygen carrying capacity until about half of the red blood cells are lost. This point identifies the "transfusion trigger" currently set at a hemoglobin concentration of about 7%. The transfusion trigger determines the point at which the organism may be in danger of becoming ischemic, a condition that must be corrected by the transfusion of blood or an oxygen carrying blood substitute. Therefore, in most cases, oxygen carrying capacity restitution is made when blood is hemodiluted, and the effects of the newly added oxygen carrying capacity are superposed to those due to the hemodiluted blood.

4.   The Underlying Physical Process and its Mathematical Analysis

The concentration of oxygen in the blood vessels can presently be routinely measured in the microcirculation by a technique developed UCSD (Torres and Intaglietta, 1993; Tsai *et al.*, 1998), which allows to determine the partial pressure of oxygen in blood and tissue. When a survey of pO_2 is made in the blood vessels arranged according to their order of branching we obtain a distribution as shown in Fig. 2. The salient features of this result is that oxygen exits continuously the microvessels, being a minimum in the capillaries and the tissue. Most of the oxygen appears to be delivered by arterioles and the increase of oxygen tension in venules is due to convective and diffusive shunts between arterioles and venules. The rate at which oxygen is delivered from the microscopic blood vessel is a function of blood flow, blood viscosity, blood composition, oxygen carrying capacity of blood, vessels dimensions and distribution, and the diffusion characteristics of gases. Ideally, we would develop an artificial material that is identical in all respects with blood, but this has proven not to be possible, and all blood substitutes to be developed have physical properties different from blood. Given the number of variables involved in the process of delivering blood to the tissue, the engineering method is needed to understand and predict the behavior of these material as suppliers of oxygen to the tissue. This technique gives a mathematical formulation that shows the relationship between the variables and allows to identify what are the critical parameters of the process.

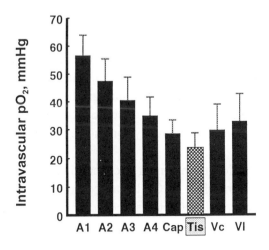

Microvessel Order

Fig. 2. Measurement of the partial pressure of oxygen in blood and tissue in the hamster window preparation (mean $\pm$ SEM). The blood vessels are grouped according to size and their position in the branching order in the microvascular network A1,2,3,4 are arterioles in descending size, Cap: capillaries, Vc: venous capillaries; Tis: tissue pO_2; and, Vl: large venules. A1 arterioles, which branch directly from small arteries, have a nominal diameter of 55 μm. The large Vl venules have 80 μm diameter. Measurements were made optically with the phosphorescence quenching technique and reported by Kerger *et al.*, 1996. The venules, which collect the blood from the tissue, have higher pO_2 than the capillaries because of the presence of flow shunts that bring arteriolar blood directly into the venules, and the diffusion of oxygen between arterioles and parallel running venules.

Analysis of the relationship between the variables that determine oxygen distribution in the microvascular network is based on mass balance considerations for a segment of a microscopic blood vessel of length dx of branching order i, where the rate of oxygen entering the upstream cross section at x is m_i, the rate of exit at $x + dx$ is $m_{i,x+dx}$ and $m_{\text{diffusion}}$ is the rate exit of oxygen through the vessel wall by diffusion:

$$m_{i,x} = m_{i,x+dx} + m_{\text{diffusion},i} \,. \tag{1}$$

The rate at which oxygen passes through a given vessel cross section is equal to the product of the volumetric flow rate of blood Q_i, and a function $F(Htc, C)$ that represents the intrinsic oxygen carrying capacity of blood due to oxygen carried by red blood cells, i.e. the hematocrit Htc, and oxygen contained by an oxygen carrier C. $S_{0,i} + n_i \text{pO}_{2,i}$ is the linearized relationship between blood oxygen expressed as a % of the maximum amount that can be dissolved in blood m_t at the prevailing partial pressure of oxygen $\text{pO}_{2,i}$ that approximates the hemoglobin oxygen dissociation curve in the branch i of length L_i ($S_{0,i}$ is oxygen saturation at zero pO_2 for each segment i, and n_i is the linearized slope of the dissociation curve for that segment). Accordingly:

$$m_i = Q_i F(Htc, C)[S_{0,i} + n_i \text{pO}_{2,i}] m_t \,. \tag{2}$$

The rate of oxygen exit by diffusion is found by solving the diffusion equation for the region of tissue surrounding the blood column. The steady state distribution of oxygen in terms of its concentration c is given by the diffusion equation for region in which there is oxygen consumption:

$$D\nabla^2 c = g_0 \tag{3}$$

where D is diffusion coefficient for oxygen, and g_0 is the rate of oxygen consumption in the vessel wall. The vessel wall is assumed to be thin, i.e. its thickness h is such that relative to vessel wall radius r_0, $h/r_0 \ll 1$. The diffusional gradients are assumed to be important primarily in this region according to recent experimental evidence, which allows to make the simplifying assumption that the geometry for the region of the vessel wall is Cartesian (rather than cylindrical).

The solution of the diffusion equation in a consuming medium for the layer cf tissue adjacent to the blood tissue interface in Cartesian coordinates, where the direction of diffusion is r (note that this is now the Cartesian coordinate), gives the concentration profile outside of the blood vessel:

$$c(r) = \frac{g_0}{2D} r^2 - \frac{g_0}{D} \delta r + \alpha \text{pO}_{2,i} \tag{4}$$

where α is the oxygen solubility of oxygen in tissue and δ is the distance from the blood tissue interface at which all the delivered oxygen has been consumed, given by:

$$\delta = \sqrt{\frac{2D\alpha \text{pO}_{2,i}}{g_0}} \,. \tag{5}$$

The concentration gradient at the blood/tissue interface is given by:

$$\left. \frac{dc}{dr} \right|_{r=0} \sim -\sqrt{\frac{2g_0 \alpha \text{pO}_{2,i}}{D}} \tag{6}$$

leading to a rate of oxygen exit per unit vessel length dx given by:

$$m_{\text{diffusion}} = \pi d_i \sqrt{2Dg_0\alpha\text{pO}_{2,i}}dx\,. \tag{7}$$

In this equation, oxygen partial pressure is the variable which in a gas is proportional to oxygen concentration. In blood, oxygen concentration is related to pO$_2$ through the oxygen saturation curve which describes the oxygen held in chemical binding. In this system, it is practical to express mass balance as a function of partial pressures rather than concentrations (molecule per unit volume) because of the changes present at the blood/vessel wall interface.

The mass balance for oxygen in vessel segment i is obtained from Eqs. (1) and (5) whereby:

$$m_{i,j} = m_{i,j} + \frac{\partial m_{i,j}}{\partial x}dx + m_{\text{diffusion}} \tag{8}$$

where the subscript j labels the differential element within the segment i. From (6) and (7), we obtain the differential equation describing the longitudinal oxygen distribution for a blood vessel of order i:

$$\frac{d\text{pO}_{2,i}}{dx} = \pi d_i \frac{\sqrt{2g_0\alpha\text{pO}_{2,i}}}{Q_i F(Htc,C)n_i m_i}\,. \tag{9}$$

Integration of Eq. (4) over the extent of the vascular segment L_i yields:

$$\text{pO}_{2,i}^{1/2} = \text{pO}_{2,i-1}^{1/2} - k_i \qquad \text{where} \qquad k_i = \frac{\pi d_i L_i}{Q_i F(Htc,C)n_i m_t}\sqrt{\frac{2g_0\alpha}{D}} \tag{10}$$

which indicates that if we know pO$_2$ at the beginning of the vascular segment, i.e. the end of the previous segment $i-1$, we can calculate pO$_2$ for the end of the ith segment. Knowing pO$_2$ at $i = 0$, i.e. the central oxygen tension, allows prediction of pO$_2$ at any subsequent order of vascular branching n (see Fig. 3) according to:

$$\text{pO}_{2,n}^{1/2} = \text{pO}_{2,0}^{1/2} - \sum_{l}^{n} k_i\,. \tag{11}$$

Blood flow in any segment is given by Poiseuille's equation:

$$Q_i = \frac{\pi d_i^4 \Delta P_i}{128\mu L_i} \tag{12}$$

where μ is blood viscosity and ΔP_i is the pressure difference between entrance and exit of the segment i. Introducing this expression into Eq. (8) and rearranging the summation we obtain:

$$K_n = \sum k_i = \frac{128\mu}{F(Htc,C)m_t}\sqrt{\frac{g_0\alpha D}{2}}\sum_{l}^{n}\frac{L_i^2}{n_i d_i^3 \Delta P_i}\,. \tag{13}$$

The summation shown in Eq. 11 shows a group of terms common to all vessel segments and including blood viscosity, blood oxygen carrying capacity and vessel wall metabolism. The second group is a summation where each term is specific to each vascular segment. The combination of these terms allows to predict capillary blood pO$_2$ for any given set of physical properties of blood.

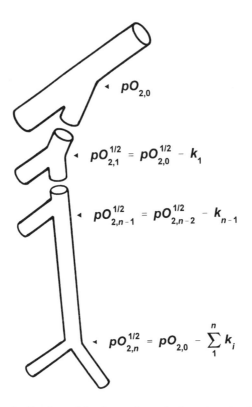

Fig. 3. Distribution of oxygen in the large blood vessels and the microcirculation determined from mass balance considerations and the solution of the diffusion equation. As blood flows through the arterial vessels, oxygen exits and passes into the tissue by diffusion. Consequently the amount of oxygen that arrives to the exchange vessels of the microcirculation is determined by the summation of the losses k_i in each vessel segment. The circulation is designed to insure that both the amount of oxygen and the partial pressure of oxygen that arrives to the microcirculation is regulated within a narrow range, and departures from this set point causes autoregulation. In the case of artificial blood, if too much oxygen is delivered the microcirculation reacts by constricting the arterioles, causing hypertension.

This analysis shows that, solely on the basis of the prevailing transport phenomena, better oxygenation results from lowering viscosity. The factor g_0 represents the vessel wall metabolism, which had not been previously recognized as a significant component in the management of oxygen distribution. The significance of this factor is that metabolism of the vessel wall is directly affected by the composition of blood and flow velocity.

The vessel wall metabolism also plays an important role, since an increase of this parameter due for instance by an inflammatory process or increased work of the vessel wall due to contraction work against blood pressure increases wall metabolism and lowers tissue oxygenation (since the increase in K_n lowers capillary pO_2).

5.   The Role of Autoregulation

The delivery of oxygen to the tissue, which is described by the capillary blood oxygen tension, is regulated by related active and passive mechanisms. As shown by the foregoing analysis, the physical properties of blood play a key role in setting the passive component of regulation through the interaction between blood viscosity, oxygen carrying capacity and position within the microvascular network of the dissociation curve for hemoglobin (i.e. the value of n at each segment i). The existence of this coupling has only recently come under scrutiny, and is of importance for blood substitutes, since their use decreases the concentration of red blood cells, the major determinant of blood viscosity. Altering blood viscosity changes blood flow velocity, thereby lowering viscosity and increasing flow velocity, which lowers diffusional oxygen exit from the arterioles and increases their oxygen tension.

Superposed to the passive delivery of oxygen determined by the physical properties of the microcirculation, in the normal tissues there is a process of active blood flow regulation and oxygen delivery that ensures that parenchymal cells receive an adequate supply of oxygen to satisfy the requirements of oxidative metabolism. Furthermore, blood flow regulatory mechanisms also prevent an overabundance of oxygen delivery to the tissues. The linkage between blood supply and oxygen demand is due to the existence of chemical vasodilator signals transmitted from the tissue cells to the resistance vessels, which relax (increase their diameter) causing blood flow to increase when tissue oxygen tension falls below critical levels. Conversely, increasing blood oxygen tension over normal levels elicits vasoconstriction.

The prevailing assumption is that both passive and active flow regulation are focused on ensuring that capillaries receive an appropriate supply of oxygenated blood. The parameter that characterizes the role of capillaries as a system for oxygenating blood is functional capillary density defined as the number of capillaries that possess red blood cell transit. This parameter is regulated in such a fashion that over-oxygenation, hypertension and hypotension lower functional capillary density with the consequence that tissue oxygen decreases.

There is also a component of autoregulation that is directly linked to flow, whereby the endothelium directly senses blood flow. The mechanism relates blood viscosity, μ, and shear rate dv/dr which determine shear stress τ_{wall} at the blood vessel wall ($\tau_{\text{wall}} \approx \mu dv/dr$). Production of the vasodilators prostacyclin and NO by the endothelium is a direct function of the shear stress generated by blood at the vascular wall. Consequently lowered blood viscosity is beneficial only if blood flow velocity increases in proportion so that τ_{wall} remains constant, otherwise if flow velocity does not increase to the required level, the net effect of lowered viscosity in terms of endothelium derived factors regulation is vasoconstriction.

The implication of these considerations is that stroma cell-free oxygen carriers may perform the task of substituting red blood cells in oxygen delivery. However, their transport properties must be tailored so that the consequent physiological reactions do not negate the intended function. This becomes apparent in predicting the consequences of substituting blood with stroma free hemoglobin solutions, which is the current direction of the technology for artificial blood.

Paradoxically, lowering blood viscosity while maintaining oxygen carrying capacity, as it occurs with most of the presently developed artificial bloods is counter productive, because as consequence of the increased flow velocity due to the lower number of red blood cells, flow velocity is increased and too much oxygen arrives to the microcirculation. This overabundance of oxygen delivery is sensed by the mechanism of "autoregulation" which is designed to maintain oxygen delivery constant, which causes constriction of the blood vessels in order to limit the oxygen supply, and hypertension. An additional adverse effect of these hemoglobin solutions is due to the shear stress at the vessel wall (i.e. the frictional force caused by the passage of blood over the blood vessel wall) which causes the release of vasodilators in proportion to the product of blood flow velocity and viscosity, and lowering both factors significantly reduces shear stress and the release of vasodilator, aggravating the vasoconstrictor effect. In fact, most of the "artificial bloods" developed so far cause hypertension.

6.   Summary and Future Developments

The mathematical model presented here describes how the physical properties of blood determine the distribution of oxygen in the circulation, and particularly the microcirculation, which is the site where blood exchanges oxygen with the tissue. Some of the variables are interdependent and inclusion of their functional relationship further refines the model. For instance, blood viscosity and hematocrit are directly related. Analysis of Eq. (13) also shows that, in general, viscosity and oxygen carrying capacity must change simultaneously if we wish to achieve normal capillary oxygen tension. Lowering blood viscosity while maintaining oxygen carrying capacity causes too much oxygen to arrive to the capillaries, a situation that is sensed by autoregulation and which leads to vasoconstriction. This equation also shows why vasoconstriction is deleterious to tissue oxygenation since a small change in vessel diameter is significantly magnified since this parameter is cubed.

The analytical development yields a very well defined result: maintenance of oxygen carrying capacity requires the maintenance of the viscosity of the mixture of blood and the oxygen carrying blood substitute at a level close to that of natural blood. Consequently, success in developing an oxygen carrying blood substitute is dependant on the availability of comparatively high viscosity infusible, biocompatible materials. This analysis illustrates the power of the engineering method in the analysis of complex phenomena and for finding the relationship between variables. In summary, the design of artificial blood may be considered as classical bioengineering endeavor, where engineering analysis is used in parallel with the knowledge of biological mechanism, which in this case are represented by our information on the range of adaptability of the circulation to a varying oxygen shown by the tolerance to hemodilution, and the reaction of the microcirculation to overabundance of oxygen.

7.   References

7.1.   *Articles*

Dodd, R. Y., The risk of transfusion transmitted infection, *N. Eng. J. Med.* **327**, 419–420 (1992).

Intaglietta, M., Johnson, P. C. and Winslow, R. M., Microvascular and tissue oxygen distribution, *Cardiovasc. Res.* **32**, 632–643 (1996).

Intaglietta, M., Whitaker Lecture 1996: Microcirculation, biomedical engineering and artificial blood, *Ann. Biomed. Eng.* **25**, 593–603 (1997).

Kerger, H., Torres Filho, I. P., Rimas, M., Winslow, R. M. and Intaglietta, M., Systemic and subcutaneous microvascular oxygen tension in conscious Syrian golden hamsters, *Am. J. Physiol.* **267** (Heart. Circ. Physiol. **37**) H802–810 (1995).

Torres Filho, I. P. and Intaglietta, M., Microvessel pO$_2$ measurements by phosphorescence decay method, *Am. J. Physiol.* **265** (Heart. Circ. Physiol. **34**) H1434–H1438 (1993).

Tsai, A. G., Friesenecker, B., Mazzoni, M. C., Kerger, H., Buerk, D. G., Johnson, P. C. and Intaglietta, M., Microvascular and tissue oxygen gradients in the rat mesentery, *Proc. Natl. Acad. Sci. USA* **95**, 6590–6595 (1998).

7.2.   *Books*

Tuma, R. R., White, J. V. and Messmer, K., eds. The Role of Hemodilution in Optimal Patient Care. Zuckeschwerdt Verlag GmbH, Munich (1989).

Winslow, R. M., Vandegriff, K. D. and Intaglietta, M., eds. Blood Substitutes. Physiological Basis of Efficacy. Birkhäuser, Boston, (1995).

Winslow, R. M., Vandegriff, K. D. and Intaglietta, M., eds. Blood Substitutes. New Challenges. Birkhäuser, Boston (1996).

Winslow, R. M., Vandegriff, K. D. and Intaglietta, M., eds. Advances in Blood Substitutes. Industrial Opportunities and Medical Challenges. Birkhäuser, Boston, (1997).

Proc. Natl. Acad. Sci. USA
Vol. 95, pp. 6590–6595, June 1998
Engineering

Microvascular and tissue oxygen gradients in the rat mesentery

(arterioles/vessel wall/endothelium/oxidative metabolism)

Amy G. Tsai[*][†], Barbara Friesenecker[*][‡], Michelle C. Mazzoni[*][§], Heinz Kerger[*][¶], Donald G. Buerk[‖],
Paul C. Johnson[*], and Marcos Intaglietta[*]

*Department of Bioengineering, University of California, San Diego, La Jolla, CA 92093-0412; and ‖Departments of Physiology and Bioengineering, and the Institute for Environmental Medicine, University of Pennsylvania, Philadelphia, PA 19104-6086

Communicated by Yuan-Cheng B. Fung, University of California, La Jolla, CA, March 27, 1998 (received for review May 14, 1997)

ABSTRACT   One of the most important functions of the blood circulation is O_2 delivery to the tissue. This process occurs primarily in microvessels that also regulate blood flow and are the site of many metabolic processes that require O_2. We measured the intraluminal and perivascular pO_2 in rat mesenteric arterioles *in vivo* by using noninvasive phosphorescence quenching microscopy. From these measurements, we calculated the rate at which O_2 diffuses out of microvessels from the blood. The rate of O_2 efflux and the O_2 gradients found in the immediate vicinity of arterioles indicate the presence of a large O_2 sink at the interface between blood and tissue, a region that includes smooth muscle and endothelium. Mass balance analyses show that the loss of O_2 from the arterioles in this vascular bed primarily is caused by O_2 consumption in the microvascular wall. The high metabolic rate of the vessel wall relative to parenchymal tissue in the rat mesentery suggests that in addition to serving as a conduit for the delivery of O_2 the microvasculature has other functions that require a significant amount of O_2.

Blood entering capillaries is only 50% saturated (1), thus one-half of the O_2 gathered by the lung exits the circulation before arrival in the capillaries. This rate of O_2 loss from the arteriolar network has been found and documented in different species and tissues at rest (2–7). It was noted in one study that this loss is an order of magnitude greater than expected from simple diffusion (8). Possible sinks for the O_2 exiting before the capillaries include O_2 shunting from arterioles to parallel venules (9), periarteriolar tissue consumption (10), and arteriolar-capillary O_2 diffusional shunting (11). The contribution of arterio-venous shunting appears to be negligible during normal conditions (9), and although arterio-capillary diffusional shunting has been demonstrated (11), the conditions under which it might occur, i.e., an arteriole crossing a capillary network, are not common in the mesentery. It therefore is important to examine in detail the loss of O_2 from arterioles to surrounding tissue. The outward flux of O_2 from blood is governed by the law of diffusion and defined by O_2 gradients between blood and the surrounding tissue. O_2 delivery to the tissue surrounding the microvessels has not been extensively studied because of the lack of methods that can measure pO_2 in both blood and tissue. Recently, an optical method has been developed that makes such a study feasible (12).

The objective of our study was to use noninvasive phosphorescence quenching microscopy to determine the radial pO_2 profiles *in vivo* for the periarteriolar tissue of the rat mesentery. With this method it was possible to determine the contribution of microvascular wall metabolism to the precap-

illary O_2 exit. Intravascular and extravascular measurements were carried out by using the same pO_2 measuring technique, and validation of tissue fluxes was obtained by O_2 mass balance between O_2 entering and exiting vascular segments, and the O_2 diffusional fluxes were determined by the measured O_2 gradients.

MATERIALS AND METHODS

Phosphorescence Quenching Microscopy. Pd-phosphorescence quenching microscopy, based on the relationship between the decay rate of excited Palladium-mesotetra-(4-carboxyphenyl)porphyrin (Porphyrin Products, Logan, UT) bound to albumin and the pO_2 according to the Stern-Volmer equation, was used to measure pO_2 in the microcirculation (13). The method was used previously in microcirculatory studies to determine blood pO_2 levels in different tissue under different conditions (14–18). In our system, a xenon strobe (EG & G, Salem, MA; decay constant of 10 μsec, frequency of 30 Hz; peak wave length of 420 nm) excites the phosphorescence by epi-illumination of a tissue area for 3 sec. Phosphorescence emission from the target tissue area passes through an adjustable rectangular optical slit and light filter (630-nm cutoff) and is captured by a photomultiplier (EMI, 9855B; Knott Elecktronick, Munich, Germany). A digital oscilloscope (Tektronix, 2434) averages 90 photomultiplier decay signals, and the resulting smoothed curve is stored to a computer. Decay curves are analyzed off-line, by using a standard single exponential least-squares numerical fitting technique, and the resultant time constants are applied to the Stern-Volmer equation to calculate pO_2, where k_Q, the quenching constant and τ_0, the phosphorescence lifetime in the absence of O_2 measured at pH = 7.4 and T = 37°C are 325 mmHg^{-1} sec^{-1} and 600 μs, respectively. The phosphorescence decay caused by quenching at a specific pO_2 yields a single decay constant (12), and *in vitro* calibration has been demonstrated to be valid for *in vivo* measurements.

As a consequence of the two-dimensional, single-layer planar characteristic of the vasculature in the exteriorized mesentery preparation, our phosphorescence exponential decay signal is not contaminated by signals from vessels above and below the focus plane and usually is fitted with a high level of correlation ($r > 0.92$) by a single exponential. Modeling of the decay curves obtained from our measurements as multiple exponentials does not alter the results obtained with a single exponential fit. Our measurements, based on minimal and

---

†To whom reprint requests should be addressed. e-mail: agtsai@ucsd.edu.

‡Present address: Department of Anesthesia and Intensive Care Medicine, The Leopold-Franzens-University of Innsbruck, Innsbruck, Austria.

§Present address: Alliance Pharmaceutical Corp., San Diego, CA 92121.

¶Present address: Institute for Anesthesiology and Operative Intensive Care, Mannheim, Heidelberg University, Mannheim, Germany.

uniform light exposure, do not show evidence of changes in phosphorescence intensity within the tissue near the vessel wall. Furthermore, we found no change in the measured pO_2 distribution when using windows slits of widths varying from 7.5 to 20 μm for characterizing the gradient when the center line of the window was assumed to be the location of the measurement, indicating that the window provides a consistent average of the decay curves within the window.

Interstitial pO_2 Measurement. The albumin-bound probe passes into the interstitium according to the exchange of albumin from blood to tissue (19). The resulting accumulation of albumin-bound dye within the tissue, which may contain up to 10% of the total albumin in the organism, allows measurement of tissue and intravascular pO_2 at high resolution with the same technique. The reflection coefficient of albumin in different vascular networks varies, changing the equilibration time between intravascular and extravascular dye/albumin. Because measurement of pO_2 is possible anywhere Palladium-porphyrin albumin-bound complex is located given an adequate signal-to-noise ratio (15–17), in our model tissue pO_2 measurements could be made within 20 min after injection. Interstitial pO_2 profiles were determined by measuring pO_2 in the tissue at specific distances from the vessel lumen.

Phosphorescence generated by the light excitation of the porphyrin probe consumes O_2. This photoactivation could be a factor affecting tissue O_2 measurements made in a slow-moving or stationary fluid. To determine the extent of this photoactivation on our interstitial measurements, we performed *in vitro* measurements in sealed tubes. We estimated O_2 consumption by the technique in sealed 75-mm long hematocrit tubes filled with Pd-meso-tetra(4-carboxyphenyl-)porphyrin bound to albumin solution (0.1 mg/ml) saturated to pO_2 of 37 mmHg with the same system used for animal experiments. The concentration of probe used is approximately that which would be found in interstitial fluid. Albumin concentration in tissue at steady state is approximately one-third lower than in blood, therefore the probe concentration of 0.26 mg/ml plasma within the blood (= 10 mg/ml probe × 0.1 ml probe/100 g animal × 100 g animal/7.0 ml blood × 1 ml blood/0.55 ml plasma) corresponds to an interstitial fluid probe level of 0.09 mg/ml. The initial pO_2 within the tube was measured at three locations 1.5 mm apart with a 3-sec, 30 flashes/sec pulsed light. The center of the tube was masked and exposed to 45 min of pulsed illumination at a rate of 30 flashes/sec, in such a fashion that 0.5 mm of tube length was exposed to flash illumination. Immediately after the pulsed exposure, the tube content was mixed by a nylon bar sealed in the tube, which was made to move by gravity along the length of the tube for 5 sec in each direction during a 5-min period. pO_2 measured in the tube at the three locations decreased from 37 to 31 mmHg. Therefore 81,000 light flashes lowered pO_2 within the exposed tube segment by (6 mmHg × 75 mm/0.5 mm × 8.1 × 10^4) = 0.01 mmHg/flash. The decrease in O_2 at the excitation spot was initially a factor of 2 higher than the rate determined for the entire flashing period. O_2 initially is consumed solely at the excitation spot, establishing a concentration gradient whereby consumption is lowered by diffusion of O_2 from surrounding areas into the excitation spot.

We used 90 flashes (3 sec) for each measurement. We estimate that the free fluid in the mesentery in which albumin-bound probe is present is at most 20% of the total tissue volume; the rest of the tissue is not occupied by the probe and constitutes a reservoir of O_2. Therefore the maximal reduction in pO_2 during each determination, assuming the highest rate found for O_2 consumption by the flash, is about 0.02 mmHg/flash × 90 flashes × 0.2 = 0.4 mmHg.

We also examined the accuracy of the tissue pO_2 measurements with the phosphorescence method *in vivo* by simultaneous continuous measurements with approximately 5-μm diameter recessed tip gold cathode microelectrodes (20). Mea-

surements were made in avascular tissue areas of the hamster skin fold preparation (15–17). In these studies, the tissue was isolated from extraneous sources of O_2 by superfusing with a Krebs Henseleit solution (1.3 × 10^{-1} M NaCl, 1.2 × 10^{-3} M Mg SO_4, 2.0 × 10^{-3} M $CaCl_2$, 4.7 × 10^{-3} M KCl, 2.2 × 10^{-2} M $NaHCO_3$) bubbled with 100% N_2. A maximum divergence of 2% was found between the methods over the tissue pO_2 range of 5–40 mmHg. Extended flashing over a period of up to 1 min did not produce a detectable change in the microelectrode measurement. The congruence between microelectrode and phosphorescence quenching microscopy measurements demonstrates that excitation of the porphyrin probe in the tissue is minimal and does not affect tissue pO_2 in our model. It should be noted that those investigators (15–17) using the same technique reported that capillary pO_2 in the hamster skin fold preparation was 29 mmHg, tissue pO_2 was 23 mmHg, and the gradient at the capillary wall was 4 mmHg (21). This finding is incompatible with the assumption that the pO_2 decrease on the order of 20 mmHg found near arterioles is caused by O_2 consumption by the dye in the tissue, because capillaries have 100-fold lower O_2 delivery capacity than the arterioles in this study, which would result in a near-zero pericapillary tissue pO_2 measurement.

Mesentery Preparation. The mesentery of male Wistar rats (250–350 g) was exteriorized, and the upper and lower surfaces of the tissue were sandwiched between two essentially gas-impermeable and transparent barriers, allowing O_2 to exit from the vasculature only in the plane of the tissue as shown diagrammatically in Fig. 1. After anesthetizing the animal (Nembutal, 40 mg/kg i.m. into left hind limb), the right femoral vein and artery are cannulated for: (*i*) injection of Pd-phosphorescence probe and supplemental anesthesia (5–10 mg/kg), and (*ii*) monitoring blood pressure. The animal is placed on a microscope stage with a circulating water heater (37°C). The mesentery is exteriorized through an epigastric incision, partially extracted, and kept moist with dripped buffered Krebs Henseleit solution, pH corrected by bubbling a gas mixture of 5% CO_2 and 95% N_2. It is viewed by draping it over a thin circular glass platform. Before the experiment the

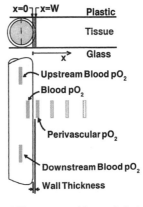

Fig. 1. (*Upper*) The geometry of the exteriorized mesentery preparation. Mesentery is sandwiched between O_2 impermeable upper and lower barrier of plastic and glass. Circle represents blood vessel with vessel wall. O_2 diffuses in the planar, x, direction, $x = 0$ is the blood side of the tissue/blood interface. W is the thickness of the vessel wall. (*Lower*) The positions of the optical window (shaded rectangles) are shown schematically. Upstream and downstream pO_2 along with vessel diameter, blood velocity with a given separation distance is used to calculate the convective O_2 losses. O_2 mass balance is studied by comparing O_2 loss along and diffusional loss out of the vessel. pO_2 levels detected in blood and at increments in tissue, illustrated by the staggered window in x-direction are used to construct pO_2 profiles. Window size was a 5 μm × 20 μm rectangle.

6592   Engineering: Tsai *et al.*

Proc. Natl. Acad. Sci. USA 95 (1998)

tissue is covered with a thin strip of transparent plastic film (Saran, Dow Corning) to prevent gas exchange [diffusion coefficient of 6.2×10^{-13} cm^3 O$_2$/(cm $\times$ mmHg $\times$ sec)] and desiccation. The heated drip without the gas is continued onto the plastic to maintain tissue at 37°C during the study. A temperature probe (Physitemp, Clifton, NJ) was placed on the plastic and positioned above the mesenteric window under study to monitor tissue temperature during the experiment.

Experimental Protocol. The animal was positioned on an inverted microscope (IMT-2, Olympus, New Hyde Park, NY) equipped with a 40× objective (WPlan FL40x, 0.7 water). Measurements began 20 min after injection of phosphorescence probe (10 mg/ml, 0.1 ml/100 g). Transillumination (halogen lamp, 12 V, 100 W) was used to measure vessel diameter (22) and blood flow velocity (23), followed by intravascular and perivascular pO$_2$ determinations using phosphorescence quenching microscopy. Transillumination measurements were made with a maximum of 2–3 V power to the lamp to reduce possible light toxicity and photoactivation of the phosphorescence probe. All pO$_2$ measurements were performed after extraneous light from the transillumination lamp and the room were eliminated. Each intravascular pO$_2$ measurement was repeated after perivascular pO$_2$ determinations to assure constant delivery conditions were maintained. Vessels whose intravascular pO$_2$ readings differed by more than 5 mmHg were not included in the study. Measurements were obtained in vessels with sharp focus and not in close proximity to other vessels (>250 μm separation). The optical measuring windows (5 μm × 20 μm) were placed as shown in Fig. 1 relative to the vasculature under study. Extravascular pO$_2$ profiles were obtained by progressively displacing the optical measuring window away from the blood/tissue interface. All measurements were performed within a 30-min period. The resolution of the optical window does not allow for pO$_2$ measurements within the vascular wall.

Vessel Wall Gradient. In this model, an O$_2$ gradient between blood and perivascular tissue greater than that dictated by simple diffusion is indicative of O$_2$ consumption by the vessel wall when the calculated diffusion flux shows mass balance with the decrease of O$_2$ content in the flowing blood. Thus the vessel wall gradient, the difference between intra- and perivascular pO$_2$, was used to determine the rate of vessel wall oxidative metabolism and later used in determining the mass balance in vessel segments. The perivascular pO$_2$ profile was used to calculate tissue oxidative metabolism. The pO$_2$ measured in tissue adjacent to the O$_2$ source (about 4 μm from the blood-tissue interface) could, in principle, be contaminated by the higher intensity phosphorescence decay signal from blood. The net effect of this contamination, if it exists, would result in a higher perivascular pO$_2$ reading and thus underestimate the measured vessel wall gradient.

Theoretical Simulation and Analysis. Our system consisting of a thin tissue whose thickness is much less than the distance between adjacent vessels, can be modeled in terms of diffusion from a linear source (the blood vessels) with thickness equal to that of the tissue (assumed to be identical to the separation between the impermeable barriers). This process is described by Fick's law of diffusion for O$_2$ into a semi-infinite region extending away from the source in the x direction, where the consumption rate is g and the diffusivity is D. The governing equation at steady state is:

$$\alpha D \, (d^2p/dx^2) - g = 0,\qquad [1]$$

where p = oxygen tension and α = solubility constant for O$_2$. We assume that there is no O$_2$ gradient perpendicular to the tissue plane and that the mesentery is a homogenous tissue with a constant rate of O$_2$ consumption.

Tissue and Vessel Wall Metabolic Rate. With boundary conditions of a constant O$_2$ source p_W at the vessel wall/tissue

interface $x = W$ and a no flux of O$_2$ at a penetration distance δ from vessel wall/tissue interface, the solution of Eq. **1** is:

$$p(x) = K_0 \, x^2 + K_1 \, x + K_2.\qquad [2]$$

The coefficients are $K_O = g/(2D\alpha)$, $K_1 = -(g\delta)/(D\alpha)$, and $K_2 = p_W - [(gW)/(D\alpha)][(W/2)-\delta]$; W, is the location of the vessel/tissue interface. Curve fitting of the perivascular pO$_2$ profiles to the second-order polynomial solution was used to determine the tissue consumption rate from the coefficient K_o. The distance of O$_2$ penetration from the vessel wall/tissue interface, δ, is determined from K_1.

A finite difference numerical approximation of Eq. **1** is used to calculate the steady-state profile for the perivascular tissue, consisting of two metabolically active regions representing the vascular wall and the tissue proper. By using the consumption rate of tissue obtained with curve fitting and the distance of O$_2$ penetration, the rate of vessel wall metabolism is iterated until the measured intra- and perivascular difference is attained. The boundary and regional conditions are: (*i*) At blood side of blood/tissue interface, $x = 0$, O$_2$ tension is equal to intravascular pO$_2$, $p = p_i$; (*ii*) At interface between vessel wall p_w and tissue p_T, $x = W$, O$_2$ flux is continuous $D\alpha \, (dp_w/dx) = D\alpha \, (dp_T/dx)$; (*iii*) At penetration distance, $x = \delta$, O$_2$ level is constant, $dp/dx = 0$; (*iv*) g_W, O$_2$ consumption rate in region comprising the vessel wall ($0 < x \le W$); and (*v*) g_T, O$_2$ consumption rate in region beyond the vessel wall ($x > W$) to the penetration depth ($x = \delta$).

Mass Conservation Analysis of O$_2$ Transport. The law of mass conservation stipulates that the amount of O$_2$ lost from a vascular segment must be equal to the diffusional O$_2$ flux, determined by the perivascular pO$_2$ gradient. We studied longitudinal and radial O$_2$ losses from nonbranching vessel segments to determine the extent of O$_2$ loss in our animal model. The equation for the rate of longitudinal O$_2$ loss along the length of a microvessel is:

Change in longitudinal O$_2$ flux

$$= (Q[O_2])_{\text{upstream}} - (Q[O_2])_{\text{downstream}},\qquad [3]$$

where $Q = v\pi r^2$ and $[O_2] = C_b \, [fHCT] \, SO_2$. The parameters are: Q, blood flow; v, blood flow velocity, corrected for velocity profile shape (24); r, vessel radius; $[O_2]$, O$_2$ content; C_b, binding capacity of blood; $[fHCT]$, ratio of microvascular to systemic hematocrit, a function of vessel diameter (25) (physically dissolved O$_2$ is neglected because of low solubility of O$_2$ in plasma); SO_2, hemoglobin O$_2$ saturation, intravascular pO$_2$ measurements are converted to SO_2 (26).

Longitudinal O$_2$ exit must be matched by the rate of O$_2$ exit by diffusion and O$_2$ consumption by the vessel wall. The area through which O$_2$ diffuses out of the blood vessel into the tissue is assumed to be the plane perpendicular to the impermeable barriers. The remaining portion of the blood vessel, V_W, in contact with the impermeable barriers is assumed to consume O$_2$ at the experimentally determined rate g_w. However this portion of the vessel wall does not contribute to the diffusive flux into the tissue. This rate of O$_2$ exit by diffusion and O$_2$ consumption by the vessel wall is expressed as

Diffusive O$_2$ flux and O$_2$ wall consumption

$$= DA\alpha[p_i - p_o]/dx + V_w g_w;\qquad [4]$$

where D, diffusivity (1.70×10^{-5} cm^2/sec) (27); A, surface area; α, O$_2$ solubility constant (2.14×10^{-5} ml O$_2$/cm^3 mmHg) (27); p_i, intravascular pO$_2$; p_o, perivascular pO$_2$; dx, vessel wall thickness, approximated as 10% of vessel diameter (2.3 μm); V_W, volume of vessel wall; g_w, O$_2$ consumption rate of vessel wall determined by the theoretical analysis of measured tissue pO$_2$ profiles.

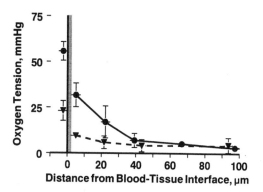

FIG. 2. Large drop in pO₂ across the blood/tissue interface. PO₂ profiles from blood into tissue for high (●) and low (▲) intravascular pO₂. Results were grouped according to intravascular pO₂ levels. Intravascular pO₂ measurements were made flush to the intraluminal vessel wall with the center of the window slit 2.5 μm from the vessel wall. Perivascular pO₂ begins with the window flush with the abluminal vessel wall. The solid horizontal bars at each data point represent the width of the rectangular slit. The vascular wall is depicted as the shaded area. Individual measurements from different profiles are grouped as function of distance from intraluminal wall (mean ± SD).

Curve fitting to the data was performed by first transforming the data to allow for linear least-squares regression. Gauss-Jordan method for matrix inversion then was used to determine the coefficients from the transformed data (Table Curve 2-D, Jandel, San Rafael, CA). Results are presented mean ± standard error unless otherwise stated.

RESULTS

PO₂ profiles were obtained in 12 animals from 31 arterioles with sharp focus and not in close proximity to other vessels (>250-μm separation). Results were grouped according to intravascular pO₂ levels ($p_i > 50$; $40 < p_i \leq 50$; $30 < p_i \leq 40$; and, $20 < p_i \leq 30$ mmHg). Profiles for two groups are shown in Fig. 2 to demonstrate the pronounced fall between intravascular ($x = 0$) and perivascular ($x = 5$ μm) pO₂, a distance spanning the vessel wall, which includes endothelium and smooth muscle. The finding of a steep fall in pO₂ near the

vessel wall was common to all groups. Fig. 3 shows a more closely sampled pO₂ profile, which exhibits the same characteristics as those more sparsely sampled profiles used in the analysis.

The theoretical profile for diffusion from a constant source into a consuming medium was constructed to determine whether measured pO₂ profiles corresponded to a region with a single metabolic rate, implying that vessel wall O₂ consumption is equal to that in the tissue proper, or whether the vessel wall and tissue have different metabolic rates (Eq. 2). The measured tissue pO₂ profiles were fitted to the theoretical solution for the tissue region beyond 2.5 μm of the blood/tissue interface to determine the tissue O₂ consumption rate and penetration. Average g_T and δ for all groups were 2.4 × 10⁻⁴ ± 0.5 ml O₂/(cm³ tissue sec) and 93.6 ± 11.1 μm, respectively. Least-squares regression correlation coefficients of the curve fit to the profiles ranged from $r = 0.87$ to 0.99, being all statistically significant $P < 0.01$. Fig. 4 shows the mean pO₂ profile from the pO₂ > 50 mmHg group and the curve fit used to determine the tissue O₂ consumption rate.

The two-compartment model was used to determine the vessel wall O₂ consumption rate relative to the tissue. We found the average vessel wall metabolic rate needed to achieve the intravascular pO₂ levels given the tissue pO₂ profiles was 277.8 ± 45.0 (range 141–323) times that of the tissue metabolic rate or 3.9 ml O₂/(cm³ tissue min).

If the O₂ consumption of the vessel wall is as great as the calculation above suggests there should be a corresponding loss of O₂ longitudinally. Therefore we measured the rate of O₂ exit both longitudinally and radially in 12 vessel segments of five animals. Longitudinal saturation drop was 2.4 ± 0.3%/100 μm in vessels with average diameter of 23.2 ± 2.9 μm and blood flow velocity of 1.5 ± 0.3 mm/sec. Intravascular pO₂ was 43.3 ± 3.7 mmHg and the intraperivascular difference was 18.1 ± 1.7 mmHg.

Loss of O₂ along each individual vessel segment (Eq. 3) was found to be within 3.3 ± 19.6% of their diffusional losses (Eq. 4), confirming that the rate of O₂ exit is caused by the extremely large O₂ gradient in the tissue compartment that includes the vessel wall. The average convective loss of 2.4 ± 0.5 × 10⁻⁵ ml O₂/cm² sec and diffusive loss of 2.2 ± 0.2 × 10⁻⁵ ml O₂/cm² sec in all of the vessels studied agreed within 8.6%. Calculation of O₂ loss from averaged parameters does not yield

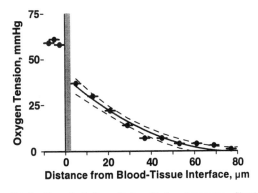

FIG. 3. Example of pO₂ profile found in the rat mesentery. Closely sampled pO₂ measurements allow for a detailed view of tissue pO₂ profile. Solid line is the curve fitting of the pO₂ profile within the tissue, dotted lines are the 95% confidence intervals of the data. Intravascular pO₂ averaged 59.3 ± 1.5 mmHg (horizontal bar), shown left of the blood tissue-interface. The wall was taken to be 2.5 μm thick, and each data point represents the pO₂ within the sampling window whose width is represented by the horizontal bars.

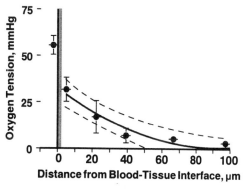

FIG. 4. Graph shows curve fitting to the tissue profile with tissue having one consumption rate. An average g_T of 2.4 × 10⁻⁴ ± 0.5 ml O₂/(cm³ tissue sec) and δ of 93.6 ± 11.1 μm were determined from the coefficients. Least-squares regression correlation coefficient of the curve fit to the perivascular portion of the pO₂ profiles is $r = 0.98$, and the dotted line represents the 95% confidence intervals. When the two-compartment model is used to fit the tissue pO₂ profile with the intravascular pO₂ of 55.6 mmHg, the vessel wall consumption rate was found to be 141 times the tissue rate. Data is grouped axially and presented as mean ± SD ($n = 6$ profiles).

6594   Engineering: Tsai *et al.*   *Proc. Natl. Acad. Sci. USA 95* (1998)

mass balance because the parameters that determine O_2 fluxes are not related in an additive form.

DISCUSSION

This study shows that there is a substantial O_2 loss from the blood as it passes through the arteriolar network in the rat mesentery. This was first reported by Duling and Berne (1) using recessed microelectrodes in the hamster cheek pouch and subsequently in other laboratories (2–7) using microelectrode and spectrophotometric measurement techniques.

In addition to the large O_2 loss, we found a large pO_2 gradient across a tissue region that includes the arteriolar vessel wall. The presence of this gradient can account for the large rate of O_2 loss from microvessels because, in accordance with the law of mass balance, our results require that the large rate of O_2 exit from arterioles be driven by a large O_2 gradient. The steep O_2 gradient across the wall could be caused by a large amount of O_2 consumption or an increased diffusional resistance within the vessel wall (lower diffusion constant and/or permeability). However, in the latter case the resulting steep gradient would lead to significantly lower O_2 exit, and therefore a large difference between longitudinal O_2 loss and tissue O_2 flux and a large disparity in mass balance.

The low levels of pO_2 in the tissue (and therefore large gradients) could be caused by O_2 consumption by the technique because the fluid containing the probe in the tissue is virtually stationary during the period of measurement. Intravascular measurements would not be affected by O_2 consumption because the blood flow would replace the consumed O_2. Simultaneous measurements of tissue pO_2 levels using our phosphorescence decay system and recessed microelectrodes found no statistical change in pO_2 during our 3-sec measurement period or during prolonged flashing. Moreover when using the phosphorescence technique to measure tissue pO_2 in the hamster skin fold chamber we found values between 23 and 25 mmHg (7, 15–17), similar to the value of 26 mmHg reported by Endrich *et al.* (28), who used surface platinum multiwire electrodes in the same tissue. Our values obtained using the phosphorescence technique are higher than those reported with microelectrode measurements in the hamster cremaster muscle, which ranged from 11 to 17 mmHg (29, 30). Therefore our tissue pO_2 measurements are within the range of previous measurements made by other investigators using different techniques and are not unduly lowered by photoactivation of the phosphorescence probe.

Our *in vitro* measurements of the amount of O_2 consumption per flash described in *Materials and Methods* also corroborated the electrode findings that O_2 quenching by photoactivation of the probe is not a factor affecting our measurements. The equivalent lowering of tissue pO_2 was found to be about 0.4 mmHg. This number is indicative of the maximum possible error in the measurement; however, the actual error is much smaller because as O_2 is consumed it also is replenished by diffusion from the arteriole. Tissue O_2 depletion by flash illumination in the area of measurement (about 140 μm in diameter, 50 μm thick) is calculated to be 1.4×10^{-12} ml O_2/sec, whereas the arterioles supply O_2 at a rate of 4.1×10^{-9} ml O_2/sec, or a rate that is about 3,000 times greater than that at which O_2 is consumed by the probe during the measurement, indicating that pO_2 measurements are not affected by probe O_2 consumption near the vessel wall.

The spatial resolution of our technique is not sufficient to measure the pO_2 profile within the vessel wall; however, we can accurately measure the pO_2 difference across the vessel wall. To measure pO_2 as close as possible to the blood-tissue interface, the optical windows were as narrow as possible, compatible with the need for adequate signals. The tissue pO_2 profile also was determined with a window that was 10 μm by 20 μm, double the width used in this study, and there was no

difference in the pO_2 measurements even in regions of steep gradients, indicating that in these experiments pO_2 is averaged within the window.

In our mass balance calculations, we approximated vessel wall thickness to be 10% of the average vessel diameter. We found that vessel wall thickness of the vessels studied ranged from 1.3 to 7.5 μm, the magnitude greatly depending on the position of the nucleus of endothelial cells and smooth muscle cells. Wall thickness tends not to be uniform in these vessels, and we used an average wall thickness based on the lumen size. Varying the vessel wall thickness from 2 to 3 μm in our mass balance calculations shifted the balance between O_2 loss along each vessel segment and the diffusional flux to deviate by -5.6% and 17.0%, respectively.

Our measurements of the pO_2 gradient in the mesentery itself provides an estimate of O_2 consumption rate of mesenteric tissue of $2.4 \pm 0.5 \times 10^{-4}$ ml O_2/cm^3 tissue sec, which is about four times higher than the value for loose connective tissue of 5.6×10^{-5} ml O_2/cm^3 tissue sec determined from the renal capsule of the goat (31), a comparison between the same tissue type but from different size animals. O_2 consumption relative to the body size is much higher than in the larger mammals. By using the empirical relationship of mammals size and O_2 consumption, assuming that connective tissue also can be similarly scaled, we estimate that the O_2 consumption rate of the rat to be on the order of 3–4 times higher than the goat (32), and therefore similar to the one found in this study.

Investigations in other tissues and models that used different measurement techniques also found evidence of a high rate of vessel wall metabolism. When endothelium is removed from the dog hind limb O_2 consumption decreases by 34% (33). Measurements in larger arterial vessels using microelectrodes show that the ratio between the O_2 consumption rate and the diffusivity increases nearly 10-fold as the electrode is advanced from the abluminal side toward the blood/vessel interface (34). Therefore given a constant diffusivity the rate of O_2 consumption increases significantly near the blood/vessel interface. However, the O_2 consumption rate suggested by that study was significantly lower than the one we measured, a discrepancy that may be, in part, because of the resolution limitation of 10–15 μm microelectrode tips. Investigations of O_2 consumption by endothelial cells cultured from vascular tissue found a metabolic rate "as much as 5,000-fold" over the rate in organized tissues (35).

The endothelial lining of blood vessels has, until recently, been regarded as simply a barrier between the blood and the parenchyma, greatly restricting passage of macromolecules while allowing rapid exchange of gases and crystalloids. Mitochondria density is not particularly pronounced within endothelium (36) and cannot solely account for the high respiratory rate of the vessel wall suggested by our findings. However, endothelial cells are the site of chemical synthesis and metabolic processes that require O_2 (renin, prostaglandins, collagen, conversion of angiotensin I to II, degradation of bradykinin and prostaglandins, and the clearance of lipids and lipoprotein) (37). The endothelium is a principal mediator in many homeostatic (38) and disease processes (39, 40) and has been considered a widely influential "organ" (41). Moreover, endothelial cells have an active actin/myosin-based contractile system that also may consume O_2 (42, 43). In the blood/tissue interface, endothelium overlays and interacts with smooth muscle throughout the vasculature with the exception of the capillaries, giving origin to the regulatory mechanism that modulates the vessel diameter and controls tissue perfusion. The metabolic cost of this array of endothelial function has not been addressed previously. It is worth noting that because of its location, the endothelium has the highest O_2 availability of all tissues.

Previous studies on microvascular O_2 distribution relied on a spectrophotometric method for intravascular O_2 determination and electrode measurements to characterize tissue pO_2 gradients. Those studies were carried out either longitudinally or radially, but only rarely in both directions with the same technique. Lack of direct evidence for steep gradients at the vessel wall led to the concept that the O_2 diffusion constant for the arteriolar wall may be an order of magnitude larger than that of water to account for the large rate of O_2 loss (8).

Microelectrode studies by Duling *et al.* (2) in pial vessels of the cat revealed both a longitudinal gradient along the vessel and transmural gradient across the vessel wall. They found a difference between blood pO_2 and vessel surface pO_2, which ranged from 27 mmHg in the largest (234 μm diameter) to 6 mmHg in the smallest (22 μm diameter) arterioles. O_2 consumption by the wall was estimated to be 2.8×10^{-2} ml O_2/sec g tissue, which is similar to our finding. Yaegashi *et al.* (44) in their study of pO_2 distribution in mesenteric microcirculation, which used O_2-sensitive fluorescence-coated 3-μm silica gel beads embedded in a silicone rubber membrane 5 μm thick suspended an average 38 μm above the mesenteric surface, found a longitudinal pO_2 gradient three times smaller than the value reported by ourselves and others. Their study did not find the wall pO_2 gradient seen in our study. Because of the separation of the O_2-sensitive membrane and the tissue surface, this method may have limited resolution. From their measurements, they estimated O_2 consumption rate in the rat mesentery to be 8.2×10^{-6} ml O_2/cm^3 tissue sec, which is 30 times lower than our finding.

The metabolic rate of endothelium may vary among organs and vessels within the same organs as they all are subjected to different local conditions and influences. Our finding of a high metabolic rate in the wall of mesenteric arterioles may help to explain in part the large O_2 loss observed in other arteriolar networks in other tissues but this requires further study.

In summary, this report documents the existence of steep O_2 gradients at the wall of arterioles in the rat mesentery. The steepness of the O_2 gradient is compatible with the hypothesis that the arteriolar vessel wall is a region of high O_2 consumption. This finding in the rat mesentery suggests that O_2 delivery to parenchymal tissue is, in part, determined by the vessel wall metabolism in the microcirculation.

This study was supported by U. S. Public Health Service–National Heart, Lung and Blood Institute Grant HL-48108.

1. Duling, B. R. & Berne, R. M. (1970) *Circ. Res.* **27**, 669–678.
2. Duling, B. R., Kuschinsky, W. & Wahl, M. (1979) *Pflügers Arch.* **383**, 29–34.
3. Pittman, R. N. & Duling, B. R. (1977) *Microvasc. Res.* **13**, 211–224.
4. Ivanov, K. P., Derii, A. N., Samoilov, M. O. & Semenov, D. G. (1982) *Pflügers Arch.* **393**, 118–120.
5. Swain, D. P. & Pittman, R. N. (1989) *Am. J. Physiol.* **256**, H247–H255.
6. Buerk, D. G., Shonat, R. D., Riva, C. E. & Cranstoun, S. D. (1993) *Microvasc. Res.* **45**, 134–148.
7. Torres Filho, I. P., Kerger, H. & Intaglietta, M. (1996) *Microvasc. Res.* **51**, 202–212.
8. Popel, A. S., Pittman, R. N. & Ellsworth, M. L. (1988) *Am. J. Physiol.* **256**, H921–H924.
9. Sharan, M. & Popel, A. S. (1988) *Math. Biosci.* **91**, 17–34.
10. Kuo, L. & Pittman, R. N. (1990) *Am. J. Physiol.* **259**, 1694–1702.
11. Ellsworth, M. L. & Pittman, R. N. (1990) *Am. J. Physiol.* **258**, H1240–H1243.
12. Vanderkooi, J. M., Maniara, G., Green, T. J. & Wilson, D. F. (1987) *J. Biol. Chem.* **262**, 5476–5482.
13. Torres Filho, I. P. & Intaglietta, M. (1993) *Am. J. Physiol.* **265**, H1434–H1438.
14. Helmlinger, G., Yuan, F., Dellian, M. & Jain, R. K. (1997) *Nat. Med.* **3**, 177–182.
15. Kerger, H., Saltzman, D. J., Menger, M. D., Messmer, K. & Intaglietta, M. (1996) *Am. J. Physiol.* **270**, H827–H836.
16. Kerger, H., Torres Filho, I. P., Rivas, M., Winslow, R. M. & Intaglietta, M. (1995) *Am. J. Physiol.* **267**, H802–H810.
17. Kerger, H., Saltzman, D. J., Gonzales, A., Tsai, A. G., van Ackern, K., Winslow, R. M. & Intaglietta, M. (1997) *Anesthesiology* **86**, 372–386.
18. Shonat, R. D., Richmond, K. N. & Johnson, P. C. (1995) *Rev. Sci. Instrum.* **66**, 5075–5084.
19. Parker, J. C., Perry, M. A. & Taylor, A. E. (1984) in *Edema*, eds. Staub, N. C. & Taylor, A. E. (Raven, New York), pp. 7–27.
20. Whalen, W. J., Riley, J. & Nair, P. (1965) *J. Appl. Physiol.* **23**, 789–794.
21. Intaglietta, M., Johnson, P. C. & Winslow, R. M. (1996) *Cardiovasc. Res.* **32**, 632–643.
22. Intaglietta, M. & Tompkins, W. R. (1973) *Microvasc. Res.* **5**, 309–313.
23. Intaglietta, M., Silverman, N. R. & Tompkins, W. R. (1975) *Microvasc. Res.* **10**, 165–179.
24. Pittman, R. N. & Ellsworth, M. L. (1986) *Microvasc. Res.* **32**, 371–388.
25. Kanzow, G., Pries, A. R. & Gaehtgens P. (1982) *Int. J. Microcirc. Clin. Exp.* **1**, 67–79.
26. Ulrich, P., Hilpert, P. & Bartels, H. (1963) *Arch. Gesamte Physiol.* **277**, 150–165.
27. Middleman, S. (1972) in *Transport Phenomena in the Cardiovascular System.* (Wiley Interscience, New York), pp. 53 and 131.
28. Endrich, B., Goetz, A. & Messmer, K. (1982) *Int. J. Microcirc. Clin. Exp.* **1**, 81–99.
29. Gorczynski, R. J. & Duling, B. R. (1978) *Am. J. Physiol.* **235**, H505–H515.
30. Klitzman, B., Popel, A. S. & Duling, B. R. (1982) *Microvasc. Res.* **25**, 108–131.
31. Lentner, C. (1986) in *Geigy Scientific Tables: Biochemistry, Metabolism of Xenobiotics, Inborn Errors of Metabolism, and Pharmacogenetics and Ecogenetics* (Ciba-Geigy, West Caldwell, NJ), Vol. 4, 8th Ed., p. 87.
32. Schmidt-Nielsen, K. (1990) in *Animal Physiology* (Cambridge Univ. Press, Cambridge), 4th Ed., pp. 193–194.
33. Curtis, S. E., Vallet, B., Winn, M. J., Caufield, J. B., King, C. Z., Chapler, C. K. & Cain, S. M. (1995) *J. Appl. Physiol.* **79**, 1351–1360.
34. Buerk, D. G. & Goldstick, T. K. (1982) *Am. J. Physiol.* **243**, H948–H958.
35. Bruttig, S. P. & Joyner, W. L. (1983) *J. Cell. Physiol.* **116**, 173–180.
36. Oldendorf, W. H., Cornford, M. E. & Jann Brown, W. (1977) *Ann. Neurol.* **1**, 409–417.
37. Seccombe, J. F. & Schaff, H. V. (1994) *Vasoactive Factors Produced by the Endothelium: Physiology and Surgical Implications* (CRC, Boca Raton, FL).
38. Cain, B. S., Meldrum, D. R., Selzman, C. H., Cleveland, J. C., Meng, X., Sheridan, B. C., Banerjee, A. & Harken, A. H. (1997) *Surgery* **122**, 516–526.
39. Ryan, U. S. & Rubanyi, G. M. (1992) *Endothelial Regulation of Vascular Tone* (Dekker, New York).
40. Luscher, T. L. (1995) *The Endothelium in Cardiovascular Disease: Pathophysiology, Clinical Presentation, and Pharmacotherapy* (Springer, New York).
41. Davies, M. G. & Tripathi, S. C. (1993) *Ann. Surg.* **218**, 593–609.
42. Boswell, C. A., Majno, G., Joris, I. & Ostrom, K. A. (1992) *Microvasc. Res.* **43**, 178–191.
43. Gotlieb, A. I. & Wong, M. K. K. (1988) in *Endothelial Cells*, ed. Ryan, U. S. (CRC, Boca Raton, FL), pp. 31–101.
44. Yaegashi, K., Itoh, T., Kosaka, T., Fukushima, H. & Morimoto, T. (1996) *Am. J. Physiol.* **270**, H1390–H1397.

CHAPTER 5

ANALYSIS OF CORONARY CIRCULATION: A BIOENGINEERING APPROACH

GHASSAN S. KASSAB

Department of Bioengineering, University of California, San Diego

1. Introduction

The coronary blood circulation supplies the heart with oxygen and nutrients and removes its waste products. The hemodynamics of the coronary circulation reveal a number of interesting phenomena such as phasic arterial inflow and venous outflow, spatial and temporal flow heterogeneity, the existence of significant vascular compliance and zero-flow pressure, and autoregulation to mention just a few (see recent review by Hoffman and Spaan, 1992). In order to understand such a complex system, it is necessary to analyze each component separately before synthesizing the whole system. In order to initiate the analysis, we consider the simplest possible situation: steady state blood flow in the diastolic, maximally vasodilated state of the coronary vasculature. The effects of systole, vasoactivity and time-varying boundary conditions will be added to the model once a foundation is established using the engineering approach.

In a bioengineering approach to coronary blood flow analysis, one should use the vascular geometry and branching pattern, mechanical properties of the coronary vessels (arteries, capillaries and veins), and rheology of blood in the coronary vasculature, apply the basic laws of physics to write down the governing equations and specify the appropriate boundary conditions, and solve the boundary-value problems. To date, only two organs have been subjected to such detailed analysis. One organ is that of the cat lung (Zhuang *et al.*, 1983; Fung, 1997) and the other is the Spinotrapezius muscle in the rat (Schmid–Schoenbein, 1986). Recently, a full set of data describing the branching pattern and vascular geometry of the entire porcine coronary vasculature, from arteries to capillaries and veins, in the diastolic, maximally vasodilated state has been obtained in our laboratory (Kassab *et al.*, 1993a, 1994, 1994b, 1997a).

In this chapter, I will briefly review the morphometric data of the coronary vasculature and illustrate some of its hemodynamic applications. I will also consider the mechanical properties of the coronary vessels and demonstrate the interaction of blood flow and vessel elasticity in formulating a new pressure-flow relationship for the coronary blood vessels.

2. Innovations in Morphometry of Coronary Vasculature

We have added four new innovations to morphometry: the Diameter-Defined Strahler ordering system for assigning the order numbers of the vessels, the distinction between series and parallel vessel segments, the connectivity matrix to describe the asymmetric branching pattern of vessels, and the longitudinal position matrix to describe the longitudinal position of daughter vessels along the length of their parent vessels. These innovations were used to study the anatomy of the coronary vasculature in the pig and have yielded the *first complete* set of morphometric data on the coronary arteries (Kassab *et al.*, 1993a, 1997a), capillaries (Kassab and Fung, 1994) and veins (Kassab *et al.*, 1994b) in health as well as that of arterial remodeling in right ventricular hypertrophy (Kassab *et al.*, 1993b). Now that the porcine coronary morphometric data base is complete, we are beginning to demonstrate some of its applications (Kassab *et al.*, 1994a, 1997b and Kassab and Fung, 1995; see accompanying manuscript).

3.   Anatomy of the Coronary Vasculature

The heart muscle is perfused via two major coronary arteries: the right coronary artery (RCA) and the left common coronary artery (LCCA). In the porcine, LCCA is a short segment (1–2 mm) which bifurcates into the left anterior descending (LAD) artery and left circumflex (LCx) artery. Figure 1 shows a cast of the LAD artery made from silicone elastomer while Fig. 2 shows examples of the branching pattern of intramural arterioles and venules (Kassab *et al.*, 1993a and Kassab *et al.*, 1994b).

There are two routes by which coronary venous flow returns to the heart. In one route, blood flows from the great cardiac vein, the posterior vein of the left ventricle, the posterior interventricular vein, the oblique vein of Marshal, and anterior cardiac vein into the coronary sinus on the epicardial surface and empties into the right atrium. In another route, blood flows through the smallest cardiac veins of Thebesius to the endocardial surface and drains directly into the heart chambers (predominantly into the right ventricle). The branching pattern of the porcine coronary arteries and endocardial veins are tree-like (Kassab *et al.*, 1993, 1994b), but the coronary capillary blood vessels have a non-tree-like topology (Kassab and Fung, 1994). Arcades are found at the epicardial surface connecting the sinusal veins

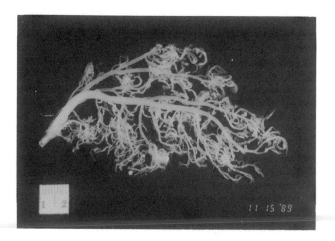

Fig. 1. Cast of porcine left anterior descending artery. Scale is 1 cm. From Kassab *et al.* (1993a), by permission.

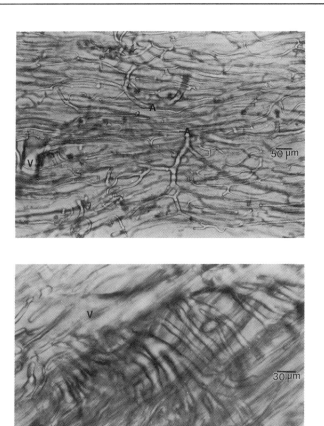

Fig. 2. Photomicrographs of arteriole (A) and venule (B) in the porcine left ventricle. From Kassab *et al.* (1993a, 1994b), by permission.

and at the endocardial surface connecting Thebesian veins (Kassab *et al.*, 1994b). There are also connections between the sinusal and Thebesian veins.

4. Mathematical Description of Coronary Arterial and Venous Trees

First, the capillary blood vessels are defined as vessels of order zero. The smallest arterioles supplying blood to the capillaries are assigned an order number of 1. The smallest venules draining the capillaries are assigned an order number of −1. When two arterioles of order 1 meet, the confluent vessel is given an order number 2 if its diameter exceeds the diameters of the order 1 vessels by an amount specified by a set of formulas or diameter criterion (Kassab *et al.*, 1993a), or remain as order number 1 if the diameter of the confluent is not larger than the amount specified by the formulas. When an order 2 artery meets another order 1 artery, the order number of the confluent is 3 if its diameter is larger by an amount specified by the diameter criterion, or remains at 2 if its diameter does not increase sufficiently. This process is continued until all arterial segments are arranged in increasing diameter and assigned the order numbers $1, 2, 3, \ldots, n, \ldots$ Similarly, the veins are assigned

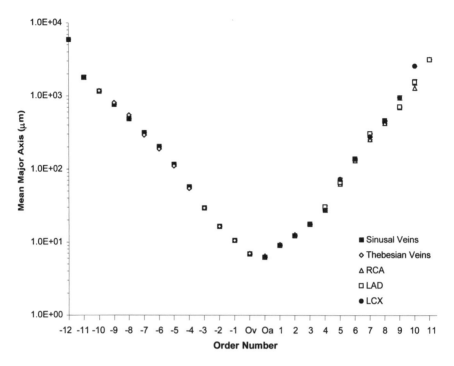

Fig. 3. Relationship between the mean length of the major axes of vessel elements in successive orders of vessels and order number of vessels in the porcine RCA, LAD, LCx, Thebesian and Sinusal Veins. From Kassab *et al.* (1993a, 1994b), by permission.

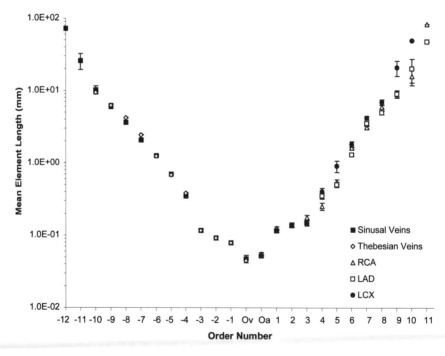

Fig. 4. Relationship between mean length of vessel elements in successive orders of vessels and order number of vessels in the porcine RCA, LAD, LCx, Thebesian and Sinusal Veins. From Kassab *et al.* (1993a, 1994b), by permission.

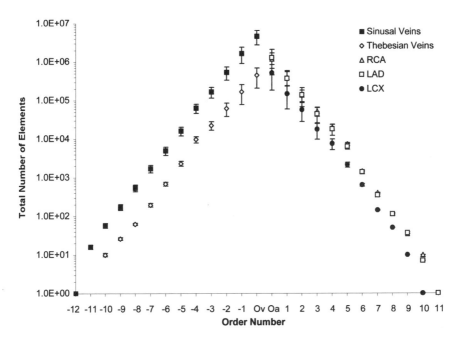

Fig. 5. Relationship between mean total number of vessel elements in successive orders of vessels and order number of vessels in the porcine RCA, LAD, LCx, Thebesian and Sinusal Veins. From Kassab *et al.* (1993a, 1994b), by permission.

the order $-1, -2, -3, \ldots, -n, \ldots$ Once the entire coronary arterial and venous trees are assigned order numbers, we define those vessel segments of the same order connected in series as elements. Figures 3, 4 and 5 show the relationship between the diameter or major axes length, element length and number of vessels and order number, respectively.

The connectivity of the various elements is presented in the form of a matrix, the component of which in row n and column m are the ratio of the total number of elements of order n sprung from elements of order m, divided by the total number of elements in order m. We call the results the *connectivity matrix* (Kassab *et al.*, 1993a, 1994b). Finally, the longitudinal position matrix of the coronary vasculature is determined, the component of which in row m and column n is the fractional longitudinal position of the vessels of order m which spring directly along the length of vessels of order n. We call the results the *longitudinal position matrix* (Kassab *et al.*, 1997a). In summary, the order number, diameter, length, connectivity matrix, longitudinal position matrix and fractions of the vessels connected in series were measured for all orders of vessels of the right coronary artery (RCA), left anterior descending artery (LAD), left circumflex artery (LCx), coronary sinus and Thebesian veins.

5.   Mathematical Description of Capillary Network

We have previously emphasized that the capillary vessels have a non-tree-like branching pattern (Kassab and Fung, 1994). As mentioned above, we designated the capillaries as blood vessels of order number zero; we further designated those capillaries fed directly by arterioles as C_{0a}, those drained into venules as C_{0v}, and those capillary vessels connected to C_{0a} and C_{0v} as C_{00} (Kassab and Fung, 1994). The capillaries are connected in patterns geometrically identified as Y, T, H or HP (hairpin) and anastomosed through transverse capillary cross-connections (C_{cc}). The C_{cc} vessels may connect adjacent capillaries or capillaries originating from different arterioles.

6.   Hemodynamic Applications of Morphometric Data

6.1.   *Analysis of total cross-sectional area and blood volume*

The data on the diameters, lengths, and number of elements can be used to compute the total cross-sectional area (CSA) and blood volume in the coronary arteries or veins. The total CSA, A_n, is equal to the product of the area of each element and the total number of elements:

$$A_n = (\pi/4)D_n^2 N_n \qquad (1a)$$

for arteries. The veins have a noncircular cross-sectional area which can be approximated by an ellipse with major and minor axis a and b, respectively (Kassab *et al.*, 1994b). Hence, the CSA for the veins can be expressed as

$$A_n = (\pi/4)a_n^2(b/a)_n \,. \qquad (1b)$$

The total blood volume in all elements of a given order, V_n, are given by

$$V_n = A_n L_n \,. \qquad (2)$$

The cumulative volume of the whole coronary artery or vein is the summation of V_n over n from 1 to N where N is the order of the largest vessel.

6.2.   *Pressure and flow distributions*

6.2.1.   *Arterial tree*

The branching pattern of the coronary arterial tree is prescribed by the connectivity and longitudinal position matrices while the vascular geometry is prescribed by the morphometric data (diameters and lengths). After the branching pattern and vascular geometry of the asymmetric network is generated, a hydrodynamical analysis can be performed (Kassab *et al.*, 1997a, 1997b). The Reynolds' and Womersley's numbers are small in most coronary vessels; hence, the physical laws governing blood flow reduce to Poiseuille's Law which applies for an analysis of steady flow of blood in the coronary vasculature in diastole. We shall initially use Poiseuille's Law to compute the pressure drop across individual vessels, which assumes that the vessels are rigid (the distensibility of the vessels will be considered later). Poiseuille flow, the volumetric flow Q_{ij}, in a vessel between any two nodes, represented by i and j, is given in terms of the pressure differential, ΔP_{ij}, and vessel conductance, G_{ij}, by

$$Q_{ij} = (\pi/128)\Delta P_{ij} G_{ij} \qquad (3)$$

where $\Delta P_{ij} = P_i - P_j$, $G_{ij} = D_{ij}^4 \mu_{ij}/L_{ij}$ and D_{ij}, L_{ij} and μ_{ij} are the diameter, length and viscosity, respectively, between nodes i and j. We assume that the coefficient of viscosity is 4.0 *cp* in vessels of order 11 to 5 and decreases linearly with order number to a value of 2.0 *cp* in the pre-capillary arteriole (Zhuang *et al.*, 1983). There are two or more vessels that emanate from the *j*th node anywhere in the tree with the number of vessels converging at the *j*th node being *mj*. By conservation of mass we must have

$$\sum_{i=1}^{mj} Q_{ij} = 0 \tag{4}$$

where the volumetric flow into a node is considered positive and flow out of a node is negative for any branch. From Eqs. (3) and (4) we obtain a set of linear algebraic equations in pressure for M nodes in the network, namely

$$\sum_{i=1}^{mj} [P_i - P_j] G_{ij} = 0. \tag{5a}$$

The set of equations represented by (5a) reduce to a set of simultaneous linear algebraic terms for the nodal pressures once the conductances are evaluated from the geometry, and suitable boundary conditions are specified. The boundary conditions are as follows:

P(at the Sinus of Valsalva) = 100 mmHg.

P(at the first bifurcation of the capillary network) = 26 mmHg. (5b)

In matrix form, this set of equations is

$$GP = G'P' \tag{6}$$

where G is the $n \times n$ matrix of conductances, P is a $1 \times m$ column vector of the unknown nodal pressures, and $G'P'$ is the column vector of the boundary pressures times the conductances of their attached vessels. The solution to Eq. (6) is obtained in the form of a column vector of the nodal pressures throughout the arterial network. The pressure drops, as well as the corresponding flows, are then computed.

Once the hemodynamics of the network have been determined, Poiseuille's hypothesis can be reexamined. Poiseuille's Law applies only when the flow has a low Reynolds' and Womersely's numbers. The former is defined by the formula UD/ν where U is the mean velocity of flow, D is the blood vessel lumen diameter and ν is the kinematic viscosity of blood. The latter is defined as $(D/2)(\omega/\nu)^{1/2}$, where w is the circular frequency of pulsatile flow and is computed for a heart rate of 110 cycles/minute. We have previously shown that the Reynolds' number is less than 120 for all orders of vessels and hence justifies the steady state assumption (Kassab *et al.*, 1997b). Womersley's number, on the other hand, is less than 1 for all orders less than 9 (Kassab *et al.*, 1997b). Hence, the inertia of blood should be taken into account for the first several largest orders. Equation (3) may also be corrected for loss due to bifurcation by replacing m with the "apparent" viscosity which depends on the Reynolds number. To account for the effect of change of kinetic energy along the stream, at each junction of vessels of order n to a vessel of order $n+1$, a static pressure drop equal to $1/2[\rho U^2{}_{n+1} - \rho U^2{}_n]$ should be added according to the well-known Bernoulli's equation (Fung, 1997) where U_n is the mean velocity of flow in the vessel of order n.

6.2.2. *Capillary network*

We have previously stressed that the topological structure of the coronary arteries and intramyocardial veins are tree-like, but the coronary capillary blood vessels have a non-tree-like topology (Kassab et al., 1993a; Kassab and Fung, 1994; Kassab et al., 1994). The analysis of blood flow in the capillaries needs a different network analysis from that presented above. The capillaries not only branch but also cross-connect along their lengths (Kassab and Fung, 1994). The presence of cross-connections in the myocardial capillaries may make the pressure and flow distributions in the capillary bed more uniform and hence must be taken into account when modeling its hemodynamics. The arterial analysis presented is valid for the flow into the capillary network. The capillary network is then simulated based on its geometry, branching pattern, distensibility and blood rheology.

At the capillary dimension, the particulate nature of the blood cells becomes important and the blood properties become non-Newtonian. The viscosity in Poiseuille's law [Eq. (3)] is no longer constant and should be considered as an apparent viscosity, μ_{app}. An empirical formula for the apparent viscosity which accounts for the non-Newtonian properties of blood is (Schmid-Schonbein, 1988)

$$\mu_{app} = [k_1 + k_2(U/D)^{-1/2}]^2 \tag{7}$$

where U is the mean velocity of blood and D is the diameter of the capillary vessel. The constants k_1 and k_2 depend on vessel diameter, hematocrit and shear rate. The constants have been experimentally determined in rigid glass tubes (Lingard, 1979) and *in vivo* viscosity measurements by Lipowsky *et al.* (1978).

6.2.3. *Venous tree*

The morphology and connectivity of the venous trees have been presented by Kassab *et al.* (1994b, 1997a). The venous trees are connected to the capillaries as specified by their connectivity matrix. The tree simulations developed above will be applicable to the venous trees with its appropriate vascular connectivity, longitudinal position of bifurcations, vascular geometry, distensibility and boundary conditions. Hence, the hemodynamics of the entire coronary vasculature will be synthesized to determine the pressure-flow relationship of the coronary circulation.

Let us consider some of the hemodynamic modifications needed for the coronary venules and veins because of their unique geometry. Unlike coronary arterioles and arteries which have cylindrical cross-sections, the venules have approximately elliptical cross-sections. If the cross-section of a vein is approximated by an ellipse, then, relative to a set of rectangular Cartesian coordinates x, y with origin located at the center, the parametric equations of the ellipse with a semi-major axis a and a semi-minor axis b are

$$x = a\cos\theta, \qquad y = b\sin\theta. \tag{8}$$

The parameters relevant to the flow can be derived from the Navier–Stokes equation. For a steady longitudinal flow of a Newtonian viscous fluid in a long cylindrical tube of elliptic cross-section subjected to a constant pressure gradient, dp/dx. In analogy to the exact solution of flow in a circular cylinder, the velocity profile

$$u = 2U[1 - (x/a)^2 - (y/b)^2] \tag{9}$$

satisfies the Navier-equation and the boundary condition that $u = 0$ on the elliptic wall described by Eq. (8). U is the mean velocity over this section. With Eq. (9), the Navier–Stokes equation yields

$$dP/dx = -4\mu U[(a^2 + b^2)/(a^2 b^2)] \tag{10}$$

where μ is the coefficient of viscosity of the fluid, and x is the length along the longitudinal axis of the tube. Then the volume rate of flow is

$$Q = \text{Area}\, U = \pi a b U = \pi/4\mu L[(a^3 b^3)/(a^2 + b^2)]dP/dx. \tag{11}$$

The resistance to flow is given by:

$$\text{Resistance} = 4\mu L/\pi[(a^2 + b^2)/(a^3 b^3)] \tag{12}$$

where L is the length of the tube. The formulas (9)–(12) show that a and the ratio b/a are the most important parameters of venous blood flow in which the Womersley number is less than one and have been measured in detail (Kassab *et al.*, 1994).

7.   Distensibility of Coronary Vessels

Vessel distensibility data are necessary because the elasticity of blood vessels of an organ is an important determinant of the pressure-flow relationship of blood flow through the organ. Pressure affects blood vessel diameter which, in turn, controls pressure distribution. Mathematically, the hemodynamic equations consist of an equation describing blood motion, and an equation for blood vessel deformation. The interaction of the two equations enters through the boundary conditions.

We have recently obtained data on the distensibility of the coronary vessels in the form of a pressure-diameter relation (unpublished data). Our data show two important features of distensibility of coronary arteries: (1) that the relationship is linear in the physiological range of pressures and (2) the compliance is small; i.e. the coronary epicardial arteries are relatively rigid in the diastolic state of the heart.

8.   Steady Laminar Flow in an Elastic Tube

With the distensibility of the blood vessels known, the mechanics of the blood vessel is coupled to the mechanics of blood flow to yield a pressure-flow relation for each vessel segment. This can be demonstrated for the cylindrical coronary arteries as follows: assume that the tube is long and slender, that the flow is laminar and steady, that the disturbances due to entry and exit are negligible, and that the deformed tube remain smooth and slender. These assumptions permit the use of Poiseuille's law for a Newtonian fluid that can be state as

$$dP/dx = (128\mu/\pi D^4)Q \tag{13}$$

where P is the pressure, x is the axial coordinate, Q is the volume-flow rate and D, and μ are the diameter, and viscosity, respectively. In a stationary, nonpermeable tube Q is a constant throughout the length of the tube. The tube diameter is a function of x because of the elastic deformation. Our data shows that, in the physiological pressure range, the elastic deformation can be described by the equation

$$D - D^* = \alpha(P - P^*) \tag{14}$$

where D is the diameter at a given intravascular pressure P, D^* is the diameter corresponding to a pressure P^* and α is the compliance constant of the vessel. Using Eq. (14), we have

$$dP/dx = dP/dDdD/dx = 1/\alpha dD/dx. \tag{15}$$

On substituting Eq. (14) into Eq. (13) and rearranging terms, we obtain

$$D^4 dD = (128\mu\alpha Q/\pi)dx. \tag{16}$$

Since the right-hand side term is a constant independent of x, we obtain the integrated result

$$D^5(x) = (640\mu\alpha Q/\pi)x + D^5(0). \tag{17}$$

The integration constant can be determined by the boundary condition, at the entry section of the vessel, that when $x = 0$, $D(x) = D(0)$. Putting $x = L$, at the exit section of a vessel, in Eq. (17) yields

$$D^5(L) - D^5(0) = 640\mu\alpha QL/\pi. \tag{18}$$

We now seek an approximate expression of Eq. (18) when $D(L) - D(0)$ is small; i.e. the vessel compliance is small. Letting $D(L) = D(0)+\varepsilon$, expanding the left hand side of Eq. (18) in power series of ε, and retaining only terms up to ε^2, we obtain the approximation

$$[D(L) - D(0)]\{1 + 2[D(L) - D(0)]/D(0)\} = (128\mu\alpha LQ)/(\pi D^4(0)). \tag{19}$$

Using Eq. (14) first at $x = L$ and then at $x = 0$ and subtracting, we have

$$D(L) - D(0) = \alpha[P(L) - P(0)]. \tag{20}$$

Combining (19) and (20), and writing D_0 for $D(0)$, we obtain

$$\Delta P + (2\alpha/D_0)\Delta P^2 = (128\mu LQ/\pi D_0{}^4) \tag{21}$$

where $\Delta P = P(L) - P(0)$. The solution to Eq. (21) takes the form

$$\Delta P_n = [-D_n + (D_n{}^2 + 8\alpha_n\Delta P_{pn}D_n)^{1/2}]/4\alpha_n \tag{22}$$

where ΔP_p is the Poiseuille's pressure drop as given by the right hand side of Eq. (21) and applies to each arterial vessel of order n. The pressure drop in Eq. (22) can be plotted as a function of the compliance α for the various orders of arterial vessels as shown in Fig. 6. It can be seen that when the compliance is zero (rigid vessel), the pressure drop corresponds to that given by Poiseuille's equation. However, when the compliance is nonzero, the pressure drop is smaller than that given by Poiseuille's equation and varies for different orders of vessels.

In the case of a non-Newtonian blood, as in the capillary vessels, the hydrodynamic law [Eq. (13)] can be combined with the elasticity [Eq. (14)] and rheology of blood [Eq. (7)] relationships to yield

$$dP/dx = (128/\pi)[k_1 + k_2[\pi(\alpha(P - P^*) + D^*)^3/4Q]^{1/2}]^2[\alpha(P - P^*) + D^*]^{-4}Q. \tag{23}$$

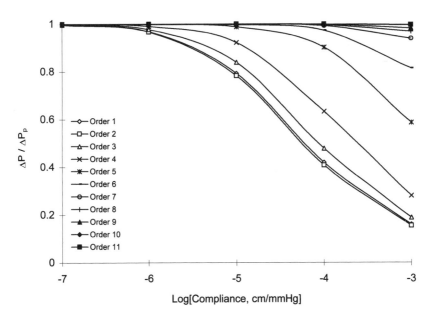

Fig. 6. Relationship between normalized pressure drop and logarithm of compliance constant for various orders of the left common coronary artery.

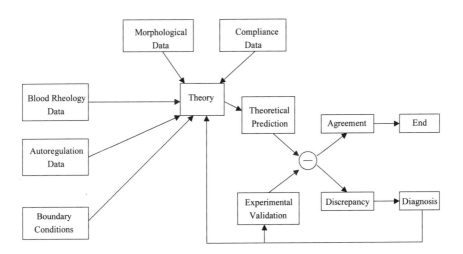

Fig. 7. A block diagram showing the interaction between input data (morphological, compliance, etc.), theoretical predictions and experimental validations.

This is the governing equation for non-Newtonian viscous blood flow in an elastic vessel at steady state conditions which can be integrated for specific boundary conditions (Kassab *et al.*, 1999).

9.   Integration of Theory and Experiment

The interaction of the anatomy, elasticity, vasoactivity, tissue/vessel interaction, theoretical analysis and experimental validation is presented in Fig. 7. When the theoretical and experimental results are compared, one may find that they do agree. In that case the process ends successfully and one gains confidence in the theory, which can then be used to predict the behavior of the physiological system. On the other hand, one may find that they disagree. Then it is necessary to examine carefully the cause of the discrepancy. With the diagnosis, one may wish to improve the experiment, or the theory, or both. With the improved theory and experiment, the process is repeated. The iterative process ensues until there is agreement between theory and experiment. This second alternative is what usually happens and provides real opportunity for learning and discovery.

10.   Concluding Remarks

A mathematical model of coronary circulation should be constructed based on the physical laws governing blood flow, the set of measured data on anatomy and elasticity of the coronary blood vessels, data on muscle/vessel interaction and vasoactivity, rheology of blood and the appropriate boundary conditions. This yields a predictive model that incorporates some of the factors controlling coronary blood flow. The virtue of the model will be determined through experimental validation. The model of normal hearts will serve as a physiologic reference state. Pathological states can then be studied in relation to changes in model parameters that alter coronary perfusion. This chapter illustrates the use of physical principles, with the help of anatomy and mechanical properties, to explain and predict the physiology of the coronary circulation in quantitative terms.

In this chapter, I hoped to demonstrate the extend to which an engineering approach can yield precise information about the blood circulation in the heart. The theory connects a large number of physical, morphometric, and rheological variables. Without such an approach, it would be very difficult to correlate all of these variables by conventional, empirical methods. In the present research, we are building our theory on continuum mechanics, and using measured geometric and elasticity data. *Ad hoc* hypotheses are kept to a minimum. Hence, the theory is basic and the agreement between the theoretical predictions and experimental results will yield conviction in the usefulness of the engineering approach.

References

1. Fung, Y. C., Biomechanics: Circulation, Springer Verlag, New York (1997).
2. Hoffman, J. I. E. and Spann, J. A. E., Pressure-flow relations in the coronary circulation, *Physiol. Rev.* **70**, 331–390 (1990).
3. Kassab, G. S. and Fung, Y. C., The pattern of coronary arteriolar bifurcations and the uniform shear hypothesis, *Ann. Biomed. Eng.* **23**, 13–20 (1995).
4. Kassab, G. S. and Fung, Y. C., Topology and dimensions of the pig coronary capillary network, *Am. J. Physiol.* **267** (Heart Circ. Physiol. 36), H319–H325 (1994).
5. Kassab, G. S., Rider, C. A., Tang, N. J. and Fung, Y.-C. B., Morphometry of pig coronary arterial trees, *Am. J. Physiol.* **265** (Heart Circ. Physiol. 34), H350–H365 (1993a).
6. Kassab, G. S., Lin, D. and Fung, Y. C., Consequences of pruning in morphometry of coronary vasculature, *Ann. Biomed. Eng.* **22**, 398–403 (1994a).

7. Kassab, G. S., Lin, D. and Fung, Y. C., Morphometry of the pig coronary venous system, *Am. J. Physiol.* **267** (Heart Circ. Physiol. 36), H2100–H2113 (1994b).

8. Kassab, G. S., Pallencaoe, E. and Fung, Y. C., The longitudinal position matrix of the pig coronary artery and its hemodynamic implications, *Am. J. Physiol.* **273** (Heart Circ. Physiol. 42) H2832–H2842 (1997a).

9. Kassab, G. S., Berkley, J. and Fung, Y. C., Analysis of pig's coronary arterial blood flow with detailed anatomical data, *Ann. Biomed. Eng.* **25**, 204–217 (1997b).

10. Kassab, G. S., Imoto, K., White, F. C., Rider, C. A., Fung, Y.-C. B. and Bloor, C. M., Coronary arterial tree remodeling in right ventricular hypertrophy, *Am. J. Physiol.* **265** (Heart Circ. Physiol. 34), H366–H375 (1993b).

11. Kassab, G. S., K. N. Le and Y. C. Fung, A hemodynamic analysis of coronary capillary blood flow based on anatomic and distensibility data. *Am. J. Physiol.* 277 (Heart Circ. Physiol. 46): H2158–H2166, (1999).

12. Oh, B.-H., Volpini, M., Kambayashi, M., Murata, K., Rockman, H., Kassab, G. S. and Ross, J. R., *Circulation* **86**, 1265–1279 (1992).

13. Schmid-Schoenbein, G. W., A theory of blood flow in skeletal muscle, *J. Biomech. Eng., Trans. ASME* **110**, 20–26 (1986).

14. Zhuang, F. Y., Fung, Y. C. and Yen, R. T., Analysis of blood flow in cat's lung with detailed anatomical and elasticity data, *J. Appl. Physiol.: Respirat. Environ. Exercise Physiol.* **55**, 1341–1448 (1983).

Annals of Biomedical Engineering, Vol. 25, pp. 204–217, 1997
Printed in the USA. All rights reserved.

0090–6964/97 $10.50 + .00
Copyright © 1997 Biomedical Engineering Society

Analysis of Pig's Coronary Arterial Blood Flow with Detailed Anatomical Data

GHASSAN S. KASSAB, JEFF BERKLEY, and YUAN-CHENG B. FUNG

Department of Bioengineering, University of California–San Diego, La Jolla, CA

Abstract—Blood flow to perfuse the muscle cells of the heart is distributed by the capillary blood vessels via the coronary arterial tree. Because the branching pattern and vascular geometry of the coronary vessels in the ventricles and atria are nonuniform, the flow in all of the coronary capillary blood vessels is not the same. This nonuniformity of perfusion has obvious physiological meaning, and must depend on the anatomy and branching pattern of the arterial tree. In this study, the statistical distribution of blood pressure, blood flow, and blood volume in all branches of the coronary arterial tree is determined based on the anatomical branching pattern of the coronary arterial tree and the statistical data on the lengths and diameters of the blood vessels. Spatial nonuniformity of the flow field is represented by dispersions of various quantities (SD/mean) that are determined as functions of the order numbers of the blood vessels. In the determination, we used a new, complete set of statistical data on the branching pattern and vascular geometry of the coronary arterial trees. We wrote hemodynamic equations for flow in every vessel and every node of a circuit, and solved them numerically. The results of two circuits are compared: one *asymmetric model* satisfies all anatomical data (including the mean *connectivity matrix*) and the other, a *symmetric model,* satisfies all mean anatomical data except the connectivity matrix. It was found that the mean longitudinal pressure drop profile as functions of the vessel order numbers are similar in both models, but the asymmetric model yields interesting dispersion profiles of blood pressure and blood flow. Mathematical modeling of the anatomy and hemodynamics is illustrated with discussions on its accuracy.

Keywords—Heart, Connectivity matrix, Hemodynamic matrix, Hemodynamics, Dispersion profile, Fractal dimension.

INTRODUCTION

In a rational coronary blood flow analysis, the circuit must agree with the anatomical data on the branching pattern and geometry of the coronary vasculature, and the

Acknowledgment—We thank Ms. Edith Pallencaoe and Ms. Aymi Schatz for their excellent technical expertise. This research was supported by the National Heart, Lung, and Blood Institute Training Grants HL-07089 and HL-43026, and by the National Science Foundation Grant BCS-89-17576.

Address correspondence to Ghassan S. Kassab, Department of Bioengineering, University of California–San Diego, 9500 Gilman Drive, La Jolla, CA 92093-0412, U.S.A.

(Received 22Sep95, Revised 14Feb96, Revised 29Apr96, Accepted 22May96)

basic hemodynamic equations and boundary conditions must be satisfied. To date, such an analysis has been applied to the rabbit and cat omentum (5,8,13), the rat and cat mesenteries (20,22,26), the bat wing (2,23,25), the mucosal circulation of rabbit small intestine (21), the human bulbar conjunctiva (5), the cat and dog lungs (6,9,32), the spinotrapezius muscle in the rat (27,28), and the pial arterial system of the rat (12). Recently, a full set of data describing the geometric properties of the porcine coronary arterial and venous trees and the coronary capillary network have been completed in our laboratory (15,17,18). Order number, diameter, length, connectivity matrix, and fractions of the vessels connected in series were measured for all orders of vessels of the right coronary artery (RCA), left anterior descending (LAD) artery, and left circumflex (LCX) artery. An application of this set of data to coronary hemodynamics is illustrated herein.

One of the central issues in coronary physiology is the high spatial nonuniformity of the coronary blood flow distribution [see critical reviews by Hoffman (10), Hoffman and Spaan (11), and Spaan (29)]. Spatial distribution of the blood flow and blood volume into the coronary capillaries has obvious physiological significance, because the nutrition of the heart muscle depends on the blood flow in the capillaries. Thus, the dispersion of flow in order 1 vessels feeding the capillaries is an important index of the coronary blood flow of each heart. Dispersion of the perfusion of the heart muscle at the capillary level depends on both the distribution of flow rates over the capillaries and the spatial distribution of capillaries in the muscle. Our hypothesis is that the distribution of flow rates over the capillaries is related to the branching network of coronary vessels. Our objective is to construct a mathematical model of the left common coronary artery (LCCA) based on the measured statistical data of vascular geometry and connectivity (17), and use it to analyze the distribution of flow rates over the capillaries.

On setting up the hemodynamic equations, we note that blood flow depends on two dimensionless numbers: the Reynolds number that affects instability, turbulence, separation, entry and exit disturbances, and resistance; and the Womersley number that affects transient response, phase shift, dynamic amplification, resonance, damping and reflection, and transmission at branching nodes. These effects are larger if these numbers are much larger than 1. They are smaller if these numbers are much smaller than 1. In the coronary arteries, the Womersley's numbers are smaller than 1, tending to 0.01 at the capillary inlets. Thus, it is a good approximation to treat the coronary blood flow as quasisteady. Reynolds numbers are smaller than 1 in coronary arteries of order 5 or smaller, and are of the order of 100 in the largest coronary arteries. In most coronary arteries, the flow can be approximated by Poiseuillean flow; only in the larger coronary arteries is a correction to the Poiseuillean resistance needed. Thus, we use Poiseuille's law as first approximation, and make corrections for higher Reynolds and Womersley numbers later.

Coronary arteries and veins are topologically tree-like (18,19). Coronary capillaries are network-like (16). Geometry of the arterial tree is specified by a modified Strahler system of branching, the statistical data of the lengths and diameters of the vessels of all orders, and the connectivity matrix C_{mn}, whose elements in row m and column n is the ratio of the total number of elements of order m that spring directly from parent elements of order n, divided by the total number of elements of order n. A realistic model of vascular circuit set up for numerical analysis must have a connectivity that is in agreement with the real, measured, anatomical connectivity matrix. Two vascular circuits are specified for a hemodynamic analysis of coronary flow distribution. One, called the *asymmetric model*, satisfies all the statistical anatomical data, including the means and standard deviations of the diameters and lengths of vessels, and the mean *connectivity matrix*. The other, called a *symmetric model*, satisfies all of the mean anatomical statistical data, but not the standard deviations and connectivity matrix. In fact, in the symmetric model, we replace the real, measured connectivity matrix with an idealized diagonal one. We shall show that, when both circuits are subjected to given inlet and outlet pressures, the longitudinal mean pressure profiles (nodal pressure plotted as a function of the order number of the blood vessels) are almost the same in both models. However, the dispersion profiles (the plot of the ratio of the standard deviation of a variable divided by its mean value) of the pressure and flow are very different in these two models.

In this study, we analyze blood flow in the coronary arterial tree with inflow pressure specified at the left coronary Valsalva sinus and a constant blood pressure at the first order bifurcation of the coronary capillary bed. If we added the coronary venous tree and capillary network and analyzed the entire heart, then we should be able to determine the distribution of capillary blood flow and pressure without any *ad hoc* hypothesis. However, the additional labor required would be considerable. Shortened analysis

of the arterial tree does illustrate quite well the way to incorporate the anatomical data in hemodynamic analysis.

METHODS

Application of Morphometric Data of the Vascular Tree

To establish a mathematical model of the tree-like arteries for the purpose of hemodynamic analysis, a complete set of morphometric data of pig coronary arteries has been obtained (19). For the purpose of mathematical modeling, three innovations in the morphometry were introduced. The first innovation is a rule for assigning the order numbers of the vessels on the basis of diameter ranges. Capillary blood vessels are defined as vessels of order 0. The smallest arterioles supplying blood to the capillaries are assigned an order of 1. When two arterioles of order 1 meet, the confluent vessel is given an order 2 if its diameter exceeds the mean diameters of the order 1 vessels by an amount specified by a formula (19) to be given herein, or remain as order 1 if the diameter of the confluent is not larger than the amount specified by the formula. When an order 2 artery meets another order 2 or order 1 artery, the order number of the confluent is 3 if its diameter is larger than the mean value of order 2 by an amount specified by the formula, or remain at 2 if its diameter does not increase sufficiently. This process is continued until all arterial segments are arranged in increasing diameter and assigned the order numbers 1, 2, 3, ... n, ... Let D_j denote the mean diameter of the vessels of order j, Δj denotes the standard deviation of the diameters of order j, whereas the subscript j denotes the order number of the vessel. Then, the formula referred to herein specifies that the dividing point of the diameters of vessels of the order j and those of vessels of order $j + 1$ lies midway between $D_j + \Delta_j$ and $D_{j+1} - \Delta_{j+1}$. An iteration process is required to determine the order number of all vessels of a given tree; but the process has been found to converge rapidly. This system of assigning order numbers to vessels is called the *diameter-defined Strahler System.*

The second innovation of Kassab et al. (15,19) is to introduce a clarifying terminology. Each blood vessel between two successive points of bifurcation is called a *segment*. If several segments of a given order j are connected in series, then they are lumped together into a unit called an *element* of order j. Thus, the total number of elements of order j is smaller than the total number of segments of order j.

Finally, the third innovation of Kassab et al. (19) is to describe asymmetric branching of the coronary vessels by a connectivity matrix C_{mn} whose element in row m and column n is the ratio of the total number of elements of order m that spring directly from parent elements of order n divided by the total number of elements of order n.

The LCCA bifurcates into the LAD and LCX. Morphometric data on the LAD and LCX have been presented in Kassab *et al.* (19). To obtain the morphometric data of the LCCA, we combined the statistical data of the LAD and LCX in each respective order. Table 1 shows the data on the diameters, lengths, and number of elements in each order of coronary vessels of the pig. Table 2 shows the data on the connectivity of the LCCA.

The process of obtaining the statistical morphometric data from a tree is unique, given that the same set of rules are followed every time. The process of constructing a circuit from the statistical morphometric data is, however, not unique. An infinite number of circuits can be constructed that are consistent with a given set of statistical data. Because blood flow can be analyzed only for a definite vascular circuit, we must consider many circuits when morphology is specified by statistical data.

Herein, we shall analyze two kinds of circuits. One, called the asymmetric model, simulates the full set of statistical morphometric data, including the mean connectivity matrix. The other, called the symmetric model, simulates all the mean morphometric data, but not the connectivity matrix and standard deviations.

A Symmetric Model of the Circuit

This is an analytical model that simulates the mean statistical data of Table 1 (assumes all standard deviations are 0), but replaces the connectivity matrix of Table 2 by a diagonal matrix whose nonvanishing components in row m and column $m + 1$ are the branching ratios of the number of the elements of order m divided by the number of elements of order $m + 1$. These numbers are shown in Table 1. Physically, it is equivalent to assuming that all the vessel elements in any order are in parallel, and the blood pressures at all of the junctions between specific orders of vessels are equal (15). In this simplified circuit, the flow in each element of order n obeys Poiseuille's formula (see Fung, Ref. 6, p. 11)

$$\dot{q}_n \frac{\pi D_n^4}{128 \mu_n L_n} (P_{n+1} - P_n) \tag{1}$$

in which $\dot{q}_n$ is the flow in ml/sec, D_n is vessel lumen diameter, μ_n is the coefficient of viscosity of blood, L_n is the length of the vessel element, P_n is the pressure of blood at the exit end of the vessel element, and P_{n+1} is that at the entry end of the vessel element, all in consistent units. There are N_n elements of order n in parallel. If the total flow is $\dot{Q}_T$, then $\dot{q}_n$ in each vessel of order n is $\dot{Q}_T/N_n$. Equation (1) then yields the pressure drop. The pressure at the Valsalva sinus, P_{11} at $n = 11$ being given (*e.g.*, an inlet diastolic coronary artery pressure of 100 mm Hg), we can compute $P_{10}, P_9, \ldots, P_{0a}$ (0a refers to an arteriolar capillary) in turn. Using the mean data given in Table 1, we obtain the pressure profile under the assumptions that the pressure at the first bifurcation of the capillary bed, P_{0a}, is a constant with a value of 26 mm Hg. The Oa capillaries are assigned a mean diameter and length of 6.2 and 52.0 μm, respectively (16). We assume that the coefficient of viscosity is 4 centipoise (0.004 Pa · sec) in vessels of orders 11 to 5, and decreases linearly with order number to 1.5 centipoise in the capillaries.

An Asymmetric Model of the Circuit

This model simulates the morphometric data of Tables 1 and 2. Figure 1 (left) shows a schematic of a branching pattern of the trunk of the LCCA that is consistent with its connectivity matrix given in Table 2. In turn, each of the branches arising from the trunk also give rise to branches consistent with the connectivity matrix and so on down to the arterial capillaries (order 0a). Figure 1 (right) shows a schematic of how an order 5 arteriole arising from the trunk of the LCCA [Fig. 1 (left)] would branch down to order 0 vessels. Each element shown in Fig. 1 may represent one or more elements in parallel, as will be determined later. The number of possible pathways for each element of Fig. 1 is shown in Fig. 2.

A realistic analysis consistent with the morphometric data must also incorporate the dispersion of diameters and lengths of various orders. In a real flow, the inflow from the aortic sinus is nonuniformly distributed to the parallel vessels of order 10, because all order 10 elements do not

TABLE 1. Mean diameters, lengths and number of vessel elements in each order of vessels in pig left common coronary artery.

Order	Diameter ± SD μm	Diameter Range μm	Length ± SD mm	Number ± Propagated Error	Branching Ratio
1	9.0 ± 0.73	8.0–10.3	0.115 ± 0.066	530,045 ± 321,512	2.63
2	12.1 ± 1.1	10.4–14.3	0.136 ± 0.088	201,823 ± 106,256	3.17
3	17.1 ± 1.7	14.4–21.3	0.149 ± 0.094	63,723 ± 28,657	2.44
4	30.1 ± 5.4	21.4–43.1	0.276 ± 0.163	26,120 ± 8,350	2.99
5	65.2 ± 11.6	43.2–95.7	0.508 ± 0.351	8,739 ± 1,468	4.23
6	138 ± 24.7	95.8–206	1.42 ± 1.04	2,066 ± 203	4.08
7	290 ± 30.1	207 –368	3.68 ± 2.04	506 ± 39	2.99
8	460 ± 42.7	369 –572	5.58 ± 3.47	169 ± 10	3.52
9	778 ± 154	573 –936	11.8 ± 10.4	48 ± 2	6.00
10	1695 ± 481	937 –2549	23.9 ± 20.1	8	8.00
11	3276	3276 –3276	52.9	1	

TABLE 2. Connectivity matrix of pig left common coronary artery. An element (m,n) in mth row and nth column is ratio of total number of elements of order m that spring directly from parent elements of order n divided by total number of elements of order n. Values are means ± SE.

	1	2	3	4	5
0	3.17 ± .117	.675 ± .080	.148 ± .081	0	0
1	.144 ± .031	2.04 ± .070	.630 ± .116	.071 ± .071	0
2		.094 ± .027	2.24 ± .102	1.50 ± .374	.084 ± .031
3			.074 ± .036	2.14 ± .294	.325 ± .060
4				.143 ± .097	2.40 ± .107
5					.205 ± .053

	6	7	8	9	10	11
0	0	0	0	0	0	0
1	0	0	0	0	0	0
2	.094 ± .019	.023 ± .010	0	0	0	0
3	.075 ± .014	.087 ± .020	.092 ± .029	.030 ± .030	0	0
4	.412 ± .037	.318 ± .035	.326 ± .051	.303 ± .102	.167 ± .167	0
5	2.51 ± .055	1.67 ± .087	1.37 ± .130	1.17 ± .233	.667 ± .333	0
6	.173 ± .020	2.42 ± .065	1.72 ± .135	1.55 ± .262	1.28 ± .565	0
7		.132 ± .021	2.08 ± .083	1.45 ± .275	2.28 ± .969	2
8			.104 ± .027	2.78 ± .237	2.71 ± .918	3
9				.119 ± .051	4.00 ± 1.15	8
10					.091 ± .091	6
11						0

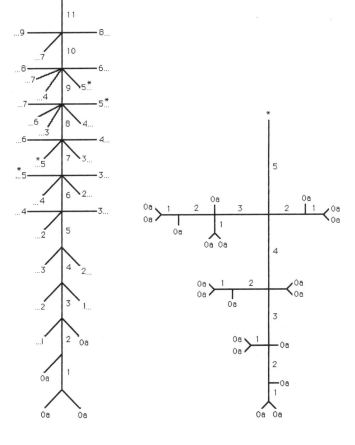

FIGURE 1. (Left) A schematic diagram of the trunk of the LCCA and its immediate branches as specified by its connectivity matrix. The "..." denotes that each branch arising from the trunk gives rise to further branches whose branching pattern is also consistent with the connectivity matrix and so on down to the arterial capillaries (order 0a). (Right) A schematic diagram of a 5th-order arteriole demonstrating all possible pathways to the capillaries as prescribed by the connectivity matrix. Asterisk shows the origin of the order 5 arteriole along the trunk of the LCCA (left).

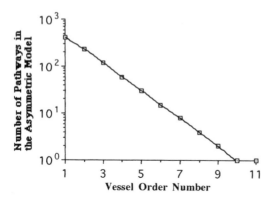

FIGURE 2. Relation between the total number of pathways and the order number in the asymmetric tree model.

have the same diameter and length and offsprings. Figure 3 shows the measured probability frequency functions for the diameters of the first several orders of vessels. The frequency functions of the diameters of successive orders of vessels have no overlap in the range of diameters in our diameter-defined Strahler System. The mean, standard deviation, and range of diameter distributions are summarized in Table 1. Figure 4 shows the measured probability frequency distribution functions (histograms) of the natural logarithm of the element length of the first several orders of vessels. Because the logarithm of the lengths are approximately normally distributed, if follows that the lengths are approximately log-normally distributed. Measured diameter and length distributions for all orders [including the 0a vessels (16)] are used in hemodynamic computations. Consequently, the elements of each order and pathway shown in Fig. 1 correspond to one or more elements in parallel whose diameter and length are specified by the appropriate histograms.

Parallel elements, by definition, have the same diameters and lengths and, thus, the same conductance. Endpoints of the parallel elements are also connected to iden-

tical pressure boundary conditions. Distribution of vessel sizes defines different conductances for vessels having different spatial positions or pathways in the network. For example, orders 11 or 10 have a single pathway and, thus, a single respective conductance. Order 9 has two pathways in the network ($11,10,9$ and $11,9$) and two equivalent conductances, respectively. Order 8, on the other hand, has four different pathways in the circuit ($11,10,9,8$; $11,10,8$; $11,9,8$; and $11,8$) and four equivalent conductances, respectively. The number of pathways for each element of the circuit of Fig. 1 is shown in Fig. 2. To clarify the pathways, let us use the symbol $N_{ijk...rst}$ to denote the number of parallel elements in pathway $ijk \ldots rst$. Let $C(t,s)$ be a component of the connectivity matrix given in Table 2, and $N_T(t)$ be the total number of elements of order t given in Table 1. Thus, the total number of elements of a given order are distributed among the various pathways as follows:

Order 11: 1 pathway
$$N_{11} = N_T(11)$$
Order 10: 1 pathway
$$N_{11,10} = N_T(10)$$
Order 9: 2 pathways

$$N_{11,10,9} = N_T(9)\left[\frac{N_{11,10}\,C(9,10)}{N_{11,10}\,C(9,10) + N_{11}\,C(9,11)}\right]$$

and

$$N_{11,9} = N_T(9)\left[\frac{N_{11}\,C(9,11)}{N_{11,10}\,C(9,10) + N_{11}\,C(9,11)}\right].$$

Order 8: 4 pathways

$$N_{11,10,9,8} = N_T(8)\left[\frac{N_{11,10,9}\,C(8,9)}{\begin{array}{c}N_{11,10,9}\,C(8,9) + N_{11,10}\,C(8,10) + \\ N_{11,9}\,C(8,9) + N_{11}\,C(8,11)\end{array}}\right]$$

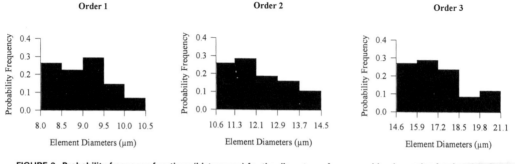

FIGURE 3. Probability frequency functions (histograms) for the diameters of coronary blood vessels of orders 1 to 3.

Order 1

Order 2

Order 3

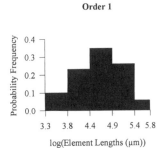

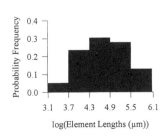

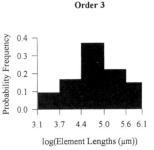

FIGURE 4. Probability frequency functions (histograms) for the natural logarithm of the lengths of coronary blood vessels of orders 1 to 3.

$$N_{11,10,8} = N_T(8)\left[\frac{N_{11,10}\,C(8,10)}{N_{11,10,9}\,C(8,9) + N_{11,10}\,C(8,10) + N_{11,9}\,C(8,9) + N_{11}\,C(8,11)}\right]$$

$$N_{11,9,8} = N_T(8)\left[\frac{N_{11,9}\,C(8,9)}{N_{11,10,9}\,C(8,9) + N_{11,10}\,C(8,10) + N_{11,9}\,C(8,9) + N_{11}\,C(8,11)}\right]$$

and

$$N_{11,8} = N_T(8)\left[\frac{N_{11}\,C(8,11)}{N_{11,10,9}\,C(8,9) + N_{11,10}\,C(8,10) + N_{11,9}\,C(8,9) + N_{11}\,C(8,11)}\right]$$

In general,

$$N_{ijk...rst} = N_T(t)\left[\frac{N_{ijk...rs}\,C(t,s)}{N_{ijk...rs}\,C(t,s) + ... + N_{ij}C(t,j) + N_i\,C(t,i)}\right]. \quad (2)$$

Equations for a hydrodynamical analysis of flow in the circuit can be set up as follows. Let the nodes of the circuit be numbered serially. The pressure at a node number j is denoted by P_j. A vessel element joining nodes i and j is denoted by vessel ij, whose diameter is D_{ij}, and length is L_{ij}. If there were N vessel elements connecting nodes i and j, then these elements will be denoted by N_{ij}. Now, as it has been explained in the *Introduction,* the flow in each vessel element is assumed to obey Poiseuille's law [*i.e.,* the rate of volume flow in the vessel connecting the two nodes, i and j, is given in terms of the pressure differential, ΔP_{ij}, vascular geometry (D_{ij} and L_{ij}), and blood viscosity (μ_{ij})] by

$$\dot{q}_{ij} = \left(\frac{\pi}{128\,\mu_{ij}}\right)\frac{D_{ij}^4}{L_{ij}}\Delta P_{ij}. \quad (3)$$

The total flow between these two nodes, $\dot{Q}_{ij}$, is the sum of the flow in the N parallel elements:

$$\dot{Q}_{ij} = N_{ij}\,\dot{q}_{ij}. \quad (4)$$

Thus, by Eq. 3

$$\dot{Q}_{ij} = \left(\frac{\pi}{128}\right)Geq_{ij}\Delta P_{ij} \quad (5)$$

where

$$Geq_{ij} = \frac{N_{ij}D_{ij}^4}{\mu_{ij}L_{ij}} \quad (6)$$

$$\Delta P_{ij} = P_i - P_j. \quad (7)$$

Figure 1 shows that there may be m_j vessels converging at the jth node. By the law of conservation of mass, we must have

$$\sum_{i=1}^{m_j}\dot{Q}_{ij} = 0, \quad (8)$$

where the volumetric flow into a node is considered positive and that out of a node is negative for any branch. From Eqs. 5 and 8, we obtain a set of linear algebraic equations in pressure for all of the nodes in the network, M in number, namely

$$\sum_{i=1}^{m_j}[P_i - P_j]\,Geq_{ij} = 0, \quad (j = 1, 2,..., M). \quad (9)$$

The set of equations represented by Eq. 9 is reduced to a set of simultaneous linear algebraic equations with the nodal pressures as unknowns. The boundary conditions are, as specified in the symmetric case,

P (at the Sinus of Valsalva) = 100 mm Hg.
P (at the first bifurcation of the capillary network) = 26 mm Hg. (10)

In matrix form, the set of M Eqs. 9 and 10 is

$$G_{eq}P = G_{eq}'P', \quad (11)$$

TABLE 3. Asymmetric model: the computed mean blood pressure at the exit section, mean blood flow per vessel, and mean blood velocity over a vessel cross section in successive orders of coronary arterial elements. The data are averages from one hundred runs of the asymmetric model program.

Vessel Order Number	Pressure (mmHg) Mean ± SD	Flow per Vessel (mL/s) Mean ± SD	Velocity (mm/s) Mean ± SD	Relative Dispersion (SD/Mean, %)	
				Flow Mean ± SD	Pressure Drop Mean ± SD
11	99.4 ± 0.042	1.07 ± 1.22E-1	128 ± 8.97	0	0
10	98.9 ± 0.49	1.04E-1 ± 9.43E-3	46.1 ± 4.17	0	0
9	98.3 ± 0.59	1.84E-2 ± 1.59E-3	38.7 ± 3.36	7.26 ± 3.70	27.8 ± 17.5
8	97.6 ± 0.65	4.99E-3 ± 4.68E-4	30.1 ± 2.87	9.63 ± 4.94	42.2 ± 24.0
7	96.4 ± 0.76	1.56E-3 ± 1.37E-4	23.6 ± 2.07	13.0 ± 5.04	50.1 ± 20.4
6	94.4 ± 1.27	4.02E-4 ± 3.43E-5	26.9 ± 2.29	20.1 ± 5.22	70.8 ± 22.6
5	90.6 ± 1.69	1.14E-4 ± 8.50E-6	34.1 ± 2.56	22.8 ± 5.04	71.9 ± 27.5
4	74.9 ± 2.76	3.83E-5 ± 2.84E-6	53.9 ± 3.98	27.9 ± 4.64	56.3 ± 14.0
3	58.8 ± 2.43	1.22E-5 ± 1.12E-6	53.3 ± 4.90	31.9 ± 5.63	49.5 ± 7.34
2	44.4 ± 1.92	4.42E-6 ± 3.43E-7	38.5 ± 2.98	46.0 ± 7.08	51.8 ± 6.45
1	32.7 ± 1.08	1.38E-6 ± 1.39E-7	21.7 ± 2.17	61.8 ± 11.3	62.6 ± 8.12

where G_{eq} is the $M \times M$ matrix of equivalent conductances (M is ~850 for the LCCA), P' is a $1 \times M$ column vector of the unknown nodal pressures, and $G_{eq}' P'$ is the column vector of the boundary pressures times the conductances of their attached vessels. We solved Eq. 11 by the Numerical Algorithms Group Subroutine (F04AEF-NAG Fortran Library Routine) that exploits the sparsity of the matrix in obtaining solutions for the large system of equations efficiently.

It has been mentioned that the statistical means, standard deviations, and histograms of the diameters and lengths of the elements were derived from the measured diameters and lengths of vessels, respectively. Since we have the raw data, we used them directly as input to the computation to avoid an intermediate step of constructing a random number table satisfying the mean and standard deviations of the published table. We numbered the entries to our data file, for each order, in positions 1, 2, 3, . . . , n, where n is the total number of measurements for that order. If the total number of elements in the computational model exceeds n, then the program uses the datasets repeatedly in sequence until the required number of elements are obtained. This algorithm is simple and as accurate as the raw data itself. To account for the randomness by which the measured data are arranged in sequence, as well as to use the entire distributions, we run the program 100 times by varying the starting point in the sequence of each order for a total of 11 orders.

Despite the sophistication of the asymmetric model, it still does not satisfy *all* of the statistical data measured previously (19). For example, only one tree topology, corresponding to the mean connectivity matrix, is considered which ignores the standard deviations of the connectivity matrix. Furthermore, the connectivity matrix shows a small number of vessels of order n branching from vessels of order n that the present model does not take into account. Finally, the asymmetric circuit is not a bifurcating tree model and cannot satisfy the statistics of the *segment-to-element ratios* (S/E) reported in Kassab *et al.* (19). The present asymmetric model includes the assumption that

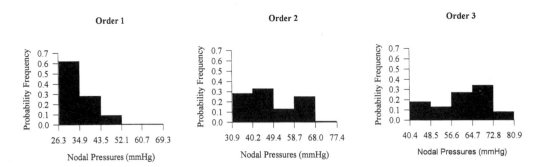

FIGURE 5. Probability frequency functions (histograms) for nodal pressures at the exit sections to the blood vessels of orders 1 to 3 in the asymmetric circuit simulating the statistical morphometric data of the coronary arterial tree of the pig, including the connectivity matrix for the case in which the capillary pressure has a constant value of 26 mm Hg.

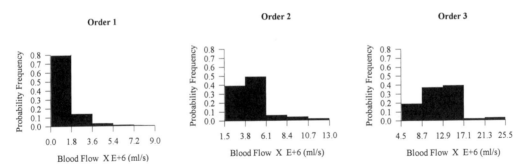

FIGURE 6. Probability frequency functions (histograms) for blood flow per element of blood vessels of orders 1 to 3 in the asymmetric circuit simulating the statistical morphometric data of the coronary arterial tree of the pig, including the connectivity matrix for the case in which the capillary pressure has a constant value of 26 mm Hg. Values of flow have been multiplied by a factor of E+6 (10^6).

the S/E = 1 for all orders of vessels that is not corroborated by experimental measurements (15). The present analysis also includes the assumption that a number of elements are grouped in parallel definition of the equivalent conductance G_{eq}, which simplifies the problem considerably.

RESULTS

The solution to Eq. 11 for the asymmetric model circuit was obtained in the form of a column vector of nodal pressures throughout the asymmetric arterial network. Statistical features of the solution of the asymmetric circuit, averaged over 100 runs of the program, are given in Table 3. Other interesting features are presented in graphs. Figures 5 and 6 show the nodal pressure and flow distributions, from a single run of the program, for the first several orders of vessels, respectively. A constant pressure of 26 mm Hg is the assumed boundary condition at the capillary network (outlet of vessels of order 0a). Distributions for the other orders shown in Figs. 5 and 6 were computed by the asymmetric model. Note the rapid change in the skewness of the distributions as the order number increases. In the vessel of order 11, the largest coronary artery, the pressure at the entry is that of the Sinus of Valsalva, and the frequency function is a Δ-function.

The heterogeneity of hemodynamic variables can be related to the asymmetry of the branching pattern by constructing a hemodynamic matrix, analogous to the connectivity matrix. The element (m,n) in the mth row and nth column of the hemodynamic matrix is the mean value of hemodynamic variables of order m that spring directly from parent elements of order n. Tables 4 to 6 show the nodal pressure matrix, pressure drop matrix, and flow matrix from a single run of the asymmetric model, respectively.

A simple analytic solution to Eq. 1 for the symmetric

model was also determined. Figures 7 to 9 show the mean values of longitudinal pressure distribution, the pressure drop per vessel element, and the coronary blood flow per vessel in the symmetric and asymmetric LCCA models, respectively. Pressure boundary conditions at the first capillary bifurcations were assumed to be a constant with a value of 26 mm Hg. Note that because only the pressures at the inlet and outlets are specified for the circuits, by Eq. 10, the total flow of the asymmetric and symmetric model circuits may be unequal, as is shown at order 11 in Fig. 9. In fact, it can be seen that the asymmetric tree carries more flow for the same pressure drop than the symmetric tree.

Once the hemodynamics of the network have been determined, the basic hypotheses leading to Poiseuille's law were examined. Stokes' derivation of Poiseuille's formula assumes a steady laminar flow of a Newtonian fluid in a circular cylinder of infinite length. It is well known (7) that, in finite tubes, these assumptions are better approximated if the values of the Reynolds number and Womer-

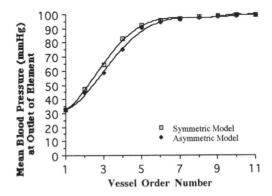

FIGURE 7. Relationship between blood pressure at the outlet of a blood vessel and the order number of the blood vessel for the arterial branches of the symmetric and asymmetric models of the LCCA. Curves are of 5th-order polynomials.

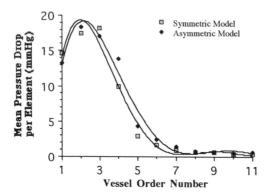

FIGURE 8. Relation between pressure drop per vessel element and the order number of the elements for the arterial branches of the symmetric and asymmetric models of the LCCA. Curves are of 5th-order polynomials.

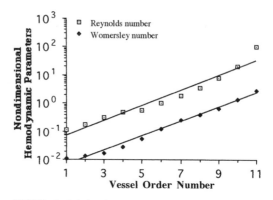

FIGURE 10. Relation between Reynolds and Womersley's numbers and order number of the elements for the arterial branches of the LCCA. Data are fitted by exponential functions.

sley number are much less than unity. Reynolds number is UD/v, where U is the mean velocity of flow, D is the blood vessel lumen diameter, and v is the kinematic viscosity of blood. The Womersley number is defined as $(D/2)(\omega/v)^{1/2}$, where ω is the radian frequency of pulsatile flow (taken to be 110 cycles/min). Figure 10 shows the relation between Reynolds and Womersley numbers and the order number of coronary blood vessels, respectively. They are not very small in all vessels. Thus, the accuracy of Poiseuille's formula needs to be discussed and is given in the *Discussion*.

Data on the diameters, lengths, and number of elements can also be used to compute the blood volume V_{np} in all blood vessels of a given order n and pathway p:

$$V_{np} = \left(\frac{\pi}{4}\right) D_{np}^2 L_{np} N_{np}, \qquad (12)$$

where D_{np}, L_{np}, and N_{np} are the diameter, length, and number of elements for a given order n and pathway p, respectively. The mean *cumulative volume* of arterial tree of order N is the summation of V_{np} over n from 1 to N, averaged over the various pathways p, and is shown in Fig. 11 for the LCCA. Because of the statistical variations of D_{np}, L_{np}, and N_{np} in different trees, the cumulative volumes of trees with different N's depend on the trees.

The computed heterogeneity of blood flow, pressure drop, and blood volume can be expressed in terms of their relative dispersion (RD = SD/mean). Figure 12 shows the variations of the relative dispersions of flow and pressure drop per element, and the cumulative blood volume and the order number of LCCA vessels.

DISCUSSION

Coronary circulation is the means by which the heart receives its nutrients and removes its wastes. The cardiac

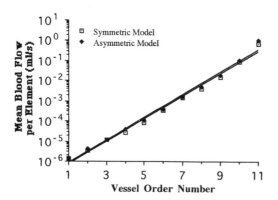

FIGURE 9. Relation between blood flow per vessel element and the order number for the arterial branches of the symmetric and asymmetric models of the LCCA. Data are fitted by exponential functions with use of the least-squares method.

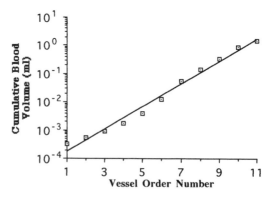

FIGURE 11. Relation between cumulative blood volume and order number for the arterial branches of the LCCA. Data are fitted by an exponential function.

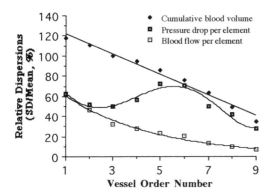

FIGURE 12. Relation between %flow dispersion, %cumulative blood volume dispersion and %pressure drop dispersion (SD/mean × 100), and the order number of the elements for the arterial branches of the asymmetric model of the LCCA. Data are fitted by an exponential function, a linear function, and a 5th-order polynomial, respectively.

muscle, unlike skeletal muscle, is only capable of aerobic respiration. Therefore, for the heart to contract continually, it must be well perfused. Many mechanisms have evolved to regulate the coronary circulation to ensure a well-perfused heart. The regulating mechanisms can be classified into three categories: physical, metabolic, and neural. In the physical category are the length and cross-sectional areas of the coronary blood vessels, the pattern by which these vessels are organized, the viscosity of the blood, the perfusion pressure, and the tissue pressure outside the blood vessels. In the present study, we have considered hemodynamic analysis based on the branching patterns, diameters, and lengths of coronary blood vessels because a full set of statistical data about these parameters has just become available. A hemodynamic analysis is done for two specific circuits: the asymmetric model satisfies the measured mean connectivity matrix and the distributions of the diameter and length data, whereas the symmetric model does not.

Distributions of blood pressure and blood flow in coronary blood vessels are shown in Figs. 5 and 6 from a single run of the asymmetric model. Hemodynamic data for the asymmetric tree, averaged from 100 runs of the model, are shown in Table 3 and Figs. 7–9 and 12. It is interesting that the symmetric model yields similar mean longitudinal blood pressure profile, pressure drop per element, and blood flow per element to that of the asymmetric model (as shown in Figs. 7–9). The greatest hemodynamic differences of these two models, however, are revealed by Figs. 5, 6, and 12. Histograms shown in Figs. 5 and 6 are those of the asymmetric model; for the symmetric model, the probability frequency function would be a Δ-function for every order. Dispersions in flow, pressure, and volume shown in Fig. 12 are those existing in the asymmetric model. For the symmetric model, the dispersions are 0 for all orders of the tree.

In both models, most of the pressure drop occurs across the first four orders of vessels (Figs. 7 and 8). This suggests that the tree topology may not play as important a role as the vascular geometry in dictating the shape of the pressure profile. Consider Eq. 1 that defines the pressure drop for the various order n vessels of a symmetric tree

$$\Delta P_n = \frac{128\mu_n L_n}{\pi N_n D_n^4} Q_T. \tag{13}$$

Suppose we define $R_D(n) = D_{n+1}/D_n$, $R_L(n) = L_{n+1}/L_n$, $R_N(n) = N_n/N_{n+1}$, and $R\mu(n) = \mu_{n+1}/\mu_n$ as the diameter, length, number, and viscosity ratios, respectively. Then, an expression for a pressure drop ratio follows from eq. 13 as

$$\frac{\Delta P_n}{\Delta P_{n+1}} = \frac{R_D^4(n)}{R_\mu(n) R_L(n) R_N(n)}. \tag{14}$$

However, since the values of the viscosity and length ratios are close to unity for the first several orders, it can be seen that the pressure drop is dictated by the ratio $R_D^4(n)/R_N(n)$. Thus, if the diameter ratio to the fourth power is decreasing faster than the increase in the number ratio, then a large pressure drop will occur. This is precisely the case at order 4 vessels, as can be seen in Fig. 7.

TABLE 4. The computed mean nodal pressure (mmHg) matrix of pig left common coronary artery. An element (m,n) in mth row and nth column is the mean nodal pressure at the exit section of order m that spring directly from parent elements of order n. Values are means ± SD from a single run of the asymmetric model.

	2	3	4	5	6	7	8	9	10	11
1	34.1 ± 6.41	36.6 ± 7.96	41.2 ± 8.29	0	0	0	0	0	0	0
2		47.8 ± 11.0	49.4 ± 7.59	61.1 ± 6.36	64.6 ± 7.50	57.7 ± 13.4	0	0	0	0
3			60.5 ± 11.0	68.4 ± 8.87	65.8 ± 14.8	71.4 ± 10.1	66.8 ± 7.73	85.3 ± 2.33	82.8	0
4				74.5 ± 10.0	71.8 ± 9.26	74.5 ± 7.51	84.0 ± 4.74	81.7 ± 3.07	96.1	0
5					88.1 ± 4.70	94.0 ± .991	92.6 ± 1.70	95.2 ± .278	96.1	0
6						92.9 ± 2.64	93.0 ± 2.26	95.6 ± 1.32	98.2	0
7							96.4 ± .637	96.3 ± .715	97.9	98.3
8								97.4 ± .363	98.4	98.4
9									98.0	98.7
10										99.1

TABLE 5. The computed mean pressure drop (mmHg) matrix of pig left common coronary artery. An element (m,n) in mth row and nth column is the mean pressure drop per element of order m that spring directly from parent elements of order n. Values are means ± SD from a single run of the asymmetric model.

	2	3	4	5	6	7	8	9	10	11
1	14.1 ± 7.46	24.4 ± 12.3	33.2 ± 9.21	0	0	0	0	0	0	0
2		13.2 ± 7.32	24.9 ± 10.5	28.3 ± 8.25	28.4 ± 8.49	38.7 ± 12.8	0	0	0	0
3			13.8 ± 7.31	20.9 ± 8.18	27.2 ± 12.9	25.0 ± 10.6	30.8 ± 7.34	12.8 ± 1.96	0	0
4				14.9 ± 12.5	21.3 ± 7.48	22.0 ± 7.40	13.5 ± 5.16	16.5 ± 2.81	16.3	0
5					4.97 ± 2.34	2.40 ± 1.20	4.97 ± 1.71	2.93 ± .574	2.98	0
6						3.56 ± 2.19	4.57 ± 1.75	2.58 ± 1.62	.874	0
7							1.19 ± .139	1.89 ± .414	1.11	1.04
8								.808 ± .065	.689	.999
9									1.07	.701
10										.318

These observations are similar to the epicardial pressure measurements reported by Chilian et al. (3,4), Kanatsuka et al. (14), and Tillmanns et al. (30). Direct comparison with these studies, however, warrants caution because measurements were made in different species, with a different degree of vasodilation and with the additional effect of cardiac contraction. Moreover, pressure measurements were made only in the epicardial vessels, unlike the pressure calculated in the present study that takes into account the entire vascular tree.

Heterogeneity of hemodynamic variables was related to the asymmetry of the branching pattern of the asymmetric model by introducing several hemodynamic matrices. The pressure matrix, pressure drop matrix, and flow matrix summarize the mean values of the respective pressure, pressure drop, and flow rate distributions obtained in a given order n as shown in Tables 4 to 6, respectively. It can be seen that, as the order of elements descending from vessels of order n increases (i.e., elements of order . . . , $n - 3$, $n - 2$, $n - 1$ in a given column of the respective hemodynamic matrix), the exit nodal pressure increases, the pressure drop per element decreases, and the flow rate per element increases. These findings imply that a smaller vessel arising from a given element has a smaller exit pressure, a larger pressure drop, and a smaller flow rate than a larger vessel arising from that same element. Although the hemodynamics of order 0a vessels (arising from orders 1, 2, and 3, respectively) have not been included in the arterial hemodynamic matrices (Tables 4 to 6), they have, however, been computed and must be used as appropriate boundary conditions into any capillary network hemodynamic analysis. Furthermore, the hemodynamic matrices will have further significance once the spatial relationship of the myocardium is related to the vessel elements and their connectivity matrix.

The shape of the total cross-sectional area (CSA) curve of all vessel elements of the same order *versus* the order number of the elements has been previously reported by Kassab et al. (19). It was shown that the CSA did not increase exponentially as the order number decreased toward one (i.e., toward the smallest arterioles). Instead, the CSA had an oscillatory variation and the CSA of order 1 arterioles was found to be only 4 times greater than that of order 11 artery (LCCA), implying that the velocity of blood does not slow down significantly in arterioles, compared with that in large coronary arteries. Table 3 shows that the mean velocity of blood flow also has an oscillatory pattern, and the mean velocity in order 1 arterioles is found to be only 6 times slower than that in the LCCA. Although, there is no complete set of experimental data on coronary flow velocities of all orders of vessels to compare our results with, some data have been reported for epicardial microcirculation. For example, Nellis et al. (24) have measured the mean flow velocity in 200 μm diameter

TABLE 6. The computed mean flow (ml/s) matrix of pig left common coronary artery. An element (m,n) in *m*th row and *n*th column is the mean blood flow per element of order *m* that spring directly from parent elements of order *n*. Values are means ± SD from a single run of the asymmetric model.

	2	3	4	5	6
1	1.33e-6 ± 5.63e-7	3.13e-6 ± 1.65e-6	4.54e-6 ± 2.61e-6	0	0
2		3.84e-6 ± 1.31e-6	5.90e-6 ± 2.06e-6	9.54e-6 ± 2.82e-6	9.09e-6 ± 3.08e-6
3			1.23e-5 ± 3.35e-6	1.76e-5 ± 6.05e-6	1.47e-5 ± 4.11e-6
4				3.84e-5 ± 8.36e-6	4.62e-5 ± 1.04e-5
5					1.20e-4 ± 1.72e-5

	7	8	9	10	11
1	0	0	0	0	0
2	6.67e-6 ± 4.65e-6	0	0	0	0
3	1.78e-5 ± 4.00e-6	1.53e-5 ± 1.48e-6	1.21e-5 ± 1.50e-6	0	0
4	4.33e-5 ± 6.40e-6	4.30e-5 ± 4.29e-6	4.02e-5 ± 9.70e-8	4.08e-5	0
5	9.74e-5 ± 2.63e-5	1.00e-4 ± 2.76e-5	1.26e-4 ± 3.16e-5	7.13e-5	0
6	4.36e-4 ± 5.10e-5	3.74e-4 ± 2.51e-5	5.20e-4 ± 6.45e-5	2.51e-4	0
7		1.61e-3 ± 1.31e-4	1.70e-3 ± 1.20e-4	1.47e-3	1.34e-3
8			4.96e-3 ± 3.59e-4	4.96e-3	5.23e-3
9				1.93e-2	1.70e-2
10					1.08e-1

epicardial arterioles to be ~21 mm/sec using fluorescent microspheres and video-image processing in the cat. This vessel caliber corresponds to our order 7 arteriole that has a computed mean velocity of 23.6 mm/sec as shown in Table 3.

As previously described, Fig. 10 shows that the Reynolds number and Womersley number of the coronary blood flow are both smaller than 1 only in vessels of orders 1 to 4. Thus, in the larger coronary blood vessels with order 5 to 11, we should correct the Poiseuille formula (Eq. 1 or 2) by multiplying the right-hand side with a function of Reynolds and Womersley numbers. Examples of such corrections for resistance and phase shift are given in Ref. 7 (p. 233). The correction should consider the effects of the flow pulsation, and the entry and exit end conditions at the points of bifurcation. Because the Reynolds number is <110 in the largest coronary arteries, turbulence is unlikely to occur because the critical Reynolds number for transition to turbulence is 2,300 or greater. Flow separation is possible only in highly stenotic conditions of atherosclerosis. These corrections will improve the numerical accuracy of the pressure and flow distributions among the vessels. But, because the major drop of blood pressure occurs in small arteries of orders 1 to 4, the corrections will not change the major features of the flow in the two models of the coronary arterial tree.

Bassingthwaighte *et al.* (1) have shown with the microsphere deposition technique that local blood flow in the myocardium is very nonuniform and that the measured nonuniformity varies with the volume of the tissue sample. They found that the relative dispersion of flow and the tissue mass obey a fractal relationship. We have shown that the morphometric data of the coronary arteries are fractal (19). Whereas we cannot calculate blood flow per

spatial compartment of varying size, we can calculate the dispersion of blood flow per element relative to cumulative blood volume, as a combination of data in Figs. 11 and 12 will show. If we can convert the cumulative blood volume to myocardial mass, then we can obtain something to compare with the results of Bassingthwaighte *et al.* (1). To do this, we plot in Fig. 13 the relationship between the cumulative arterial blood volume per its perfused myocardial mass and the myocardial region for four coronary arterial trees, namely the LCCA + RCA, the LCCA, the right ventricular branches in control, and the right ventricular branches in hypertrophy from data in Refs. 17 and 19. The myocardial mass perfused by the various arteries was determined with silicone elastomer as described previously (17). Briefly, a border was determined histologically between the elastomer-perfused and nonperfused myocardial region of each respective artery. Figure 13

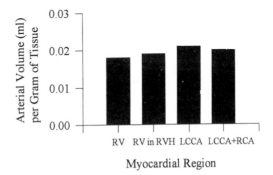

FIGURE 13. Relation between arterial blood volume per perfused myocardial mass for the myocardial regions of LCCA + RCA, LCCA, and right ventricular (RV) branches in control and right ventricular hypertrophy (RVH) (15,17).

shows a fairly constant relationship between the arterial blood volume per myocardial mass and its respective region. It is interesting that even the hypertrophic heart has a similar ratio, which indicates that the relationship is maintained during tissue remodeling in hypertrophy. The constant obtained from Fig. 13 yields 0.02 ml of arterial blood per 1 g of myocardial mass. This ratio translates into a 3.5 ml/100 g of left ventricle for a 150 g heart (with an 85 g left ventricle). Using this result, we can determine the relation between the myocardial mass and the relative dispersion of blood flow per element (as shown in Fig. 14). The relation has a fractal character with a fractal dimension of 1.27. This is similar to the results of Bassingthwaighte *et al.*'s (1) study that reported fractal dimensions of 1.20, 1.16, and 1.22 for autoregulated baboon, sheep, and rabbit hearts, respectively. Furthermore, the flow dispersion in subtrees that perfuse 1-g tissue pieces was found to be 16% (see Fig. 14), which is well within the range of 7 to 43% reported by Bassingthwaighte *et al.* (1).

Recently, VanBavel and Spaan (31) modeled the porcine coronary arterial branching pattern for the purpose of estimating flow heterogeneity. The quantitative basis for their dichotomous tree model was provided by defining and measuring the relation between diameters of parent and daughter segments at arterial nodes, as well as the relation between diameter and length of vessel segments. These relations were used to generate computer models of the coronary arterial trees for vessels <500 μm in diameter and subsequently analyzed with Strahler's ordering scheme. VanBavel and Spaan (31) reconstructed tree segment for segment and then calculated the flow and pressure drop in each segment. The flow in their simulated networks were very heterogeneous. They found that the

relationship between the level of flow heterogeneity and the perfused volume, as expressed by the number of terminal segments in a subtree, obeyed a fractal relation with a fractal dimension of 1.20.

There is no doubt that knowledge of various factors controlling coronary blood flow and its distribution is important in human health and disease, and today our understanding of those mechanisms is incomplete. The contribution of the present study is to yield a predictive model that incorporates some of the factors controlling coronary blood flow. The virtue of the model will be determined through experimental validation. The model of normal hearts will serve as a physiological reference state. Pathological states can then be studied in relationship to changes in model parameters that alter coronary perfusion.

REFERENCES

1. Bassingthwaighte, J. B., R. B. King, and S. A. Roger. Fractal nature of regional myocardial blood flow heterogeneity. *Circ. Res.* 65:578–590, 1989.
2. Chen, I. H. A mathematical representation for vessel networks. II. *J. Theor. Biol.* 104:647–654, 1983.
3. Chilian, W. M., C. L. Eastham, and M. L. Marcus. Microvascular distribution of coronary vascular resistance in beating left ventricle. *Am. J. Physiol.* 251:H779–H788, 1986.
4. Chilian, W. M., S. M. Layne, E. C. Klausner, C. L. Eastham, and M. L. Marcus. Redistribution of coronary microvascular resistance produced by dipyridamole. *Am. J. Physiol.* 256:H383–H390, 1989.
5. Fenton, B. M., and B. W. Zweifach. Microcirculatory model relating geometrical variation to changes in pressure and flow rate. *Ann. Biomed. Eng.* 9:303–321, 1981.
6. Fung, Y. C. Biodynamics: Circulation. New York: Springer-Verlag, 1984, pp. 10–12.
7. Fung, Y. C. Biomechanics: Motion, Flow, Stress, and Growth. New York: Springer-Verlag, 1990, pp. 229–234.
8. Gross, J. F., M. Intaglietta, and B. W. Zweifach. Network model of pulsatile hemodynamics in the microcirculation of the rabbit omentum. *Am. J. Physiol.* 226:1117–1123, 1974.
9. Haworth, S. T., J. H. Linehan, T. A. Bronikowski, and C. A. Dawson. A hemodynamic model representation of the dog lung. *J. Appl. Physiol.* 70:15–26, 1991.
10. Hoffman, J. I. Heterogeneity of myocardial blood flow. *Basic Res. Cardiol.* 90:103–111, 1995.
11. Hoffman, J. I., and J. A. Spaan. Pressure-flow relations in the coronary circulation. *Physiol. Rev.* 70:331–390, 1990.
12. Hudetz, A. G., K. A. Conger, J. H. Halsey, M. Pal, O. Dohan, and A. G. B. Kovach. Pressure distribution in the pial arterial system of rats based on morphometric data and mathematical models. *J. Cereb. Blood Flow Metab.* 7:342–355, 1987.
13. Intaglietta, D., R. Richardson, and W. R. Tompkins. Blood pressure, flow, and elastic properties in microvessels of cat omentum. *Am. J. Physiol.* 221:922–928, 1971.
14. Kanatsuka, H., K. G. Lamping, C. L. Eastham, M. L. Marcus, and K. C. Dellsperger. Coronary microvascular resistance in hypertensive cats. *Circ. Res.* 68:726–733, 1991.
15. Kassab, G. S. Morphometry of the Coronary Arteries in the

$$RD = 15.9 \ [Mass]^{-0.27}; \ R^2 = 0.97$$

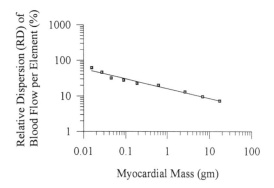

FIGURE 14. Fractal relation between myocardial mass and relative dispersion of blood flow. Data are fitted by a power law function with use of the least-squares method.

Pig. La Jolla: University of California, San Diego, Ph.D. Thesis, 1990.

16. Kassab, G. S., and Y. C. Fung. Topology and dimensions of pig coronary capillary network. *Am. J. Physiol.* 267(*Heart Circ. Physiol.* 36):H319–H325, 1994.

17. Kassab, G. S., K. Imoto, F. C. White, C. A. Rider, Y. C. Fung, and C. M. Bloor. Coronary arterial tree remodeling in right ventricular hypertrophy. *Am. J. Physiol.* 265(Part 2):H366–H375, 1993.

18. Kassab, G. S., D. H. Lin, and Y. C. Fung. Morphometry of pig coronary venous system. *Am. J. Physiol.* 267(*Heart Circ. Physiol.* 36):H2100–H2113, 1994.

19. Kassab, G. S., C. A. Rider, N. J. Tang, and Y. C. Fung. Morphometry of pig coronary arterial trees. *Am. J. Physiol.* 265(*Heart Circ. Physiol.* 34):H350–H365, 1993.

20. Lee, J. S., and S. Nellis. Modeling study on the distribution of flow and volume in the microcirculation of cat mesentery. *Ann. Biomed. Eng.* 2:206–216, 1974.

21. Levitt, D. G., B. Sircar, N. Lifson, and E. J. Lender. Model for mucosal circulation of rabbit small intestine. *Am. J. Physiol.* 237:E373–E382, 1979.

22. Lipowsky, H. H., and B. W. Zweifach. Network analysis of microcirculation of cat mesentery. *Microvasc. Res.* 7:73–83, 1974.

23. Mayrovitz, H. N., M. P. Wiedeman, and A. Noordergraaf. Analytical characterization of microvascular resistance distribution. *Bull. Math. Biol.* 38:71–82, 1976.

24. Nellis, S. H., K. L. Carroll, and M. Eggleston. Measurement of phasic velocities in vessels of intact freely beating hearts.

25. Popel, A. S. A model of pressure and flow distribution in branching networks. *J. Appl. Mech.* 47:247–253, 1980.

26. Pries, A. R., T. W. Secomb, P. Gaehtgens, and J. F. Gross. Blood flow in microvascular networks. Experiments and simulation. *Circ. Res.* 67:826–834, 1990.

27. Schmid-Schoenbein, G. W. A theory of blood flow in skeletal muscle. *J. Biomech. Eng. Trans. ASME* 110:20–26, 1986.

28. Skalak, T. C. A Mathematical Hemodynamic Network Model of the Microcirculation in Skeletal Muscle, Using Measured Blood Vessel Distensibility and Topology. University of California, San Diego (University Microfilms International No. DA8418303), Ph.D. Thesis, 1984.

29. Spaan, J. A. Coronary Blood Flow: Mechanics, Distribution, and Control. Boston: Kluwer Academic, 1991.

30. Tillmanns, H., M. Steinhausen, H. Leinberger, H. Thederan, and W. Kubler. Pressure measurements in the terminal vascular bed of the epimyocardium of rats and cats. *Circ. Res.* 49:1202–1211, 1981.

31. VanBavel, E., and J. A. Spaan. Branching patterns in the porcine coronary arterial tree: estimation of flow heterogeneity. *Circ. Res.* 71:1200–1212, 1992.

32. Zhuang, F. Y., Y. C. Fung, and R. T. Yen. Analysis of blood flow in cat's lung with detailed anatomical and elasticity data. *J. Appl. Physiol.: Respir. Environ. Exerc. Physiol.* 55:1341–1448, 1983.

Am. J. Physiol. 260 (*Heart Circ. Physiol.* 29):H1264–H1275, 1991.

CHAPTER 6

WHAT LIES BEYOND BIOINFORMATICS?

BERNHARD O. PALSSON

Department of Bioengineering, University of California, San Diego

1. Introduction

Since this material is prepared for a diverse audience, I'd like to start off my comments by briefly reviewing the central dogma of molecular biology (Fig. 1). The DNA molecule contains the inherited material that is passed from one generation to the next. DNA is a long molecule in the form of a double helix, in which each strand of the helix consists of a complementary sequence of base pairs. Well-defined segments of the DNA molecule are transcribed to an RNA molecule (a close chemical relative of DNA) that leaves the nucleus of the cell. There, this RNA transcription is "translated" into a protein molecule that comprised an amino acid sequence specified by the base pair sequence in the transcript molecule. The genetic code relates this base pair sequence to the amino acid sequence and the resulting protein molecule then carries out a particular chemical or physical function in the cell.

The order of magnitude of the information contained in a DNA molecule can be obtained from the following: the number of base pairs in the human genome is a few billion, in simple bacteria a few million, and in simple viruses, a few thousand. The number of genes found in the human genome is estimated to be on the order of 70,000 to 100,000, while some of the bacteria described below have gene numbers that are on the order of a few thousand, whereas a virus may contain half a dozen to a dozen genes.

2. Emergence of Bioinformatics

This central dogma of molecular biology has been well known for over 30 years, so what is new? The ability to generate massive amounts of compositional and structural data on these biomolecules is leading to the compilation of practically complete information about the genetic and biochemical composition of particular organisms. This has developed as a result of DNA sequencing technology, DNA chip technology, and other methods that allow us to determine the biochemical details of the schema shown in Fig. 1 on a whole organism-scale.

There are a number of small genomes that have been fully sequenced and published. The first full DNA sequence was published in the summer of 1995 for *Haemophilus influenzae*, a human bacterial pathogen (1). Since then, a growing number of small genomes have been fully sequenced, see, e.g. (Table 1) and published. There are undoubtedly many more whose sequence has been established but not published. The full sequence for higher order organisms, including human, is expected to be established within just a few years.

Central Dogma of Molecular Biology

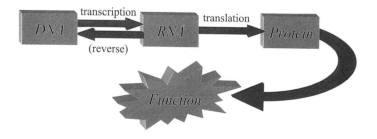

Fig. 1. The central dogma of molecular biology.

Completed Microbial Genome Annotation

Organism	Genome Size(Mb)	Number of Gene	% ORF assignment
Haemophilus influenzae	1.83	1703	83%
Mycoplasma genitalium	0.58	470	70%
Methanococcus jannaschii	1.66	1738	38%
Mycoplasma pneumoniae	0.81	677	76%
Saccharomyces cerevisiae	13.00	5800	43%
Escherichia coli K-12	4.60	4288	68%
Helicobacter pylori	1.66	1590	69%

Table 1.

Once the sequence of base pairs in the DNA molecule is known, the coding regions in the sequence are identified. Such identification leads to the location of the so-called "open reading frames" or ORFs. These are the regions of the DNA that are transcribed as shown in Fig. 1. The numbers of ORFs found in the fully sequenced small genomes are shown in Table 1. These single cellular organisms have on the order of a few thousand identified ORFs. It is interesting to note that *E. coli* has 4288 identified ORFs. This organism can grow on a single nutrient, such as glucose, along with a few salts, and generate all of its cellular components from these basic starting materials. It is also able to survive in a wide variety of environments. Thus, it can been called "the complete organic chemist." In contrast, the pathogen *H. influenzae*, has 1748 identified ORFs. This organism requires many nutrients to survive, and can only grow in certain microenvironments.

Once the ORFs have been identified, one then tries to assign function to the gene products that they encode. This task is more difficult than identifying the location of the ORFs. Such assignments are possible by looking at similarities to known genes with known functions. For some classes of genes, this task is easier than others. For instance, the metabolic enzymes have been known for a long time, and finding functional assignment to the corresponding ORFs is relatively easy. On the other extreme, there are ORFs with

Evolution of Bioinformatic Databases

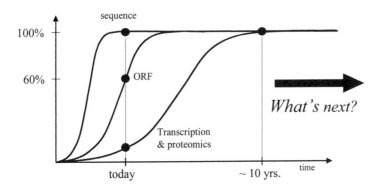

Fig. 2. Where is Bioinformatics leading us? For some model organisms, the DNA sequence, the functional assignments of open reading frames (ORFs), and the use of gene products under different circumstances (proteonomics) is headed towards completion. This figure was prepared with *E. coli* K-12 in mind.

sequences that bear no similarity to any protein of known function. The percentage of the ORFs that have been assigned function are shown in Table 1. For *E. coli*, about two-thirds of the ORFs have assignments, whereas in *H. influenzae*, the assignments are now at over 80%.

3.   Pressing Questions

Let us put this situation in perspective. We are effectively establishing complete "part catalogs" of simple cells. Taken as a function of time, these events can be represented by Fig. 2 using *E. coli* K-12 as an example. The DNA sequence is known, and we have assigned function to about 2/3 of the ORFs. Which of these genes are expressed under different conditions can be determined by a variety of techniques, giving basis for the so-called proteonomics . Therefore, we have the expectation that within a decade or so, we will have extensive, if not complete knowledge about bacteria like *E. coli* in both genetic and biochemical terms. Although the outline above gives a narrow view of bioinformatics through small genome sequencing, we can generally state that we are headed towards effectively knowing all the details of the events depicted in Fig. 1 for a selected number of organisms.

That brings us to the pressing question of what is next? What lies beyond bioinformatics? The DNA sequencing efforts that I just outlined lead to the definition of the genotype of a cell, e.g. a list of the genes contained in the organisms and their individual function. The term phenotype is used to describe the function or behavior of the cell under particular conditions. With the complete genetic makeup of a growing number of organisms becoming available, the pressing question arises as to how the phenotype can be predicted from the genotype. Some of the current thinking has been that there is a linear, or one-to-one relationship, between genes and cellular functions (2). Thus, we have the breast cancer gene, genes that confer chemical dependencies, the baldness gene, and so on.

The expectation then, is that the identification of these genes will lead to the possibility of manipulating the corresponding physiological functions through genetic alterations, or by affecting the function of their gene products. However, experience with an increasing number of knockout mice shows that the removal of particular genes leads to null phenotypes, and the knock-out animal develops "normally" without the gene product derived from the currently deleted gene.

Thus, the relationship between the genotype and the phenotype is complex and highly nonlinear, and cannot be predicted from simply cataloging and assigning gene functions to genes found in a genome (3). Genomics will give us detailed and complete compositional information, but not the dynamic and systemic characteristics that determine the physiological functions of living systems. The establishment of the genotype-phenotype relationship will lead to an entirely new field of study, one that I propose we call *Phenomics*.

4.   The Axioms of Phenomics

Relating the genotype to the phenotype is a challenging task. In attempting to formulate a strategy for meeting this challenge, it appears to me that two fundamental statements can be made from which we can build.

AXIOM #1: The function of a gene product is physico-chemical in nature. The metabolic gene products have well defined catalytic functions that are described using enzyme kinetics. This statement also holds true for the biomechanical properties of cytoskeletal elements, the function of cellular motors, the DNA binding functions of transcription factors, the kinases involved in signal transduction, and so on. Thus, the function of gene products must ultimately be described by the governing physico-chemical laws. This observation in turn, calls for the analysis of gene products by the appropriate kinetic theory, transport theory, thermodynamic laws, and biomechanical means.

AXIOM #2: Most, if not all cellular functions require the coordinated function of multiple gene products. For instance, the universal glycolytic pathway requires about twelve gene products, and signal transduction pathways have a few dozen players. Consequently, essentially all cellular functions are integrative in nature, calling for systems analysis to elucidate their integrated function. Methods to deal with complex systems are developed in many fields, although it is likely that special methods will have to be formulated for the analysis of multigeneic functions.

5.   The Concept of a Genetic Circuit

It follows from these two axioms that individual cellular functions rely on a network of cooperating gene products that can be called *genetic circuits* (Fig. 3). This term indicates that the information for the function of gene product in the network is specified by the gene, but that all of the participating gene products in a cellular function together perform as a dynamic circuit.

Genetic circuit function is greater than the sum of its constituent parts. This statement follows from their nonlinear dynamic nature, and the fact that nonlinear couplings

From Genomics to Genetic Circuits

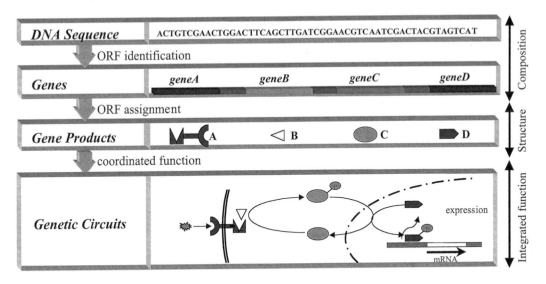

Fig. 3. From Genomics to Genetic Circuits. After the DNA sequence is known, the ORFs have been located and function assigned to them, our next challenge is to analyze the behavior of multiple gene products that together perform a cellular function. Such operations are termed genetic circuits.

lead to holistic functionalites that cannot be predicted from the study of the individual elements. The principle of superimposition does not apply mathematically nor conceptually. Therefore, it is difficult to predict changes in genetic circuit function based on changes in just one of their elements. Genetic circuits must be viewed, studied, analyzed, and manipulated as integrated entities.

6.   The Properties of Genetic Circuits

Although not all the fundamental properties of genetic circuits are presently known, some important ones can be stated:

- *Genetic circuits have many components; they are complex.* For instance, comparative analysis of the first two fully sequenced bacterial genomes has led to the identification of 256 genes, which together perform about a dozen cellular functions that constitute a minimal gene set for a modern cell (4).

- *Genetic circuits are flexible.* They are "redundant," i.e. in most cases, one can remove their components without compromising their overall function (5). For instance, many knockout mice have effectively normal phenotypes, even if the genes removed were thought to have critical roles. In addition, multiple oncogenes are typically needed for a transformation to a malignant phenotype. Further, genetic circuits are "robust" in the sense that the function of many of their elements can be altered without compromising overall function.

- **Genetic circuits have built-in controls.** That is to say, once the genes are expressed, the coordinated function of the gene products is autonomous. Genetic circuits are capable of functioning by themselves.

- **Creative functions and decision-making.** Embedded in these control structures are the capabilities to perform creative functions. Such functions include sustained oscillatory behavior and multiple steady states leading to built-in "decision making" mechanisms (6).

- **Evolutionary dynamics.** Finally, it appears that once a genetic circuit has been established, it is evolutionarily preserved. This preservation leads to unity in biology, such as the universal glycolytic and Ras-signaling pathways. However, the precise function of such genetic circuits changes over the course of evolution, and adjusts to organismic-specific needs.

7. Describing Genetic Circuit Function

In searching for methods to describe and analyze genetic circuit function, it is helpful to look at the field of metabolic modeling. Given the early elucidation of metabolic pathways and the development of enzyme kinetics, systems analysis in cellular biology is perhaps best developed for metabolic dynamics. Several methods have been developed for the analysis of integrated metabolic functions. Analysis methods such as metabolic control analysis (MCA) (7), biochemical systems theory (8), flux-balance analysis (FBA) (9, 10), and modal analysis (11) have emerged and proven useful for metabolic studies (12).

Some of these existing methods will prove important in analyzing growing genomic databases. FBA for instance, can be applied to the initial analysis of freshly sequenced genomes with the assignment of ORFs, resulting in the definition of the "metabolic genotype" of the organism. The metabolic capabilities and characteristics of this metabolic genotype can then be assessed using established FBA approaches (13). Although enzyme kinetic properties are presently not as well catalogued or "assignable" as ORFs, it is likely that bioinformatics will eventually allow similar analysis of metabolic dynamics. We can anticipate the use of existing methods to analyze the systemic kinetic behavior of newly sequenced genomes and their metabolic genotypes.

8. Genetic Circuit Analysis: Flux Balances

There are specific examples of genetic circuit analysis that have appeared. I am going to present the analysis of the genotype-phenotype relationship in metabolic circuits using a flux balance analysis (Fig. 4). One can think of this approach as analogous to Kirchhoff's laws for circuit design, in which the sum of all the currents going into a node has to equal to the sum of the currents going out.

A metabolic map, illustrated as a "node-branch" diagram in Fig. 4, has arrows or branches representing the chemical conversions catalyzed by the enzymes, and the dots represent the different metabolites. Looking at each of these metabolites, one sees what is shown in the enlarged bubble in Fig. 4. The metabolite is in the middle, and the four different types of reactions by which it is synthesized, degraded, exported, or used to meet

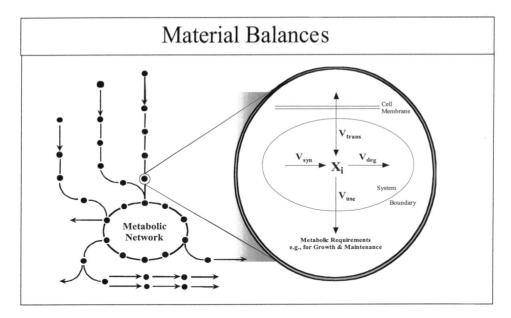

Fig. 4. Branch and node representation of metabolism and the use of flux balances to describe the steady state distribution of fluxes through the network.

metabolic requirements are also shown. Dynamic mass balances describe how the time derivative of the concentration of metabolites is equal to the sum of their formation and degradation flux. The vector "b" is the net exchange vector with the environment of the metabolic system, and it is a vector which we can measure.

The transients associated in the metabolism are relatively rapid compared to the time constants of growth, and these metabolites do not accumulate, so one can put this system of equations into a steady state, and solve the corresponding flux balance equation. The vector "b" represents the experimental data, and the matrix S the stoichiometry of the metabolic reactions taking place. Clearly, the fully sequenced genomes with assigned metabolic functions lead directly to the construction of S.

The matrix S is not square, so I cannot simply invert it and calculate the distribution of the metabolic fluxes. Therefore, for any set of inputs and outputs, the cell has choices in which way it distributes its internal fluxes. Conceptually, this is important because it represents the choices that the cell has in meeting the demands upon which it has been placed. Mathematically, the solutions to the flux balance equations are confined to a space; not every possible vector is a solution to these equations. The solution vectors are found in the null space of the stoichiometric matrix. Biologically, this space represents the capabilities of the defined metabolic genotype.

The metabolic phenotype, on the other hand, would be one particular solution in the null space. One can explore the characteristics of these phenotypes and study how they relate to the underlying genotype. One can use linear optimization to find the "best" phenotypes arising from a genotype, if objectives are stated for metabolic function. You can find the particular solution that would be the "best" metabolic phenotype, which, in biological terms, means the highest possibility for survival. We have used the objective of optimizing growth in our experimental and theoretical studies of *E. coli*. It turns out that the observed behavior of *E. coli* is consistent with this objective, and one can analyze, interpret, and even predict the phenotypic behavior of *E. coli* based on its genotype, if this is based on the optimal growth objective.

9.   The E. coli Metabolic Genotype

Based on bioinformatic data on *E. coli* genomics and biochemistry, we have synthesized its metabolic genotype (14). The number of metabolic reactions in this metabolic genotype is 720 operating on 436 metabolites. This is the integrated metabolic circuit that *E. coli* uses to synthesize all of the organic chemical structures that it needs for its own synthesis and function. With the interconnectivity of the gene products known, one can study the systems properties of the metabolic genotypes.

10.   Redundancy Characteristics of the E. Coli Metabolic Genotype

We have studied the results of changing the metabolic genotype by 1, 2, or 3 gene increments, and have determined what the effects of such deletions are on the overall functionality of *E. coli*. Figure 5 shows the effects of deleting the genes in the core metabolic pathways one at a time, based on the ability of the network to satisfy all growth requirements from glucose as the sole substrate. What this figure shows is the growth rate that these mutants can achieve relative to the wild-type, or the genotype that has all of the genes. These results are quite interesting. They show that almost two-thirds of these genes can be deleted, and the organism will still grow at essentially the same rate. It can reroute its fluxes and metabolic traffic around those deleted enzymes. There is a group of enzymes here that we call retarding, whose loss will slow the growth rate, but the organism will still grow. Then there are a few deletions that are lethal; if you remove them, the organism cannot grow on

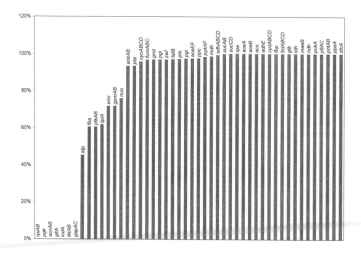

Fig. 5. Gene deletions and changes in the capabilities of the *E. coli* K-12 *in silico* metabolic genotype. This figure shows the maximal growth rates on glucose for all possible single gene deletions in the central metabolic pathways as a fraction of the growth rate obtained by the full set of genes.

glucose. Interestingly enough, some of these enzymes have isozymes, so the cell has several genes for closely related gene products that can carry out the same chemical conversion.

You may wonder what these studies mean? I call this the *in silico E. coli strain*. It is basically a computer strain that has a metabolic genotype ascribed to it which is the same as that defined for the wild type strain based bioinformatics. It has the functional properties that have been described above. It turns out that many of the mutants whose behavior we have been predicting are already known. You can go into the literature and get information about the metabolic behavior of these mutants. Remarkably, 60 out of the 65 cases presented here are accurately predicted based on this flux balance analysis (14).

Finally, I would like to paraphrase Jim Watson, with a phrase from his book *Molecular Biology of the Gene* (15),

> *"Cells obey the laws of [physics and] chemistry"*

to which we humbly add

> *... including mass energy and redox balances.*

By imposing these balances on the defined genotype we can actually analyze, interpret, and predict what cells are capable and not capable of doing.

11.   Conclusion

Bioinformatics is beginning to give us essentially complete spare-part catalogs of cells. This information *demands* functional analysis, which in turn needs system science and the physico-chemical sciences. I am predicting that genetic circuits in the sense that I have described them here, will arise as a fundamental new paradigm in cellular and molecular biology. We will not be thinking about individual genes, but instead about these circuits and how we can modify them along with the circuits that represent particular cellular functions.

I believe a new field will arise, one which I tentatively call *Phenomics*, as an analogy to genomics, which will deal with the genotype/phenotype relationship. The field of bioengineering is poised to play a key, leading role in these developments. It will be the bioengineering of the future which is going to be an effective integration of the biological sciences, the physico-chemical sciences, and the systems sciences, in ways that we have not seen before.

I would like to end on one historical note. I think we all know what happened to electrical engineering following the 1950's when it took on integrated circuits. I would like to submit that bioengineering is at a very similar point in its history. The development of genetic circuits may lead to similar influence in bioengineering, as integrated circuits did for electrical engineering.

Acknowledgments

The figures in this article were prepared by Christophe Schilling. The author was the recipient of a Fulbright and Ib Heinriksen Fellowships in 1996 during his stay at the Technical University of Denmark. Many of the thoughts outlined herein were formulated there.

References

1. Fleischmann, R. D., Adams, M. D. and White, O., Whole-genome random sequencing and assembly of haemophilus influenzae rd *Science* **269**, 496–512 (1995).
2. Palsson, B. O., What lies beyond bioinformatics? *Nature Biotechnol.* **15**(1), 3–4 (1997).
3. Strothman, R. C., The coming Kuhnian revolution in biology, *Nature Biotechnol.* **15**, 194–199 (1997).
4. Mushegian, A. R. and Koonin, E. V., A minimal gene set for cellular life derived by comparison of complete bacterial genomes, *Proc. Natl. Acad. Sci. USA* **93**, 10268–10273 (1996).
5. Edwards, J. S. and Palsson, B. O., How will bioinformatics influence metabolic engineering? *Biotechnol. Bioeng.*, 1998 (in press).
6. Reich, J. G. and Sel'kov, E. E., Energy metabolism of the cell, 2nd edition, Academic Press, New York (1981).
7. Fell, D., Understanding the control of metabolism, Portland Press, London (1996).
8. Savageau, M. A., Biochemical systems analysis. II. The steady state solutions for an n-pool system using a power-law approximation, *J. Theoret. Biol.* **25**, 370–379 (1969).
9. Varma, A. and Palsson, B. O., Metabolic flux balancing: Basic concepts, scientific and practical use, *Biol. Technol.* **12**, 994–998 (1994).
10. Bonarius, H. P. J., Schmid, G. and Tramper, J., Flux analysis of underdetermined metabolic networks: The quest for the missing constraints, *Trends Biotechnol.* **15**, 308–314 (August 1997).
11. Palsson, B. O., Joshi, A. and Ozturk, S. S., Reducing complexity in metabolic networks: Makaing metabolic meshes manageable, *Fed. Proc.* **46**, 2485–2489 (1987).
12. Heinrich, R. and Schuster, S., The regulation for cellular systems, Chapman & Hall, New York (1996).
13. Palsson, B. O., *et al.*, Metabolic flux balance analysis, *Metabolic Engineering* (ed. L. A. Papoutsakis) (1998).
14. Edwards, J. S., Keasling, J. D. and Palsson, B. O., Escherichia coli K-12 in silico: Definition of its metabolic genotype and analysis of its capabilities (submitted, 1997).
15. Watson, J. D., Molecular biology of the gene, 1st edition, Garland Publishing Inc., New York (1972).

288   *Biotechnol. Prog.* **1999**, *15*, 288–295

Toward Metabolic Phenomics: Analysis of Genomic Data Using Flux Balances

Christophe H. Schilling, Jeremy S. Edwards, and Bernhard O. Palsson*

Department of Bioengineering, University of California, San Diego, La Jolla, California 92093-0412

Small genome sequencing and annotations are leading to the definition of metabolic genotypes in an increasing number of organisms. Proteomics is beginning to give insights into the use of the metabolic genotype under given growth conditions. These data sets give the basis for systemically studying the genotype–phenotype relationship. Methods of systems science need to be employed to analyze, interpret, and predict this complex relationship. These endeavors will lead to the development of a new field, tentatively named phenomics. This article illustrates how the metabolic characteristics of annotated small genomes can be analyzed using flux balance analysis (FBA). A general algorithm for the formulation of *in silico* metabolic genotypes is described. Illustrative analyses of the *in silico Escherichia coli* K-12 metabolic genotypes are used to show how FBA can be used to study the capabilities of this strain.

During the first half of the 20th century, genetics was based on the premise that an invisible information-containing unit called a gene existed. Throughout this time period, genetics focused on the functional aspects of a gene, as little was known about its molecular structure. Since then science has unraveled the intricacies and mechanistic details of genetic information transfer and determined the structure of DNA and the nature of the genetic code, establishing DNA as the source of heredity containing the blueprints from which organisms are built. With this in mind it is of no surprise that the scientific community has devoted considerable efforts toward determining the genome of many microorganisms, and even more complex eucaryotes such as the nematode worm, the mouse, and even humans. These revolutionary endeavors have rapidly spawned the development of new technologies and emerging fields of research based around genome sequencing efforts (i.e., functional genomics, structural genomics, proteomics, and bioinformatics).

The goals of sequencing entire genomes find their roots in the initiation of the Human Genome Project (HGP). Setting the stage for modern genomics, the HGP sparked sequencing initiatives for a number of organisms from all the major kingdoms of life. Currently there are 20 completely sequenced and annotated genomes that are publicly available and published (14 Eubacteria, 4 Archaea, and 2 Eucaryote) (*1*). Small genome sequencing is becoming routine, and in the future, studies of nearly every organism will be aided by the knowledge and availability of the complete DNA sequence of their genomes (*2*).

Once presented with the entire sequence of a genome, the first step is to identify the location and size of the genes and their open reading frames (ORFs). These activities in conjunction with sequencing are performed in the domain of genomics. ORFs are determined through the use of various gene prediction programs. Once ORF identification is completed, the ORFs are searched against protein and DNA databases to establish sequence similarity to genes/proteins with known function. Upon the existence of these similarities, functional homology can then be inferred and putative gene assignments designated to ORFs. For the fully sequenced genomes approximately 45–80% of the identified ORFs have been given genetic assignments, and these percentages are increasing at a rapid pace (*3*).

We are therefore at the brink of having a complete "part catalog" of many organisms, including ourselves. The next step is to search for the blueprints that build and operate the cell given the structural knowledge of its components along with a first-hand glimpse at their function. How can we use all of this genomic and biochemical information to gain insight into the relationship between an organism's genotype and its phenotype?

DNA sequence data now need to be translated into functional information, both in terms of the biochemical function of individual genes, as well as their systemic role in the operation of multigeneic functions. The study of function is encompassed by the field of functional genomics, which can be viewed as a two-tiered approach to the study of genetic information transfer. The first level is the prediction and determination of the biochemical function of individual gene products from their sequence through comparison with other known genes and genomes. The second level, and perhaps the more challenging one (*4, 5*), involves the use of this functional information of individual genes to analyze, interpret, and predict cellular functions resulting from their coordinated activity.

The prediction of function of individual genes is at present an area of intense experimental and computational activity. It is accompanied by growing attempts to define all of the proteins expressed by an organism's genome under given conditions, or the proteome (*6*). These are useful first steps in understanding the genotype–phenotype relationship, but they can only take us so far. It is now clear that we need to develop creative approaches and technologies to use all of this information

* To whom correspondence should be adressed. Telephone: (619) 534-5668. Fax: (619) 822-0240. E-mail: palsson@ucsd.edu.

10.1021/bp9900357 CCC: $18.00   © 1999 American Chemical Society and American Institute of Chemical Engineers
Published on Web 04/24/1999

Biotechnol. Prog., 1999, Vol. 15, No. 3

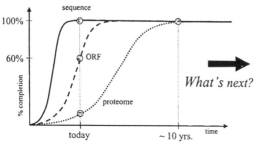

Figure 1. Genomics and *E. coli*: Currently the entire genome of *E. coli* has been sequenced, and the assignment of open reading frames is at about 65% of completion, while ongoing efforts are underway to determine the proteome. We can envision near completion of the genome annotation and a large portion of the proteome studies in the future. With the generation of all these data for various other organisms sure to follow the question remains: how can these data be used to determine the relationship between the genotype and the phenotype?

to explore and determine genome function. We must essentially take on the view of a gene that we began with over 50 years ago, wherein the focus was on functional attributes of a gene within the context of the whole organism. At that time scientists lacked any structural knowledge of a gene; however, today we have the capabilities to gain complete structural knowledge of every gene, placing us in a favorable position to decompose the complexities that link the phenotype to the genotype. Genomics will provide detailed and complete compositional information, but not the dynamic and systemic characteristics that determine the physiological functions of living systems.

From Genotype to Phenotype

As methods in genomics continue to improve in their efficiency, it has become quite obvious that we will be limited not by the availability of information (*7, 8*) but by our lack of tools available to analyze and interpret these data (Figure 1). While the process of genomics and functional genomics is essential and important in itself, it serves as a precursor to the greater goal of using genomic information to understand the phenotypic characteristics of a particular organism.

It is becoming clear that there is not a one-to-one relationship between individual genes and overall cellular functions (*9*). This is evidenced by the number of knockout mice that exhibit null phenotypes following the removal of particular genes; thus the development of the animal is more or less normal without the deleted gene. The relationship between the genotype and the phenotype is complex and highly nonlinear and cannot be predicted from simply cataloging and assigning gene functions to genes found in a genome.

Since cellular functions rely on the coordinated activity of multiple-gene products, the interrelatedness and connectivity of these elements becomes critical. The coordinated action of multiple-gene products can be viewed as a network, or a "genetic circuit" (*10*), which is the collection of different gene products that together are required to execute a particular function. Thus, if we are to understand how cellular functions operate, the function of every gene must be placed in the context of its role in attaining the set goals of a cellular function. The study of the genotype–phenotype relationship, through the use of genomic data and analysis of multigeneic functions in this manner, constitutes a field of its own;

one which may be called "phenomics". Phenomics is expected to rise as a new scientific endeavor, one that seeks to analyze, interpret, and predict the genotype–phenotype relationship from genomic data. Given all the biological data and computer power available today, the development of phenomics seems inevitable. Strategic approaches need to be formulated and methods developed which utilize the wealth of data provided by genomics and other facets of functional genomics.

Microorganisms currently offer the best test-bed for establishing *in silico* methods that can be used to analyze, interpret, and predict the genotype–phenotype relationship starting from complete genomic DNA sequences. The best physiological function from which to launch the development of this line of study is metabolism, where there already exists a history of studying its systemic properties. In addition, a major portion of genes in microbial cells encode products with metabolic functions (*11*) and the biochemical functions of most metabolic gene products are known. Thus, once the ORF assignments have been carried out, the entire metabolic map representing the stoichiometry of all the metabolic reactions taking place in the cell can be constructed. Many years of research on microbial metabolism has resulted in the development of various mathematical and analytical methods to understand the metabolic dynamics from stoichiometric and kinetic information (*12–14*). However, oftentimes the *in vivo* kinetic parameters for all the enzymes in the network are unknown, making dynamic studies and simulations extremely difficult. In fact, estimation of *in vivo* enzyme kinetics is a main problem for the area of metabolic modeling and engineering. Despite the lack of kinetic information on individual reactions, the stoichiometry of these reactions is well-known and can be used to study the theoretical capabilities and functions of metabolic networks. One approach that has recently been used for characterizing metabolic genotypes on the basis of stoichiometry is flux balance analysis (FBA) (*15–18*).

This article will illustrate the use of FBA as a method that utilizes the metabolic genotype of an organism to analyze, interpret, and predict its metabolic phenotype under particular conditions. This use provides an example of the types of system science methods that can be developed in order to realize the benefits of genomics for unraveling the genotype–phenotype relationship. A brief background and overview of FBA will be provided, and then it will be shown how to integrate genomic data to establish organism specific models. These *in silico* genomically based models can be used to predict phenotypic characteristics of organisms, including the assessment of metabolic capabilities and genetic deletions. The focus here is not on FBA itself, but rather on the use of FBA as a guiding example for methods to utilize genomic data to further scientific progress.

Fundamentals of Flux Balance Analysis

FBA is an approach that is well-suited to account for genomic detail, as it has been developed on the basis of the well-known stoichiometry of metabolic reactions. The approach is based on metabolic flux balancing in a metabolic steady state. The principles are outlined in Figure 2. The fundamentals and uses of flux balance methods for metabolic network analyses have been reviewed (*15, 16, 19*).

The history of flux balance analysis for metabolic studies is relatively short, contained mostly within the field of metabolic engineering (*20*). It has been applied

290   *Biotechnol. Prog.*, 1999, Vol. 15, No. 3

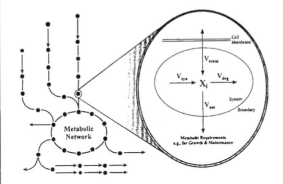

Fundamentals of Flux Balance Analysis

The fundamental principle underlying FBA is the conservation of mass. A flux balance can be written for each metabolite(X_i) within a metabolic system to yield the dynamic mass balance equations that interconnect the various metabolites.

Dynamic Flux Balances

$$\frac{dX}{dt} = S \cdot v - b$$

$$v = fn(X,...)$$

X = Metabolite Concentrations
S = Stoichiometric Matrix
v = Metabolic Reaction Fluxes
 (V_{syn}, V_{deg})
b = Net Transport out of Network
 (V_{use}- V_{trans})

Steady State Analysis

$$S \cdot v = b$$

Solve for Unknown Metabolic Fluxes v

A steady state assumption is made which produces the system of linear equations in the lower box, which simply states that the formation fluxes are balanced by degradation fluxes. This situation is analogous to Kirchhoff's current law used in electrical circuit analysis, where the sum of the currents coming in and out of a node must sum to zero.

Typically there are more fluxes in the system than metabolites. Thus, the system of equations is underdetermined and can therefore be formulated as a linear programming (LP) problem, in which one finds the optimal flux distribution that minimizes or maximizes a particular objective. Various objective function can be used such as those listed below:

- *Minimize ATP production*
- *Minimize nutrient uptake*
- *Minimize redox production*

- *Maximize metabolite production*
- *Maximize biomass production (i.e growth)*
- *Minimize the Euclidean norm of the flux vector*

Figure 2.

to metabolic networks (21) and the study of adipocyte metabolism (22). Acetate secretion from *E. coli* under ATP maximization conditions (23) and ethanol secretion by yeast (24) have also been investigated using this approach. Importantly, FBA has demonstrated its ability to quantitatively predict the behavior of *E. coli* under a variety of conditions (18, 25, 26).

Constructing *in Silico* Strains from Bioinformatics

The interconnectivity of metabolites is contained within the stoichiometric matrix. A stoichiometric matrix, **S**, is an $m \times n$ matrix where m (the number of rows) corresponds to the number of metabolites and n (the number of columns) corresponds to the number of fluxes occurring in the network. The S_{ij} element of a stoichiometric matrix corresponds to the stoichiometric coefficient of the reactant i in the reaction denoted by j. The stoichiometric matrix allows us to represent the metabolic network in a mathematical format permitting further analysis. It is an invariant property of the metabolic network that describes the architecture and topology of the system.

From annotated genomes we can directly construct the stoichiometric matrix for the entire metabolic network of an organism to be further analyzed using FBA, or other related approaches for the determination of metabolic pathways. The procedure that is used to synthesize an organism and genome-specific stoichiometric matrix and subsequent *in silico* models is outlined in Figure 3. A universal stoichiometric matrix (U-Stoma) can be constructed as a database of metabolic reactions from which organism-specific metabolic reactions can be selected.

Therefore, individual metabolic genotypes of all organisms are comprised of a subset of columns from the U-Stoma.

By capitalizing on genomics, a genome can be sequenced and ORFs assigned, and through the first level of functional genomics, we can obtain the metabolic genotype of an organism by utilizing sequence similarities and inferred homologies. This organism specific metabolic genotype can be used to select the columns of the U-Stoma representing a particular organisms metabolic profile. From the list of genes comprising the metabolic genotype we simply include all of the possible reactions that the gene products are associated with. It is also necessary in the case of well-studied organisms to include reactions, known to occur from experimental evidence and biochemical assays, for which no associated genes have been assigned in the genome. Thus, we have a simple algorithm for the construction of a genome-specific stoichiometric matrix (Gs-Stoma) once the ORF assignments have been made and existing literature surveyed.

Following this procedure, the genome-specific stoichiometric matrix can be obtained immediately for any recently sequenced and annotated genome. The Gs-Stoma will allow the metabolic capabilities of these organisms to be determined and serve as the crucial component in the prediction of metabolic phenotypes. One advantage of this approach is its ability to absorb errors in ORF assignments generated through genomics. If an assignment is incorrect or a new gene is found or determined to be biochemically present in an organism, a simple

Biotechnol. Prog., 1999, Vol. 15, No. 3

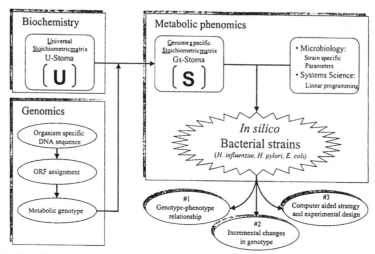

Figure 3. Synthesis of *in silico* metabolic genotypes: A schematic illustration of the algorithm used to generate genome specific stoichiometric matrices on the basis of genomics, ORF assignments, strain specific information, and methods of systems science.

addition or removal of appropriate columns in the Gs-Stoma is all that is necessary.

To perform the complete FBA additional strain-specific parameters must be determined, such as the biomass composition, maximal substrate uptake rates, and maintenance requirements, all of which relate to the basic capabilities and demands placed on the metabolic network (*18, 27*). This information along with the Gs-Stoma complete the organism specific model/*in silico* strain, which can then be applied to the methods of linear programming for further analysis.

The *E. coli in Silico* Metabolic Genotype

Using the above algorithm, an *in silico* strain of *E. coli* K-12 has been constructed (*28*) from annotated sequence data (*29*) and from biochemical information (*30*).

Briefly, the *E. coli* K-12 metabolic genotype contains 720 metabolic processes that influence 436 metabolites. There are 587 enzymes involved, of which 540 have ORF assignments in the newly sequenced genome (*29*), while the remaining 47 have not been assigned but are known to be present from biochemical data. This *in silico* strain accounts for the metabolic capabilities of *E. coli*. It includes membrane transport processes, the central catabolic pathways, utilization of alternative carbon sources, and the biosynthetic pathways that generate all the components of the biomass. In the case of *E. coli* K-12, we can call upon the wealth of data on overall metabolic behavior and detailed biochemical information about the *in vivo* genotype to which we can compare the behavior of the *in silico* strain. One utility of FBA is the ability to learn about the physiology of the particular organism and explore its metabolic capabilities without any specific biochemical data. This ability is important, considering possible future scenarios in which the only data that we may have for a newly discovered bacterium (perhaps pathogenic) could be its genome sequence.

In Silico Strains and Flux Balance Analysis

Now that we can construct "*in silico*" strains for particular organisms, what types of *in silico* studies and manipulations can be performed?

Study Variations in the Genotype. FBA can be used to interpret and predict the phenotypic affects of genetic

deletions to an organism. FBA has been used to explore the relation between the *E. coli* metabolic genotype and phenotype in an *in silico* knockout study (*28*). In this study, all possible combinations of single, double, and triple deletions of 48 core metabolic genes/gene products were analyzed for their effects on the ability of the network to produce from glucose minimal media the necessary components of the biomass and energy/maintenance requirements, which together represent the growth capabilities. To simulate these effects the fluxes through reactions associated with the genes in question are constrained to equal zero, and an objective function is set to maximize for growth. To maximize for growth the objective function typically represents a flux that drains on metabolites such as amino acids, nucleotides, lipids, and other metabolites that are incorporated into the biomass in specific ratios in accordance with the biomass composition. Based on the resulting ability of the network to maximize this objective, these 48 gene products were classified as essential, important/nonessential (retarding), and redundant. It was found that a surprising number of the metabolic gene products were determined to be redundant for growth on glucose minimal media (32 of 48 cases (67%), see Table 1), which indicated that the deletions showed no effect on the capability of the network to meet the growth demands. In the remaining 16 cases the deletion led to either a complete loss in the ability to meet the growth demands or a decrease in the efficiency of reaching these demands.

Even when multiple knockouts were performed, a surprising number of the double and triple deletions resulted in no effect on growth, 424 of 1128 cases (38%) and 3260 of 17 296 cases (19%), respectively. The predicted metabolic phenotype of 60 of 66 experimentally characterized mutants examined was accurate. This deletion study has immediate implications to the interpretation of cellular metabolic physiology. For each one of these cases it is possible to examine the precise production deficiencies that arise in the network due to the absence of the gene(s) in question. In this way one can readily use FBA to determine auxotrophic growth requirements for mutant strains of a bacteria by determining which components of the biomass cannot be produced due to alterations in the genotype.

Biotechnol. Prog., 1999, Vol. 15, No. 3

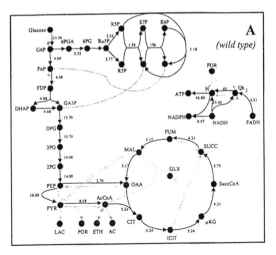

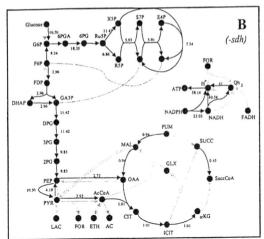

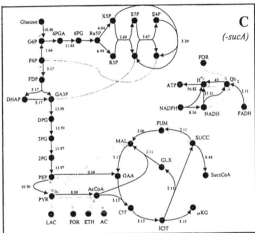

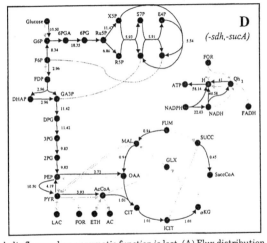

Figure 4. Flexibility in the metabolic genotype: Rerouting of metabolic fluxes when enzymatic function is lost. (A) Flux distribution for the basic gene set under aerobic conditions. Biomass yield is 0.49 g DW g^{-1} glucose. (B) Flux distribution for maximal biomass yield for *sdh* mutant under aerobic conditions. Biomass yield is 0.48 g DW g^{-1} glucose. (C) Flux distribution for maximal biomass yield for *sucA* mutant under aerobic conditions. Biomass yield is 0.49 g DW g^{-1} glucose. (D) Flux distribution for *sdh sucA* double mutant under aerobic conditions. Biomass yield is 0.48 g DW g^{-1} glucose. The solid lines represent enzymes that are being utilized, with the corresponding flux value noted. The gray lines represent enzymes that are not being utilized. The fluxes are relative to the glucose uptake rate (10.5 mmol glucose h^{-1} g^{-1} DW, 26).

Table 1. List of the Genes Predicted To Be Essential, Retarding, or Redundant under Growth on Glucose Minimal Media[a]

Essential Gene Set
  rpiAB, pgk, acnAB, gltA, icdA, tktAB, gapAC
Retarding Gene Set
  atp, fba, pfkAB, tpiA, eno, gpmAB, nuo, ackAB, pta
Redundant Gene Set
  cyoABCD, fumABC, gnd, pgl, zwf, talB, pts, pgi, aceEF, ppc, pykAF, mdh, sdhABCD, sucAB, sucCD, rpe, aceA, aceB, acs, adhE, cydABCD, fbp, frdABCD, glk, ldh, maeB, ndh, pckA, pflAC, pntAB, ppsA, sfcA

[a] Redundant genes are defined as genes whose removal leads to the ability of achieving a growth rate of 95−100% of the complete gene set. Essential genes are those for which no growth is expected. Genes whose gene products form one functional protein are considered as a single gene group (i.e., *sdhABCD*). Each set of isozymes is also clustered into one single gene group for this analysis (i.e., *pykA* and *pykF* into *pykAF*).

Prediction of Metabolic Shifts. As an extension to the previous example, FBA can be used to examine the shifts in metabolic routing when subjected to the loss of the function of a gene product through genetic deletions. Figure 4A shows the flux distribution for the central metabolic pathways as determined using the complete gene set present in *E. coli* for maximal biomass yield with glucose as the carbon source. The computed value for biomass yield (0.49 g DW g^{-1} glucose) (DW = dry weight) compares quantitatively with experimental data (31).

The ability of *E. coli* to respond to the loss of an enzymatic function (through gene mutation or inhibition of activity) can be assessed by removing a gene from the basic gene set as mentioned in the previous section. The predicted alterations in flux distribution under aerobic conditions resulting from the loss of the *sucA* and *sdh* gene products are discussed below (20), due to the availability of experimental data for both single mutants and the double mutant. Parts B and C of Figure 4 show the optimal metabolic flux distribution when the *sucA* gene and the *sdh* gene are removed from the basic gene set, respectively. The *sucA* gene codes for an essential component of the 2-oxoglutarate dehydrogenase complex, and the *sdh* gene codes for the succinate dehydrogenase enzyme.

A mutant defective in the *sdh* gene product has been shown to be able to grow on glucose minimal medium (*32*). This is also the case in the *in silico* strain, and the flux distribution for the *sdh* mutant is illustrated in Figure 4B. The metabolic energy needs of this mutant are greatly increased, and for optimal biomass synthesis, the pentose phosphate pathway must support large fluxes to generate redox potential to be used for the production of high-energy phosphate bonds. This mutation decreases the predicted biomass yield by less than 1%. The *in silico* analysis shows that the ability of the *sucA* mutant to synthesize biomass is nearly equivalent to the ability of the complete gene set, showing that the *sucA* gene product is nonessential for growth on glucose. The flux distribution for the *sucA* mutant is provided in Figure 4C. Experimental observations show that mutants in the *sucA* gene are able to grow anaerobically on glucose but are unable to grow aerobically. However, in these mutants defective in the *sucA* gene, revertants often arise that also inactivate the *sdh* gene product, forming a double mutant that is able to grow aerobically on glucose minimal medium (*32*). Figure 4D shows the *in silico* predicted metabolic flux distribution for this *sdh sucA* double mutant to be in agreement with the experimental data (*32*).

This example shows the stoichiometric flexibility in central metabolism under the conditions considered. The metabolic network has the ability to redistribute its metabolic fluxes with little change in its capabilities to support biomass synthesis, even if faced with the loss of key enzymes. More importantly, it illustrates the complexity and unresolved nature of the genotype−phenotype relationship.

Predicting Genome Scale Shifts in Gene Expression. FBA can be used to predict metabolic phenotypes under different growth conditions, such as substrate and oxygen availability, by simply constraining the appropriate fluxes. The calculated flux distributions can then be correlated to experimental information on genetic expression levels. The relation between the flux value and the gene expression levels is nonlinear. However, FBA can give qualitative (on/off) information as well as the relative importance of gene products under a given condition. On the basis of the magnitude of the metabolic fluxes, qualitative assessment of gene expression can be inferred. To demonstrate this we studied changes in the flux distributions that are predicted to occur in *E. coli* as oxygen availability is decreased. The constraint on the oxygen uptake flux was decreased from 25 mmol g^{-1} DW to 0 mmol g^{-1} DW in increments of 0.1. Changes in the predicted biomass yield and the production rate of various byproducts are traced in Figure 5. Also shown are the six distinct phases of the metabolic response to varying oxygen availability, going from completely anaerobic (phase I) to completely aerobic (phase VI). Additionally lists of the genes associated with fluxes that are predicted to be activated, increased, inactivated, and decreased upon the shift from aerobic growth (phase V) to nearly complete anaerobic growth (phase II) are provided in Table 2.

Corresponding experimental data sets are now becoming available. Using high-density oligonucleotide arrays, the expression levels of nearly every gene in *Saccharomyces cerevisiae* can now be analyzed under various growth conditions (*33*). From these studies it was shown

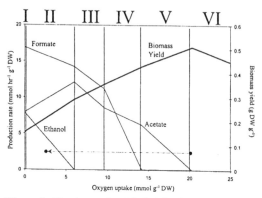

Figure 5. Six phases of metabolic behavior associated with varying oxygen availability, going from completely aerobic to completely anaerobic in *E. coli*. The glucose uptake flux was fixed under all conditions, and the resulting optimal biomass yield is indicated along with the output fluxes associated with three metabolic byproducts: acetate, formate, and ethanol. The dashed line indicates the shift from aerobic to anaerobic conditions, which provided the data for Table 2.

that nearly 90% of all yeast mRNAs are present in growth on rich and minimal media, while a large number of mRNAs were shown to be differentially expressed under these two conditions. Another recent article shows how the metabolic and genetic control of gene expression can be studied on a genomic scale using DNA microarray technology (*34*). The temporal changes in genetic expression profiles that occur during the diauxic shift in *S. cerevisiae* were observed for every known expressed sequence tag in this genome. As shown above, FBA can be used to qualitatively simulate shifts in metabolic genotype expression patterns due to alterations in growth environments (*35*). Thus, FBA can possibly serve to complement current studies in metabolic gene expression, by providing a fundamental approach to analyze, interpret, and predict the data from such experiments.

Assessing the Metabolic Capabilities and Underlying Pathway Structure in a Defined Metabolic Genotype. Often we seek to gain insight into the architecture of a metabolic network in biochemical terms. Namely, we are interested in the inherent structure and connectivity of the elements within the network. In the case of a metabolic network obeying the law of material conservation, this interest can be explored by effectively spanning the solution space (flux space) to the system of linear equations, which define the biochemical reaction network. This flux space is a subset of the null space of the stoichiometric matrix. In biochemical terms, the flux space contains all possible flux distributions through a system and hence all the possible metabolic phenotypes which the system can choose to operate.

Biologically relevant pathways can be defined by a set of basis vectors spanning the null space of the stoichiometric matrix (*36*), which can then be used to interpret the capabilities and structure of an organism's metabolic repertoire. Perhaps the most effective means by which to define the region of all admissible flux vectors is through the use of a nonnegative basis, which can be derived through concepts from convex analysis (*37*). Approaches such as these allow for the interpretation of metabolic functions and flux distributions from a pathway-based perspective as opposed to an individual reaction-based perspective (for a review, see ref 47). This approach of metabolic pathway analysis accompanied with FBA

Table 2. Changes in the Calculated Flux Levels of the 48 Genes Considered in Core Metabolism under a Shift from Aerobic to Anaerobic Conditions[a]

status	gene	aerobic (VI)	anaerobic (II)
activated	*adhE*	0.00	5.24
	pflAC	0.00	16.08
	pntAB	0.00	5.92
increased	*ackAB*	−0.52	9.86
	pta	−0.52	9.11
	pykAF	0.56	7.02
	pgi	4.53	9.94
	tpiA	−7.14	−9.57
	fba	7.25	9.61
	pfkAB	7.25	9.61
	eno	14.04	18.39
	gpmAB	14.04	18.39
	pgk	15.66	18.99
	gapAC	15.66	18.99
decreased	*ppc*	2.66	1.05
	rpiAB	2.47	0.27
	nuo	35.92	3.89
	rpe	2.80	−0.29
	tktAB	2.80	−0.29
	atp	−54.16	−5.53
	cyoABCD	40.57	4.00
	aceEF	5.33	0.40
	acnAB	5.33	0.40
	gltA	5.33	0.40
	icdA	5.33	0.40
	fumABC	5.27	0.38
	mdh	5.27	0.38
	talB	1.58	−0.08
	sucCD	3.85	−0.18
inactivated	*gnd*	5.31	0.00
	pgl	5.31	0.00
	zwf	5.31	0.00
	aceEF	8.18	0.00
	sdhABCD	4.31	0.00
	sucAB	4.31	0.00
not used	*aceA*	0.00	0.00
	aceB	0.00	0.00
	ldh	0.00	0.00
	acs	0.00	0.00
	cydABCD	0.00	0.00
	fbp	0.00	0.00
	frdABCD	0.00	0.00
	glk	0.00	0.00
	ndh	0.00	0.00
	pckA	0.00	0.00
	ppsA	0.00	0.00
	sfcA	0.00	0.00

[a] Negative flux values indicate that the reaction is proceeding in the reverse direction. To determine whether a gene was activated, increased, inactivated, or decreased, the ratio of the absolute value of the two flux values was used. The data for the aerobic and anaerobic conditions were taken for the oxygen uptake flux constrained to 20.2 and 2.0 mmol g^{-1} DW, respectively. Note for the *pts* genes there was no change in the flux level between the two conditions as this flux is a fixed input to the system.

techniques can allow one to assess the basic structural capabilities and fitness of an organism as a result of the pathways that are theoretically present within the cell. Additionally a pathway-based perspective can be used to interpret genome-scale expression data and may provide insight into the regulatory logic that should be imposed by the cell to control the function of the entire metabolic genotype. Thus, the structure of the stoichiometric matrix itself can reveal the answers to many important questions regarding the connectivity of the metabolic network and the capabilities of the system.

Design of Defined Media. An important economic consideration in large-scale bioprocesses is optimal medium formulation. FBA can be used to design such media

(*38*). Following the approach defined above, a flux-balance model for the first completely sequenced free living organism, *Haemophilus influenzae*, has been generated (*39*). One application of this model is to predict a minimal defined media. It was found that *H. influenzae* can grow on the minimal defined medium as determined from the ORF assignments and predicted using FBA. This predicted minimal medium was compared to the previously published defined media and was found to differ in only one compound, inosine. It is known that inosine is not required for growth; however, it does serve to enhance growth to a certain extent (*40*). Again the *in silico* results obtained were consistent with published *in vivo* research. These results provide confidence in the use of this type of approach for the design of defined media for organisms in which there currently does not exist a defined media.

The Future of Phenomics

As previously mentioned, genomic data are compositional in nature and contain limited information about the dynamic behavior of integrated cellular processes. However, there is a clear shift underway today from "structural genomics" to "functional genomics" (*5*), which is expanding the scope of biological investigation from a one-gene approach to a more systems-based "holistic" approach. With recent advances such as oligonucleotide chip technology (*33, 41, 42*) and DNA microarrays (*43, 44*), it is possible to take genomic sequences and begin analyzing the dynamic events which occur in gene expression as a consequence of operational shifts in genetic circuits and cellular systems. In addition to mountainous genomic sequence databases we will soon be confronted with burgeoning gene expression databases (*45*). Just as experimental technologies will be critical to growth in the post-genomic era, so too will the development of systems science based approaches and computational strategies making sense of all the information provided thus far from genomics. Thus the science of phenomics will emerge, focused on extending systems science to take advantage of genomic information as well as expression data to address gene function on a systemic scale. While we strive to improve and develop algorithms to perform tasks such as sequence similarity searching, we must also search for the "algorithms" which operate within the cell.

Here we have illustrated the use of FBA as a tool for the analysis of the massive amounts of genomic data being currently generated. This method can be used for a broad range of scientific interests, all related to the deepest goal of understanding the genotype–phenotype relationship. FBA offers an example of the type of techniques and applications that lie ahead if we are to truly profit from genomics. Future methods will need to be developed and tailored for specific uses, which move beyond the management of sequence data. While we may have the complete genetic sequence of an organism available, we still have not identified all of the ORFs within the genome, nor can we faithfully assign a predicted function to many of these ORFs (*46*). As a result of the current state of genome annotation, any tools that are based on ORF assignments must be flexible in handling continuous changes in the annotation of genomic data.

Summary

Genomics is now well-established and proceeding at a highly efficient pace; it has been estimated that as many as 100 microbial genomes will have been sequenced by the year 2000 (*7*). Meanwhile proteomics is offering us

Biotechnol. Prog., 1999, Vol. 15, No. 3

snapshots and hints toward the relationships and activities of gene products. However, the value of genomics will not be fully utilized until methods and techniques which capitalize on genomic data are developed to establish the systemic function of the identified genes, allowing us to understand the phenotypic data generated through proteomics. We now must seriously face the challenges of developing methods to deal with these massive amounts of genetic and expression data if we are to progress in our understanding of the genotype–phenotype relationship. The development of phenomics and the establishment of the relations between gene products and their systemic characteristics have now become urgent, and approaches to accomplish them must be envisioned.

References and Notes

(1) TIGR Microbial Database: http://www.tigr.org/tdb/mdb/mdb.html.

(2) Ash, C. Year of the genome. *Trends Microbiol.* **1997**, *5* (4), 135–139.

(3) Pennisi, E. Laboratory Workhorse Decoded. *Science* **1997**, *277*, 1432–1434.

(4) Fields, S. The future is function. *Nat. Genet.* **1997**, *15*, 325–327.

(5) Hieter, P.; Boguski, M. Functional genomics: It's all how you read it. *Science* **1997**, *278*, 601–602.

(6) Wilkins, M. R.; et al. Progress with proteome projects: why all proteins expressed by a genome should be identified and how to do it. *Biotechnol. Genet. Eng. Rev.* **1996**, *13*, 19–50.

(7) Blaine Metting, F.; Romine, M. F. Microbial genomics: the floodgates open. *Trends Microbiol.* **1997**, *5* (3), 91–92.

(8) Fox, J. L. Microbial genomics: Milestones mount exponentially. *Nat. Biotechnol.* **1997**, *15*, 211–212.

(9) Strothman, R. C. The Coming Kuhnian Revolution in Biology. *Nat. Biotechnol.* **1997**, *15*, 194–199.

(10) Palsson, B. O. What lies beyond bioinformatics? *Nat. Biotechnol.* **1997**, *15*, 3–4.

(11) Ouzounis, C.; Casari, G. Computational comparisons of model genomes. *Trends Biotechnol.* **1996**, *14* (August), 280–285.

(12) Heinrich, R.; Schuster, S. *The regulation of cellular systems*; Chapman & Hall: New York, 1996.

(13) Fell, D. *Understanding the Control of Metabolism*; Portland Press: London, 1996.

(14) Reich, J. g.; Sel'kov, E. E. *Energy Metabolism of the Cell*, 2nd ed.; Academic Press: New York, 1981.

(15) Varma, A.; Palsson, B. O. Metabolic Flux Balancing: Basic concepts, Scientific and Practical Use. *Biotechnol. Bioeng.* **1994**, *12*, 994–998.

(16) Bonarius, H. P. J.; Schmid, G.; Tramper, J. Flux analysis of underdetermined metabolic networks: the quest for the missing constraints. *Trends Biotechnol.* **1997**, *15* (August), 308–314.

(17) Sauer, U.; Cameron, D. C.; Bailey, J. E. Metabolic capacity of *Bacillus subtilis* for the production of purine nucleosides, riboflavin, and folic acid. *Biotechnol. Bioeng.* **1998**, *59* (2), 227–238.

(18) Pramanik, J.; Keasling, J. D. Stoichiometric Model of *Escherichia coli* Metabolism: Incorporation of Growth-Rate Dependent Biomass Compositon and Mechanistic Energy Requirements. *Biotechnol. Bioeng.* **1997**, *56*, 399–421.

(19) Edwards, J. S.; et al. Metabolic Flux Balance Analysis, in *Metabolic Engineering*; Lee, S. Y., Papoutsakis, E. T., Eds.; Springer-Verlag: New York, 1999, in press.

(20) Edwards, J. S.; Palsson, B. O. How will bioinformatics influence metabolic engineering? *Biotechnol. Bioeng.* **1998**, *58*, 162–169.

(21) Watson, M. R. A discrete model of bacterial metabolilsm. *Comput. Appl. Biosci.* **1986**, *2*, 23–7.

(22) Fell, D. A.; Small, J. A. Fat synthesis in adipose tissue. An examination of stoichiometric constraints. *J. Biochem.* **1986**, *238*, 781–786.

(23) Majewski, R. A.; Domach, M. M. Simple constrained optimization view of acetate overflow in *E. coli. Biotechnol. Bioeng.* **1990**, *35*, 732–738.

(24) Sonnleitner, B.; Kappeli, O. Growth of *Saccharomyces cerevisae* is controlled by its limited respiratory capacity: Formulation and verification of a hypothesis. *Biotechnol. Bioeng.* **1986**, *28*, 927–937.

(25) Varma, A.; Palsson, B. O. Metabolic Capabilities of *Escherichia coli*. II. Optimal Growth Patterns. *J. Theor. Biol.* **1993**, *165*, 503–522.

(26) Varma, A.; Palsson, B. O. Stoichiometric Flux Balance Models Quantitatively Predict Growth and Metabolic By-Product Secretion in Wild-Type *Escherichia coli* W3110. *Appl. Environ. Microbiol.* **1994**, *60* (10), 3724–3731.

(27) Varma, A.; Palsson, B. O. Parametric Sensitivity of Stoichiometric Flux Balance Models Applied to Wild-Type *Escherichia Coli* Metabolism. *Biotechnol. Bioeng.* **1995**, *45*, 69–79.

(28) Edwards, J. S.; Palsson, B. O. *Escherichia coli* K-12 *in silico*: Definition of its metbolic genotype and analysis of its capabilities. Submitted for publication.

(29) Blattner, F. R.; et al. The complete genome sequence of *Escherichia coli* K-12. *Science* **1997**, *277*, 1453–1474.

(30) Neidhardt, F. C., Ed. *Escherichia coli and Salmonella: Cellular and Molecular Biology*, 2nd ed.; ASM Press: Washington, DC, 1996; Vol. 1.

(31) Neidhardt, F. C.; Ingraham, J. L.; Schaechter, M. *Physiology of the bacterial cell*; Sinauer Associates, Inc.: Sunderland, MA, 1990.

(32) Creaghan, I. T.; Guest, J. R. Succinate dehydrogenase-dependent nutritional requirement for succinate in mutants of *Escherichia coli* K-12. *J. Gen. Microbiol.* **1978**, *107*, 1–13.

(33) Wodicka, L.; et al. Genome-wide expression monitoring in *Saccharomyces cerevisiae*. *Nat. Biotechnol.* **1997**, *15*, 1359–1367.

(34) DeRisi, J. L.; Iyer, V. R.; Brown, P. O. Exploring the metabolic and genetic control of gene expression on a genomic scale. *Science* **1997**, *278*, 680–686.

(35) Varma, A.; Boesch, B. W. Palsson, B. O. Stoichiometric interpretation of *Escherichia coli* glucose catabolism under various oxygenation rates. *Appl. Environ. Microbiol.* **1993**, *59* (8), 2465–2473.

(36) Schilling, C. H.; Palsson, B. O. The underlying pathway structure of biochemical reaction networks. *Proc. Natl. Acad. Sci. U.S.A.* **1998**, *95*, 4193–4198.

(37) Schuster, S.; et al. Elementary modes of functioning in biochemical networks. In *Comutation in Cellular and Molecular Biological Systems*; Cuthbertson, R., Holcombe, M., Paton, R., Eds.; World Scientific: London, 1996; pp 151–165.

(38) Xie, L.; Wang, D. Integrated approaches to the design of media and feeding strategies for fed-batch cultures of animal cells. *Trends Biotechnol.* **1997**, *15* (3), 109–113.

(39) Edwards, J. S.; Palsson, B. O. Properties of the *Haemophilus influenzae* Rd metabolic genotype. *J. Biol. Chem.*, in press.

(40) Talmadge, M. B.; Herriott, R. M. A chemically defined medium for growth, transformation, and isolation of nutritional mutants of *Hemophilus influenzae*. *Biochem. Biophys. Res. Commun.* **1960**, *2* (3), 203–206.

(41) Hoheisel, J. D. Oligomer-chip technology. *Trends Biotechnol.* **1997**, *15*, 465–469.

(42) Ramsay, G. DNA chips: State-of-the art. *Nat. Biotechnol.* **1998**, *16*, 40–44.

(43) Shalon, D.; Smith, S. J.; Brown, P. O. A DNA microarray system for analyzing complex DNA samples using two-color fluorescent probe hybridization. *Genome Res.* **1996**, *6*, 639–645.

(44) Schena, M.; et al. Microarrays: biotechnology's discovery platform for functional genomics. *Trends Biotechnol.* **1998**, *16*, 301–306.

(45) Church, G. M. Databases for gene expression. *Nat. Biotechnol.* **1996**, *14*, 828.

(46) Pollack, J. D. *Mycoplasma* genes: a case for reflective annotation. *Trends Microbiol.* **1997**, *5* (10), 413–418.

(47) Schilling, C. H.; et al. Metabolic Pathway Analysis: Basic Concepts and Scientific Applications. *Biotechnol. Prog.* **1999**, *15*, xxx.

Accepted March 10, 1999.

BP9900357

The *Escherichia coli* MG1655 *in silico* metabolic genotype: Its definition, characteristics, and capabilities

J. S. Edwards* and B. O. Palsson[†]

Department of Bioengineering, University of California, San Diego, La Jolla, CA 92093-0412

Communicated by Yuan-Cheng B. Fung, University of California, San Diego, La Jolla, CA, March 3, 2000 (received for review October 14, 1999)

The *Escherichia coli* MG1655 genome has been completely sequenced. The annotated sequence, biochemical information, and other information were used to reconstruct the *E. coli* metabolic map. The stoichiometric coefficients for each metabolic enzyme in the *E. coli* metabolic map were assembled to construct a genome-specific stoichiometric matrix. The *E. coli* stoichiometric matrix was used to define the system's characteristics and the capabilities of *E. coli* metabolism. The effects of gene deletions in the central metabolic pathways on the ability of the *in silico* metabolic network to support growth were assessed, and the *in silico* predictions were compared with experimental observations. It was shown that based on stoichiometric and capacity constraints the *in silico* analysis was able to qualitatively predict the growth potential of mutant strains in 86% of the cases examined. Herein, it is demonstrated that the synthesis of *in silico* metabolic genotypes based on genomic, biochemical, and strain-specific information is possible, and that systems analysis methods are available to analyze and interpret the metabolic phenotype.

bioinformatics | metabolism | genotype-phenotype relation | flux balance analysis

T he complete genome sequence for a number of microorganisms has been established (The Institute for Genomic Research at www.tigr.org). The genome sequencing efforts and the subsequent bioinformatic analyses have defined the molecular "parts catalogue" for a number of living organisms. However, it is evident that cellular functions are multigeneic in nature, thus one must go beyond a molecular parts catalogue to elucidate integrated cellular functions based on the molecular cellular components (1). Therefore, to analyze the properties and the behavior of complex cellular networks, one needs to use methods that focus on the systemic properties of the network. Approaches to analyze, interpret, and ultimately predict cellular behavior based on genomic and biochemical data likely will involve bioinformatics and computational biology and form the basis for subsequent bioengineering analysis.

In moving toward the goal of developing an integrated description of cellular processes, it should be recognized that there exists a history of studying the systemic properties of metabolic networks (2) and many mathematical methods have been developed to carry out such studies. These methods include approaches such as metabolic control analysis (3, 4), flux balance analysis (FBA) (5–7), metabolic pathway analysis (8–11, 69), cybernetic modeling (12), biochemical systems theory (13), temporal decomposition (14), and so on. Although many mathematical methods and approaches have been developed, there are few comprehensive metabolic systems for which detailed kinetic information is available and where such detailed analysis can be carried out (see refs. 15–17 for a few noteworthy exceptions).

To analyze, interpret, and predict cellular behavior, each individual step in a biochemical network must be described, normally with a rate equation that requires a number of kinetic

constants. Unfortunately, it currently is not possible to formulate this level of description of cellular processes on a genome scale. The kinetic parameters cannot be estimated from the genome sequence and these parameters are not available in the literature. In the absence of kinetic information, it is, however, still possible to assess the theoretical capabilities of one integrated cellular process, namely metabolism, and examine the feasible metabolic flux distributions under a steady-state assumption. The steady-state analysis is based on the constraints imposed on the metabolic network by the stoichiometry of the metabolic reactions, which basically represent mass balance constraints. The steady-state analysis of metabolic networks based on the mass balance constraints is known as FBA (7, 18, 19). This analysis differs from detailed kinetic modeling of cellular processes, in that it does not attempt to predict the exact behavior of metabolic networks. Rather it uses known constraints on the integrated function of multiple enzymes to separate the states that a system can reach from those that it cannot. Then within the domain of allowable behavior one can study the genotype-phenotype relation, such as the stoichiometric optimal growth performance in a defined environment.

In this manuscript, we have used the biochemical literature, the annotated genome sequence data, and strain-specific information, to formulate an organism scale *in silico* representation of the *Escherichia coli* MG1655 metabolic capabilities. FBA then was used to assess metabolic capabilities subject to these constraints leading to qualitative predictions of growth performance.

Materials and Methods

Definition of the *E. coli* MG1655 Metabolic Map. An *in silico* representation of *E. coli* metabolism has been constructed. We have used the biochemical literature (20), genomic information (21), and the metabolic databases (22–24). Because of the long history of *E. coli* research, there was biochemical or genetic evidence for every metabolic reactions included in the *in silico* representation, and in most cases, there was both genetic and biochemical evidence (Table 1). The complete list of genes included in the *in silico* analysis is shown in Table 1, and the metabolic reactions catalyzed by these genes can be found on the web (http://gcrg.ucsd.edu/downloads.html). The stoichiometric coefficients for each metabolic reaction within this list were used to form the stoichiometric matrix S.

Determining the Capabilities of the *E. coli* Metabolic Network. The theoretical metabolic capabilities of *E. coli* were assessed by FBA

Abbreviations: FBA, flux balance analysis; LP, linear programming; TCA, tricarboxylic acid; PPP, pentose phosphate pathway.

*Present address: Department of Genetics, Harvard Medical School, Boston, MA 02115.

[†]To whom reprint requests should be addressed. E-mail: palsson@ucsd.edu.

Table 1. The genes included in the *E. coli* metabolic genotype (21)

Central metabolism (EMP, PPP, TCA cycle, electron transport)	*aceA, aceB, aceE, aceF, ackA, acnA, acnB, acs, adhE, agp, appB, appC, atpA, atpB, atpC, atpD, atpE, atpF, atpG, atpH, atpI, cydA, cydB, cydC, cydD, cyoA, cyoB, cyoC, cyoD, dld, eno, fba, fbp, fdhF, fdnG, fdnH, fdnI, fdoG, fdoH, fdoI, frdA, frdB, frdC, frdD, fumA, fumB, fumC, galM, gapA, gapC_1, gapC_2, glcB, glgA, glgC, glgP, glk, glpA, glpB, glpC, glpD, gltA, gnd, gpmA, gpmB, hyaA, hyaB, hyaC, hybA, hybC, hycB, hycE, hycF, hycG, icdA, lctD, ldhA, lpdA, malP, mdh, ndh, nuoA, nuoB, nuoE, nuoF, nuoG, nuoH, nuoI, nuoJ, nuoK, nuoL, nuoM, nuoN, pckA, pfkA, pfkB, pflA, pflB, pflC, pflD, pgi, pgk, pntA, pntB, ppc, ppsA, pta, purT, pykA, pykF, rpe, rpiA, rpiB, sdhA, sdhB, sdhC, sdhD, sfcA, sucA, sucB, sucC, sucD, talB, tktA, tktB, tpiA, trxB, zwf, **pgl** (30), **maeB** (30)*
Alternative carbon source	*adhC, adhE, agaY, agaZ, aldA, aldB, aldH, araA, araB, araD, bglX, cpsG, deoB, fruK, fucA, fucI, fucK, fucO, galE, galK, galT, galU, gatD, gatY, glk, glpK, gntK, gntV, gpsA, lacZ, manA, melA, mtlD, nagA, nagB, nanA, pfkB, pgi, pgm, rbsK, rhaA, rhaB, rhaD, srlD, treC, xylA, xylB*
Amino acid metabolism	*adi, aldH, alr, ansA, ansB, argA, argB, argC, argD, argE, argF, argG, argH, argI, aroA, aroB, aroC, aroD, aroE, aroF, aroG, aroH, aroK, aroL, asd, asnA, asnB, aspA, aspC, avtA, cadA, carA, carB, cysC, cysD, cysE, cysH, cysI, cysJ, cysK, cysM, cysN, dadA, dadX, dapA, dapB, dapD, dapE, dapF, dsdA, gabD, gabT, gadA, gadB, gdhA, glk, glnA, gltB, gltD, glyA, goaG, hisA, hisB, hisC, hisD, hisF, hisG, hisH, hisI, ilvA, ilvB, ilvC, ilvD, ilvE, ilvG_1, ilvG_2, ilvH, ilvI, ilvM, ilvN, kbl, ldcC, leuA, leuB, leuC, leuD, lysA, lysC, metA, metB, metC, metE, metH, metK, metL, pheA, proA, proB, proC, prsA, putA, sdaA, sdaB, serA, serB, serC, speA, speB, speC, speD, speE, speF, tdcB, tdh, thrA, thrB, thrC, tnaA, trpA, trpB, trpC, trpD, trpE, tynA, tyrA, tyrB, ygjG, ygjH, **alaB** (42), **dapC** (43), **pat** (44), **prr** (44), **sad** (45), **methylthioadenosine nucleosidase** (46), **5-methylthioribose kinase** (46), **5-methylthioribose-l-phosphate isomerase** (46), **adenosyl homocysteinase** (47), **L-cysteine desulfhydrase** (44), **glutaminase A** (44), **glutaminase B** (44)*
Purine & pyrimidine metabolism	*add, adk, amn, apt, cdd, cmk, codA, dcd, deoA, deoD, dgt, dut, gmk, gpt, gsk, guaA, guaB, guaC, hpt, mutT, ndk, nrdA, nrdB, nrdD, nrdE, nrdF, purA, purB, purC, purD, purE, purF, purH, purK, purL, purM, purN, purT, pyrB, pyrC, pyrD, pyrE, pyrF, pyrG, pyrH, pyrI, tdk, thyA, tmk, udk, udp, upp, ushA, xapA, yicP, **CMP glycosylase** (48)*
Vitamin & cofactor metabolism	*acpS, bioA, bioB, bioD, bioF, coaA, cyoE, cysG, entA, entB, entC, entD, entE, entF, epd, folA, folC, folD, folE, folK, folP, gcvH, gcvP, gcvT, gltX, glyA, gor, gshA, gshB, hemA, hemB, hemC, hemD, hemE, hemF, hemH, hemK, hemL, hemM, hemX, hemY, ilvC, lig, lpdA, menA, menB, menC, menD, menE, menF, menG, metF, mutT, nadA, nadB, nadC, nadE, ntpA, pabA, pabB, pabC, panB, panC, panD, pdxA, pdxB, pdxH, pdxJ, pdxK, pncB, purU, ribA, ribB, ribD, ribE, ribH, serC, thiC, thiE, thiF, thiG, thiH, thrC, ubiA, ubiB, ubiC, ubiG, ubiH, ubiX, yaaC, ygiG, **nadD** (49), **nadF** (49), **nadG** (49), **panE** (50), **pncA** (49), **pncC** (49), **thiB** (51), **thiD** (51), **thiK** (51), **thiL** (51), **thiM** (51), **thiN** (51), **ubiE** (52), **ubiF** (52), **arabinose-5-phosphate isomerase** (22), **phosphopantothenate-cysteine ligase** (50), **phosphopantothenate-cysteine decarboxylase** (50), **phospho-pantetheine adenylyltransferase** (50), **dephosphoCoA kinase** (50), **NMN glycohydrolase** (49)*
Lipid metabolism	*accA, accB, accD, atoB, cdh, cdsA, cls, dgkA, fabD, fabH, fadB, gpsA, ispA, ispB, pgpB, pgsA, psd, pssA, **pgpA** (53)*
Cell wall metabolism	*ddlA, ddlB, galF, galU, glmS, glmU, htrB, kdsA, kdsB, kdtA, lpxA, lpxB, lpxC, lpxD, mraY, msbB, murA, murB, murC, murD, murE, murF, murG, murI, rfaC, rfaD, rfaF, rfaG, rfaI, rfaJ, rfaL, ushA, **glmM** (54), **lpcA** (55), **rfaE** (55), **tetraacyldisaccharide 4′ kinase** (55), **3-deoxy-D-manno-octulosonic-acid 8-phosphate phosphatase** (55)*
Transport processes	*araE, araF, araG, araH, argT, aroP, artI, artJ, artM, artP, artQ, brnQ, cadB, chaA, chaB, chaC, cmtA, cmtB, codB, crr, cycA, cysA, cysP, cysT, cysU, cysW, cysZ, dctA, dcuA, dcuB, dppA, dppB, dppC, dppD, dppF, fadL, focA, fruA, fruB, fucP, gabP, galP, gatA, gatB, gatC, glnH, glnP, glnQ, glpF, glpT, gltJ, gltK, gltL, gltP, gltS, gntT, gpt, hisJ, hisM, hisP, hisQ, hpt, kdpA, kdpB, kdpC, kgtP, lacY, lamB, livF, livG, livH, livJ, livK, livM, lldP, lysP, malE, malF, malG, malK, malX, manX, manY, manZ, melB, mglA, mglB, mglC, mtlA, mtr, nagE, nanT, nhaA, nhaB, nupC, nupG, oppA, oppB, oppC, oppD, oppF, panF, pheP, pitA, pitB, pnuC, potA, potB, potC, potD, potE, potF, potG, potH, potI, proP, proV, proW, proX, pstA, pstB, pstC, pstS, ptsA, ptsG, ptsI, ptsN, ptsP, purT, putP, rbsA, rbsB, rbsC, rbsD, rhaT, sapA, sapB, sapD, sbp, sdaC, srlA_1, srlA_2, srlB, tdcC, tnaB, treA, treB, trkA, trkG, trkH, tsx, tyrP, ugpA, ugpB, ugpC, ugpE, uraA, xapB, xylE, xylF, xylG, xylH, **fruF** (56), **gntS** (57), **metD** (43), **pnuE** (49), **scr** (56)*

The *in silico E. coli* MG1655 metabolic genotype used herein is available on the web: http://gcrg.ucsd.edu/downloads.html.

(5–7). The metabolic capabilities of the *in silico* metabolic genotype were partially defined by mass balance constraints; mathematically represented by a matrix equation:

$$S \cdot v = 0. \qquad [1]$$

The matrix **S** is the *mxn* stoichiometric matrix, where *m* is the number of metabolites and *n* is the number of reactions in the network. The *E. coli* stoichiometric matrix was 436 × 720. The vector **v** represents all fluxes in the metabolic network, including the internal fluxes, transport fluxes, and the growth flux. The optimal **v** vector was determined and defined the steady-state metabolic flux distribution.

For the *E. coli* metabolic network, the number of fluxes was greater than the number of mass balance constraints; thus, there was a plurality of feasible flux distributions that satisfied the mass balance constraints (defined in Eq. 1), and the solutions (or feasible metabolic flux distributions) were confined to the nullspace of the matrix **S**.

In addition to the mass balance constraints, we imposed constraints on the magnitude of each individual metabolic flux.

$$\alpha_i \leq v_i \leq \beta_i. \qquad [2]$$

The linear inequality constraints were used to enforce the reversibility/irreversibility of metabolic reactions and the max-

MICROBIOLOGY

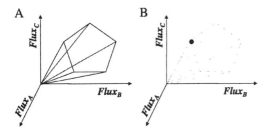

Fig. 1. The feasible solution set for a hypothetical metabolic reaction network. (*A*) The steady-state operation of the metabolic network is restricted to the region within a cone, defined as the feasible set (8). The feasible set contains all flux vectors that satisfy the physicochemical constraints (Eqs. 1 and 2). Thus, the feasible set defines the capabilities of the metabolic network. All feasible metabolic flux distributions lie within the feasible set, and (*B*) in the limiting case, where all constraints on the metabolic network are known, such as the enzyme kinetics and gene regulation, the feasible set may be reduced to a single point. This single point must lie within the feasible set.

imal metabolic fluxes in the transport reactions. The intersection of the nullspace and the region defined by the linear inequalities formally defined a region in flux space that we will refer to as the feasible set. The feasible set defined the capabilities of the metabolic network subject to the subset of cellular constraints, and all feasible metabolic flux distributions lie within the feasible set (see Fig. 1). However, every vector **v** within the feasible set is not reachable by the cell under a given condition because of other constraints not considered in the analysis (i.e., maximal internal fluxes and gene regulation). The feasible set can be further reduced by imposing additional constraints, and if all of the necessary details to describe metabolic dynamics are known, then the feasible set may reduce to a small region or even a single point (see Fig. 1).

For the analysis presented herein, we defined $\alpha_i = 0$ for irreversible internal fluxes, and $\alpha_i = -\infty$ for reversible internal fluxes. The reversibility of the metabolic reactions was determined from the biochemical literature and is identified for each reaction on the web site. The transport flux for inorganic phosphate, ammonia, carbon dioxide, sulfate, potassium, and sodium was unrestrained ($\alpha_i = -\infty$ and $\beta_i = \infty$). The transport flux for the other metabolites, when available in the *in silico* medium, was constrained between zero and the maximal level

$(0 < v_i < v_i^{max})$. However, when the metabolite was not available in the medium, the transport flux was constrained to zero. The transport flux for metabolites that were capable of leaving the metabolic network (i.e., acetate, ethanol, lactate, succinate, formate, pyruvate, etc.) always was unconstrained in the outward direction.

A particular metabolic flux distribution within the feasible set was found by using linear programming (LP). A commercially available LP package was used (LINDO, Lindo Systems, Chicago). LP identified a solution that minimized a particular metabolic objective (subject to the imposed constraints) (5, 25, 26), and was formulated as shown. Minimize $-Z$, where

$$Z = \Sigma c_i \cdot v_i = \langle \mathbf{c} \cdot \mathbf{v} \rangle. \qquad [3]$$

The vector **c** was used to select a linear combination of metabolic fluxes to include in the objective function (27). Herein, **c** was defined as the unit vector in the direction of the growth flux, and the growth flux was defined in terms of the biosynthetic requirements:

$$\sum_{all\ m} d_m \cdot X_m \xrightarrow{v_{growth}} Biomass, \qquad [4]$$

where d_m is the biomass composition of metabolite X_m (defined from the literature; ref. 28), and the growth flux is modeled as a single reaction that converts all of the biosynthetic precursors into biomass.

Results

FBA was used to examine the change in the metabolic capabilities caused by gene deletions. To simulate a gene deletion, the flux through the corresponding enzymatic reaction was restricted to zero. Genes that code for isozymes or genes that code for components of same enzyme complex were simultaneously removed (i.e., *aceEF, sucCD*). The optimal value of the objective (Z_{mutant}) was compared with the "wild-type" objective (*Z*) to determine the systemic effect of the gene deletion. The ratio of optimal growth yields (Z_{mutant}/Z) was calculated (Fig. 2).

Gene Deletions. *E. coli* MG1655 *in silico* was subjected to deletion of each individual gene product in the central metabolic pathways [glycolysis, pentose phosphate pathway (PPP), tricarboxylic acid (TCA) cycle, respiration processes], and the maximal ca-

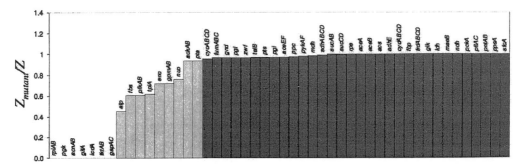

Fig. 2. Gene deletions in *E. coli* MG1655 central intermediary metabolism; maximal biomass yields on glucose for all possible single gene deletions in the central metabolic pathways. The optimal value of the mutant objective function (Z_{mutant}) compared with the "wild-type" objective function (*Z*), where *Z* is defined in Eq. 3. The ratio of optimal growth yields (Z_{mutant}/Z). The results were generated in a simulated aerobic environment with glucose as the carbon source. The transport fluxes were constrained as follows: $\beta_{glucose}$ = 10 mmol/g-dry weight (DW) per h; β_{oxygen} = 15 mmol/g-DW per h. The maximal yields were calculated by using FBA with the objective of maximizing growth. The biomass yields are normalized with respect to the results for the full metabolic genotype. The yellow bars represent gene deletions that reduced the maximal biomass yield to less than 95% of the *in silico* wild type.

pability of each *in silico* mutant metabolic network to support growth was assessed with FBA. The simulations were performed under an aerobic growth environment on minimal glucose medium.

The results identified the essential (required for growth) central metabolic genes (Fig. 2). For growth on glucose, the essential gene products were involved in the three-carbon stage of glycolysis, three reactions of the TCA cycle, and several points within the PPP. The remainder of the central metabolic genes could be removed and *E. coli in silico* maintained the potential to support cellular growth. This result was related to the interconnectivity of the metabolic reactions. The *in silico* gene deletion results suggest that a large number of the central metabolic genes can be removed without eliminating the capability of the metabolic network to support growth under the conditions considered.

Are the *in Silico* Redundancy Results Consistent with Mutant Data?

The *in silico* gene deletion study results were compared with growth data from known mutants. The growth characteristics of a series of *E. coli* mutants on several different carbon sources were examined and compared with the *in silico* deletion results (Table 2). From this analysis, 86% (68 of 79 cases) of the *in silico* predictions were consistent with the experimental observations.

How Are Cellular Fluxes Redistributed?

The potential of many *in silico* deletion strains to support growth led to questions regarding how the *E. coli* metabolic genotype deals with the loss of metabolic functions. The answer involves the degree of stoichiometric connectivity of key metabolites. For illustration, the flux redistributions to optimally support growth of a single mutant and a double mutant were investigated.

The optimal metabolic flux distribution for the *in silico* wild type was calculated (Fig. 3). The constraints used in the LP problem are defined in the figure legend. The *in silico* results suggest that optimally the oxidative branch of the PPP was used to generate a large fraction of the NADPH (66% *in silico*: 20–50% reported in the literature, ref. 29), and the TCA cycle produced NADH. The optimal flux distribution also suggested that the majority of the high-energy phosphate bonds were generated via oxidative phosphorylation and acetate secretion because of limitations of the oxygen supply.

The *in silico* gene deletion results predicted that the optimal biomass yield of the *zwf⁻* (glucose-6-phosphate dehydrogenase) *in silico* strain was slightly less than the wild type. The optimal flux distribution of the *zwf⁻ in silico* strain was calculated, and the NADPH was optimally generated through the transhydrogenase reaction and an elevated TCA cycle flux. The PPP biosynthetic precursors were generated in the nonoxidative branch. This metabolic flux rerouting resulted in an optimal biomass yield that was 99% of the *in silico* wild type.

The transhydrogenase (*pnt*) also was deleted *in silico*, creating an *in silico* double deletion mutant and eliminating an alternate source of NADPH. The double mutant still maintained growth potential. The optimal flux distribution (Fig. 2) used the isocitrate dehydrogenase and the malic enzyme to produce NADPH. The optimal biomass yield of the double mutant was 92% of the *in silico* wild type. The FBA results were consistent with the experimental observations that the *zwf⁻* strain (30) and the *pnt⁻* strain (29) are able to grow at near wild-type yields. Furthermore, the *zwf⁻ pnt⁻* double mutant strain also has been shown to grow ($\mu_{mutant}/\mu_{wild\ type} = 57\%$) (29).

Discussion

Extensive information about the molecular composition and function of several single-cellular organisms has become available. A next important step will be to incorporate the available information to generate whole-cell models with interpretative

Table 2. Comparison of the predicted mutant growth characteristics from the gene deletion study to published experimental results with single mutants

Gene	glc	gl	succ	ac	Reference
aceA	+/+		+/+	-/-	(58)
aceB				-/-	(58)
aceEF*	-/+				(60)
ackA				+/+	(61)
acn	-/-			-/-	(58)
acs				+/+	(61)
cyd	+/+				(62)
cyo	+/+				(62)
eno†	-/-	-/+	-/-	-/-	(30)
fba‖	-/+				(30)
fbp	+/+	-/-	-/-	-/-	(30)
frd	+/+		+/+	+/+	(60)
gap	-/-	-/-	-/-	-/-	(30)
glk	+/+				(30)
gltA	-/-			-/-	(58)
gnd	+/+				(30)
idh	-/-			-/-	(58)
mdh††	+/+	+/+	+/+		(63)
ndh	+/+	+/+			(59)
nuo	+/+	+/+			(59)
pfk†	-/+				(30)
pgi‡	+/+	+/-	+/-		(30)
pgk	-/-	-/-	-/-	-/-	(30)
pgl	+/+				(30)
pntAB	+/+	+/+	+/+		(29)
ppc§	±/+	-/+	+/+		(63, 64)
pta				+/+	(61)
pts	+/+				(30)
pyk	+/+				(30)
rpi	-/-	-/-	-/-	-/-	(30)
sdhABCD	+/+		-/-	-/-	(58)
sucAB	+/+		-/+	-/+	(60)
tktAB	-/-				(30)
tpi**	-/+	-/-	-/-	-/-	(30)
unc	+/+		±/+	-/-	(66–68)
zwf	+/+	+/+	+/+		(30)

Results are scored as + or − meaning growth or no growth determined from *in vivo/in silico* data. The ± indicates that suppressor mutations have been observed that allow the mutant strain to grow. In 68 of 79 cases the *in silico* behavior is the same as the experimentally observed behavior. glc, glucose; ac, acetate; gl, glycerol; succ, succinate.

*The *in vivo aceAE* strain is able to grow under anaerobic growth conditions by using the pyruvate formate lyase.

†The *in silico pfk* strain is able to grow by increasing the PPP flux ≈ 5× and using the *pps* gene product to overcome PEP deficiency.

‡The *in silico pgi* strain is unable to grow with glycerol or succinate as the carbon source because it is unable to synthesize glycogen and one carbohydrate component in the lipopolysaccharide. These are likely nonessential components of the biomass.

§The grow on glycerol and glucose is possible through the utilization of the glyoxylate bypass. Constitutive mutations in the glyoxylate bypass can suppress the *ppc* phenotype.

¶The *in silico eno* strain is able to grow by the synthesis and degradation of serine.

‖There is evidence that *fba* has an inhibitory effect on stable RNA synthesis (65). Such an inhibition cannot be predicted by FBA.

**The inability of *tpi* mutants to grow on glucose may be related to the accumulation of dihydroxyacetone phosphate, which leads to the formation of the bactericidal compound methylglyoxal (30).

††Very slow growth on glycerol and succinate.

and predictive capability. Herein, we have taken a step in that direction by using a set of constraints on cellular metabolism on the whole-cell level to analyze the metabolic capabilities of the

MICROBIOLOGY

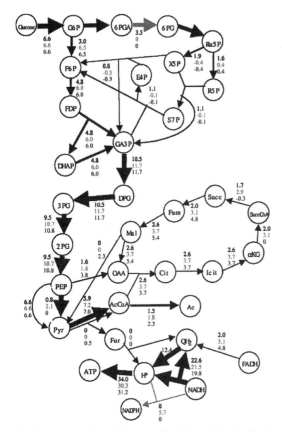

Fig. 3. Rerouting of metabolic fluxes. (Black) Flux distribution for the complete gene set. (Red) *zwf* mutant. Biomass yield is 99% of the results for the full metabolic genotype. (Blue) *zwf pnt* mutant. Biomass yield is 92% of the results for the full metabolic genotype (see text). The solid lines represent enzymes that are being used, with the corresponding flux value noted. The fluxes [substrates converted/h per g-dry weight (DW)] were calculated by using FBA with the input parameters of glucose uptake rate ($\beta_{glucose}$ = 6.6 mmol glucose/h per g-DW) and oxygen uptake rate (β_{oxygen} = 12.4 mmol oxygen/h per g-DW) (41).

extensively studied bacterium *E. coli*. We have calculated the optimal metabolic network utilization with a FBA. The *in silico* results, based only on stoichiometric and capacity constraints, were consistent with experimental data for the wild type and many of the mutant strains examined.

The construction of comprehensive *in silico* metabolic maps provided a framework to study the consequences of alterations in the genotype and to gain insight into the genotype-phenotype relation. The stoichiometric matrix and FBA were used to analyze the consequences of the loss of a gene product function on the metabolic capabilities of *E. coli*. The results demonstrated an important property of the *E. coli* metabolic network, namely that there are relatively few critical gene products in central metabolism. The nonessential genes in several organisms have been found experimentally on a genome scale (31, 32), which opens up the opportunity to critically test the *in silico* predictions. The *in silico* analysis also suggests that although the ability to grow in one defined environment is only slightly altered the ability to adjust to different environments may be diminished

(33). Therefore, the *in silico* analysis provides a methodology for relating the specific biochemical function of the metabolic enzymes to the integrated properties of the metabolic network.

The *in silico* analysis presented herein is not the typical metabolic modeling; more appropriately, the analysis can be thought of as a constraining approach. This approach defines the "best" the cell can do and identifies what the cell cannot do, rather than attempting to predict how the cell actually will behave under a given set of conditions. To accomplish this, we have used a set of physicochemical constraints for which there is reliable information available, in particular the stoichiometric properties. FBA does not directly consider regulation or the regulatory constraints on the metabolic network.

The results of FBA can be interpreted in a qualitative or a quantitative sense. At the first level we can ask whether a cell is able to grow under given circumstances and how a loss of the function of a gene product influences this ability. The results presented herein fall into this category. Quantitative predictions would hold true if the cell optimized its growth under the growth conditions considered. Therefore, when applying LP to predict quantitatively the optimal metabolic pathway utilization, it is assumed that the cell has found an "optimal solution" for survival through natural selection, and we have equated survival with growth. Although *E. coli* may grow optimally in defined media, one should not expect that optimizing growth is the governing objective of the cell under all growth conditions. For example, the regulatory mechanisms can only evolve to stoichiometric optimality in a condition to which the cell has been exposed. Furthermore, the growth behavior of mutant strains is unlikely to be optimal. However, FBA can still be used to delineate the metabolic capabilities of mutant cells based on constraining features, because both wild-type and mutant cells must obey the physicochemical constraints imposed.

The constraints on the system accurately reflect the steady-state capabilities of the metabolic network, but does the calculated optimal flux vector in the feasible set accurately reflect the behavior of the actual metabolic network? It has been shown that in a minimal media the metabolic behavior of wild-type *E. coli* is consistent with stoichiometric optimality (34). Furthermore, more detailed and critical experimental results are consistent with the hypothesis that *E. coli* does optimize its growth in acetate or succinate minimal media (33). Taken together these results call for critical experimental investigation to evaluate the hypothesis that stoichiometric and capacity constraints are the principal constraints that limit *E. coli* maximal growth. Even though growth and metabolic behavior in minimal media are consistent with FBA results, one still must determine the generality of optimal performance. The call for critical experimentation is particularly timely, given the increasing number of genome scale measurements that are now possible through two-dimensional gels (35, 36) and DNA array technology (37, 38). Furthermore, the ability to precisely remove ORFs can be used to design critical experiments (39). The *in silico* model can be used to choose the most informative knockouts and to design growth experiments with the knockouts.

At the present time, the annotation of the *E. coli* genome is incomplete, and about one-third of its ORFs do not have a functional assignment. Thus, the metabolic genotype studied here may lack some metabolic capabilities that *E. coli* possesses. The biochemical literature also was used to define the *in silico* metabolic genotype, and given the long history of *E. coli* metabolic research (20), a large percentage of the *E. coli* metabolic capabilities likely have been identified. However, if additional metabolic capabilities are discovered (40), the *E. coli* stoichiometric matrix can be updated, leading to an iterative model building process. Additionally, the *in silico* analysis can help identify missing or incorrect functional assignments by

Edwards and Palsson

144

identifying sets of metabolic reactions that are not connected to the metabolic network by the mass balance constraints.

The ability to analyze, interpret, and ultimately predict cellular behavior has been a long sought-after goal. The genome sequencing projects are defining the molecular components within the cell, and describing the integrated function of these molecular components will be a challenging task. The results presented herein suggest that it may be possible to analyze cellular metabolism based on a subset of the constraining features. Continued prediction and experimental verification will be an integral part in the further development of *in silico* strains. Deciphering the complex relation between the genotype and the phenotype will involve the biological sciences, computer science, and quantitative analysis, all of which must be included in the bioengineering of the 21st century.

We thank Ramprasad Ramakrishna, George Church, and Christophe Schilling for critical advice and input. National Institutes of Health Grant GM 57089 and National Science Foundation Grant MCB 9873384 supported this research.

1. Weng, G., Bhalla, U. S. & Iyengar, R. (1999) *Science* **284**, 92–96.
2. Bailey, J. E. (1998) *Biotechnol. Prog.* **14**, 8–20.
3. Kacser, H. & Burns, J. A. (1973) *Symp. Soc. Exp. Biol.* **27**, 65–104.
4. Fell, D. (1996) *Understanding the Control of Metabolism* (Portland, London).
5. Varma, A. & Palsson, B. O. (1994) *Bio/Technology* **12**, 994–998.
6. Edwards, J. & Palsson, B. (1999) *J. Biol. Chem.* **274**, 17410–17416.
7. Bonarius, H. P. J., Schmid, G. & Tramper, J. (1997) *Trends Biotechnol.* **15**, 308–314.
8. Schilling, C. H., Schuster, S., Palsson, B. O. & Heinrich, R. (1999) *Biotechnol. Prog.* **15**, 296–303.
9. Liao, J. C., Hou, S. Y. & Chao, Y. P. (1996) *Biotechnol. Bioeng.* **52**, 129–140.
10. Schuster, S., Dandekar, T. & Fell, D. A. (1999) *Trends Biotechnol.* **17**, 53–60.
11. Mavrovouniotis, M. & Stephanopoulos, G. (1992) *Comput. Chem. Eng.* **16**, 605–619.
12. Kompala, D. S., Ramkrishna, D., Jansen, N. B. & Tsao, G. T. (1986) *Biotechnol. Bioeng.* **28**, 1044–1056.
13. Savageau, M. A. (1969) *J. Theor. Biol.* **25**, 365–369.
14. Palsson, B. O., Joshi, A. & Ozturk, S. S. (1987) *Fed. Proc.* **46**, 2485–2489.
15. Shu, J. & Shuler, M. L. (1989) *Biotechnol. Bioeng.* **33**, 1117–1126.
16. Lee, I.-D. & Palsson, B. O. (1991) *Biomed. Biochim. Acta* **49**, 771–789.
17. Tomita, M., Hashimoto, K., Takahashi, K., Shimizu, T. S., Matsuzaki, Y., Miyoshi, F., Saito, K., Tanida, S., Yugi, K., Venter, J. C., *et al.* (1999) *Bioinformatics* **15**, 72–84.
18. Edwards, J. S., Ramakrishna, R., Schilling, C. H. & Palsson, B. O. (1999) in *Metabolic Engineering*, eds. Lee, S. Y. & Papoutsakis, E. T. (Dekker, New York), pp. 13–57.
19. Sauer, U., Cameron, D. C. & Bailey, J. E. (1998) *Biotechnol. Bioeng.* **59**, 227–238.
20. Neidhardt, F. C., ed. (1996) *Escherichia coli and Salmonella: Cellular and Molecular Biology* (Am. Soc. Microbiol., Washington, DC).
21. Blattner, F. R., Plunkett, G., 3rd, Bloch, C. A., Perna, N. T., Burland, V., Riley, M., Collado-Vides, J., Glasner, J. D., Rode, C. K., Mayhew, G. F., *et al.* (1997) *Science* **277**, 1453–1474.
22. Karp, P. D., Riley, M., Saier, M., Paulsen, I. T., Paley, S. M. & Pellegrini-Toole, A. (2000) *Nucleic Acids Res.* **28**, 56–59.
23. Selkov, E., Jr., Grechkin, Y., Mikhailova, N. & Selkov, E. (1998) *Nucleic Acids Res.* **26**, 43–45.
24. Ogata, H., Goto, S., Fujibuchi, W. & Kanehisa, M. (1998) *Biosystems* **47**, 119–128.
25. Pramanik, J. & Keasling, J. D. (1997) *Biotechnol. Bioeng.* **56**, 398–421.
26. Bonarius, H. P. J., Hatzimanikatis, V., Meesters, K. P. H., DeGooijer, C. D., Schmid, G. & Tramper, J. (1996) *Biotechnol. Bioeng.* **50**, 299–318.
27. Varma, A. & Palsson, B. O. (1993) *J. Theor. Biol.* **165**, 503–522.
28. Neidhardt, F. C. & Umbarger, H. E. (1996) in *Escherichia coli and Salmonella: Cellular and Molecular Biology*, ed. Neidhardt, F. C. (Am. Soc. Microbiol., Washington, DC), Vol. 1, pp. 13–16.
29. Hanson, R. L. & Rose, C. (1980) *J. Bacteriol.* **141**, 401–404.
30. Fraenkel, D. G. (1996) in *Escherichia coli and Salmonella: Cellular and Molecular Biology*, ed. Neidhardt, F. C. (Am. Soc. Microbiol., Washington, DC), Vol. 1, pp. 189–198.
31. Hutchison, C. A., Peterson, S. N., Gill, S. R., Cline, R. T., White, O., Fraser, C. M., Smith, H. O. & Venter, J. C. (1999) *Science* **286**, 2165–2169.
32. Winzeler, E. A., Shoemaker, D. D., Astromoff, A., Liang, H., Anderson, K., Andre, B., Bangham, R., Benito, R., Boeke, J. D., Bussey, H., *et al.* (1999) *Science* **285**, 901–906.
33. Edwards, J. S. (1999) Ph.D thesis (Univ. of California-San Diego, La Jolla).
34. Varma, A. & Palsson, B. O. (1994) *Appl. Environ. Microbiol.* **60**, 3724–3731.
35. Vanbogelen, R. A., Abshire, K. Z., Moldover, B., Olson, E. R. & Neidhardt, F. C. (1997) *Electrophoresis* **18**, 1243–1251.
36. Link, A. J., Robison, K. & Church, G. M. (1997) *Electrophoresis* **18**, 1259–1313.
37. Richmond, C. S., Glasner, J. D., Mau, R., Jin, H. & Blattner, F. R. (1999) *Nucleic Acids Res.* **27**, 3821–3835.
38. Brown, P. O. & Botstein, D. (1999) *Nat. Genet.* **21**, 33–37.
39. Link, A. J., Phillips, D. & Church, G. M. (1997) *J. Bacteriol.* **179**, 6228–6237.
40. Reizer, J., Reizer, A. & Saier, M. H., Jr. (1997) *Microbiology* **143**, 2519–2520.
41. Jensen, P. R. & Michelsen, O. (1992) *J. Bacteriol.* **174**, 7635–7641.
42. Reitzer, L. J. (1996) in *Escherichia coli and Salmonella: Cellular and Molecular Biology*, ed. Neidhardt, F. C. (Am. Soc. Microbiol., Washington, DC), Vol. 1, pp. 391–407.
43. Greene, R. C. (1996) in *Escherichia coli and Salmonella: Cellular and Molecular Biology*, ed. Neidhardt, F. C. (Am. Soc. Microbiol., Washington, DC), Vol. 1, pp. 542–560.
44. McFall, E. & Newman, E. B. (1996) in *Escherichia coli and Salmonella: Cellular and Molecular Biology*, ed. Neidhardt, F. C. (Am. Soc. Microbiol., Washington, DC), Vol. 1, pp. 358–379.
45. Berlyn, M. K. B., Low, K. B., Rudd, K. E. & Singer, M. (1996) in *Escherichia coli and Salmonella: Cellular and Molecular Biology*, ed. Neidhardt, F. C. (Am. Soc. Microbiol., Washington, DC), Vol. 2, pp. 1715–1902.
46. Glansdorff, N. (1996) in *Escherichia coli and Salmonella: Cellular and Molecular Biology*, ed. Neidhardt, F. C. (Am. Soc. Microbiol., Washington, DC), Vol. 1, pp. 408–433.
47. Matthews, R. G. (1996) in *Escherichia coli and Salmonella: Cellular and Molecular Biology*, ed. Neidhardt, F. C. (Am. Soc. Microbiol., Washington, DC), Vol. 1, pp. 600–611.
48. Neuhard, J. & Kelln, R. A. (1996) in *Escherichia coli and Salmonella: Cellular and Molecular Biology*, ed. Neidhardt, F. C. (Am. Soc. Microbiol., Washington, DC), Vol. 1, pp. 580–599.
49. Penfound, T. & Foster, J. W. (1996) in *Escherichia coli and Salmonella: Cellular and Molecular Biology*, ed. Neidhardt, F. C. (Am. Soc. Microbiol., Washington, DC), Vol. 1, pp. 721–730.
50. Jackowski, S. (1996) in *Escherichia coli and Salmonella: Cellular and Molecular Biology*, ed. Neidhardt, F. C. (Am. Soc. Microbiol., Washington, DC), Vol. 1, pp. 687–694.
51. White, R. L. & Spenser, I. D. (1996) in *Escherichia coli and Salmonella: Cellular and Molecular Biology*, ed. Neidhardt, F. C. (Am. Soc. Microbiol., Washington, DC), Vol. 1, pp. 680–686.
52. Meganathan, R. (1996) in *Escherichia coli and Salmonella: Cellular and Molecular Biology*, ed. Neidhardt, F. C. (Am. Soc. Microbiol., Washington, DC), Vol. 1, pp. 642–656.
53. Funk, C. R., Zimniak, L. & Dowhan, W. (1992) *J. Bacteriol.* **174**, 205–213.
54. Mengin-Lecreulx, D. & van Heijenoort, J. (1996) *J. Biol. Chem.* **271**, 32–39.
55. Raetz, C. R. H. (1996) in *Escherichia coli and Salmonella: Cellular and Molecular Biology*, ed. Neidhardt, F. C. (Am. Soc. Microbiol., Washington, DC), Vol. 1, pp. 1035–1063.
56. Postma, P. W., Lengeler, J. W. & Jacobson, G. R. (1996) in *Escherichia coli and Salmonella: Cellular and Molecular Biology*, ed. Neidhardt, F. C. (Am. Soc. Microbiol., Washington, DC), Vol. 1, pp. 1149–1174.
57. Lin, E. C. C. (1996) in *Escherichia coli and Salmonella: Cellular and Molecular Biology*, ed. Neidhardt, F. C. (Am. Soc. Microbiol., Washington, DC), Vol. 1, pp. 307–342.
58. Cronan, J. E., Jr. & Laporte, D. (1996) in *Escherichia coli and Salmonella: Cellular and Molecular Biology*, ed. Neidhardt, F. C. (Am. Soc. Microbiol., Washington, DC), Vol. 1, pp. 189–198.
59. Tran, Q. H., Bongaerts, J., Vlad, D. & Unden, G. (1997) *Eur. J. Biochem.* **244**, 155–160.
60. Creaghan, I. T. & Guest, J. R. (1978) *J. Gen. Microbiol.* **107**, 1–13.
61. Kumari, S., Tishel, R., Eisenbach, M. & Wolfe, A. J. (1995) *J. Bacteriol.* **177**, 2878–2886.
62. Calhoun, M. W., Oden, K. L., Gennis, R. B., de Mattos, M. J. & Neijssel, O. M. (1993) *J. Bacteriol.* **175**, 3020–3025.
63. Courtright, J. B. & Henning, U. (1970) *J. Bacteriol.* **102**, 722–728.
64. Vinopal, R. T. & Fraenkel, D. G. (1974) *J. Bacteriol.* **118**, 1090–1100.
65. Singer, M., Walter, W. A., Cali, B. M., Rouviere, P., Liebke, H. H., Gourse, R. L. & Gross, C. A. (1991) *J. Bacteriol.* **173**, 6249–6257.
66. Harold, F. M. & Maloney, P. C. (1996) in *Escherichia coli and Salmonella: Cellular and Molecular Biology*, ed. Neidhardt, F. C. (Am. Soc. Microbiol., Washington, DC), Vol. 1, pp. 283–306.
67. von Meyenburg, K., Jørgensen, B. B., Nielsen, J. & Hansen, F. G. (1982) *Mol. Gen. Genet.* **188**, 240–248.
68. Boogerd, F. C., Boe, L., Michelsen, O. & Jensen, P. R. (1998) *J. Bacteriol.* **180**, 5855–5859.
69. Karp, P. D., Krummenacker, M., Paley, S. & Wagg, J. (1999) *Trends Biotechnol.* **17**, 275–281.

MICROBIOLOGY

CHAPTER 7

TISSUE ENGINEERING OF ARTICULAR CARTILAGE

R. L. SAH

Department of Bioengineering, University of California, San Diego

The goal of this chapter is to provide an introduction to some approaches to the tissue engineering of articular cartilage. Initially, an introduction to articular cartilage in health and disease and a definition of tissue engineering are provided. Then, several selected tissue engineering therapies for articular cartilage defects are reviewed. From these therapies emerge several general ideas. (1) There are characteristic length and time scales associated with the biological, biomechanical, and biotransport processes underlying tissue engineering. (2) Quantitative measures and models help to interpret experimental results and to develop improved therapies. In this context, a number of interrelated processes involved in the development and refinement of cartilage tissue engineering therapies are analyzed in some detail. These include the repair processes leading to integration of an implant with host cartilage, the biomechanical regulation of transplanted chondrocytes, and the tissue-scale biomechanical and structural properties of articular cartilage.

1. Introduction

Articular cartilage is the hydrated connective tissue that provides a low-friction wear-resistant bearing surface in diarthrodial joints and distributes stresses to underlying bone (26, 117, 131). Adult articular cartilage consists of cells and extracellular material. The cells are called chondrocytes and make up a small percentage ($\sim$1–10% by volume) of the tissue. The extracellular material makes up most of the tissue volume and is composed primarily of fluid, collagens, and proteoglycans. The fibrillar collagens of the extracellular matrix form a meshwork that is strong in tension, while the proteoglycans contribute to the swelling and compressive properties of the tissue. The swelling propensity of proteoglycans is due to the acidic charge groups on the glycosaminoglycan (GAG) side chains, predominantly chondroitin sulfate, attached to the protein core. The balance between the swelling propensity of the proteoglycans and restraining function of the collagen meshwork is a major determinant of the biomechanical behavior of cartilage (15, 31, 116).

Unfortunately, articular cartilage and synovial joints do not always remain normal throughout life (27). Millions of people are stricken with degenerative joint disease, also called osteoarthritis, and suffer from pain and impaired joint function (141). In osteoarthritis, the articular cartilage becomes fibrillated and gradually erodes away. In addition, the bone undergoes pathologic remodeling. This includes sclerosis and cyst formation in the subchondral area and formation of cartilage-covered bone, called osteophytes, at the margin of the joint.

Damaged adult articular cartilage does not naturally undergo a successful repair process. More than two centuries ago, Hunter noted that "articular cartilage, once damaged, has a poor capacity to heal" (86). Since then, many studies have verified the limited intrinsic reparative ability of articular cartilage (27, 113). Clinical observations suggest that damaged cartilage, if left untreated, will adversely affect the joint and lead to osteoarthritic degeneration of the surrounding and opposing cartilage. Long-term studies in animals have shown gradual deterioration of not only the repair tissue that forms after an injury that penetrates through cartilage into the subchondral bone (163) but also the damaged and surrounding cartilage after an experimental cartilage laceration (63). Currently, there is no treatment that reverses the course of osteoarthritis, and the available treatment regimens only provide symptomatic relief (83, 84, 157). Such treatment includes physical therapy, weight loss programs, footwear and mechanical aids, analgesic medications, nonsteroidal antiinflammatory medications, local noninvasive therapies (heat, ice, ultrasound, and electrical stimulation), steroid injections, and surgical interventions. Surgical treatments can be divided into those that sacrifice and those that spare the articular cartilage.

The typical end-stage surgical therapy for degenerate joints is replacement with a prosthesis fabricated from metal and plastic (34, 47). Joint replacement with such prostheses have been enormously successful in providing pain relief and joint function for many individuals with debilitating arthritis. In the United States alone, hundreds of thousands of such procedures, primarily for the hip and knee, are done each year (141). However, such nonliving prostheses do exhibit wear and generate problematic debris (10, 66). Such prostheses are also not sufficiently durable for physically active individuals.

Tissue engineering has been defined as "the application of principles and methods of engineering and life sciences toward fundamental understanding of structure-function relationships in normal and pathologic mammalian tissues and the development of biological substitutes to restore, maintain, or improve tissue functions" (166). The major factor that distinguishes tissue engineering therapies from others is the primary involvement of living cells — the engineered implant either includes living cells or is designed to modulate the function of endogenous cells (60). Since the early 1990's, the field of tissue engineering has emerged and become the subject of review articles (103, 132), books (16, 104, 129, 138, 139), conference symposia (1), as well as a new journal and society, both entitled *Tissue Engineering*.

Tissue engineering of articular cartilage has received a great deal of interest, both in the scientific literature (114) and the popular press (169). A living tissue replacement may provide a long-term solution for damaged cartilage and arthritic joints. A tissue engineering approach to treating cartilage damage has potentially important intrinsic advantages over the modern generation of metal-and-plastic joint prostheses. A tissue engineering approach may minimize the generation of nonresorbed wear debris, and thus not elicit a host immune response. Such an approach also provides a material that can adapt biologically to changing physiological demands.

2.   Tissue Engineering Therapies for Articular Cartilage Defects

The concept of replacing a damaged joint, or portions of it, with engineered living tissue is not new. For example, muscle can be induced *in vivo* to form vascularized bone grafts in the shape of a femoral head or mandible (95). Currently, cartilage tissue engineering therapies are focused on the treatment of localized articular cartilage defects. It should be noted that these therapies are not directed toward treating end-stage arthritis, where the majority of opposing joint surfaces are typically damaged. On the other hand, the diagnosis and treatment of smaller and localized defects may lessen the incidence of end-stage arthritis. These therapies include the transplantation of autogeneic or allogeneic tissues, cells, scaffolds, chemicals, or combinations of these substances [Fig. 1, reviewed in (28, 135, 148)]. The tissues that have been used alone, and in combination, include cartilage, bone, marrow, perichondrium, periosteum, and even an induced fracture callus.

The cellular component of engineered tissues has typically been derived from the tissues listed above. Some cells are used in tissue engineering therapies soon after they are isolated from tissues. Other cells are expanded in number to amplify cellular effects or select for certain types of cells. The phenotype of a cell is defined by its biological behavior. The normal phenotype of articular chondrocytes includes expression of type II collagen and the large proteoglycan, aggrecan (26). Aggrecan is found in a number of load-bearing tissues. On the other hand, expression of type I collagen indicates a dedifferentiated (fibroblastic) chondrocyte phenotype, while expression of type X collagen indicates a hypertrophic (bone-forming) phenotype. Chondrocytes are known to change their phenotype when cultured in monolayer at low density (19); this is the culture method used commonly for many types of cells. Chondrocytes maintain their phenotype better when cultured in monolayer at high density (100) or in three-dimensional gels (19, 79). Dedifferentiated chondrocytes can also redifferentiate into chondrocytes when cultured in three-dimensional gels (19). The cells used in tissue engineering therapies for cartilage range from differentiated and dedifferentiated chondrocytes (23, 67) to mesenchymal stem cells, which are capable of differentiating into the cells of cartilage and bone (35). Ultimately, such cells probably need to express the articular chondrocyte phenotype to form normal cartilage tissue.

The materials used in cartilage tissue engineering therapies include biological substances

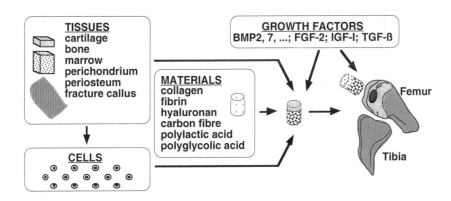

Fig. 1. General approaches to cartilage tissue engineering.

such as fibrin clot, collagen, and hyaluronan (a component of cartilage and synovial fluid) as well as nonbiological materials such as carbon fiber, polylactic acid, and polyglycolic acid. These materials degrade to varying degrees and over various time scales into different chemical products; these products may have effects on the engineered tissue (143). Biological materials may be degraded through pathways that are normally regulated in tissues, whereas nonbiological materials may be degraded through these or other pathways. Some materials are designed to have a specific chemical structure, for example, to enhance and modulate ingrowth of reparative cells. Other materials are designed to release specific chemicals, such as the growth factors [e.g. bone morphogenetic protein-2 (159)], also to stimulate reparative cells. Still other materials, such as bone fragments, may provide mechanical support.

There is a spectrum of biomechanical and biological approaches to tissue engineering of articular cartilage. Some tissue-engineered constructs are designed to restore mechanical function immediately after implantation; these constructs usually include underlying bone. Other constructs are cells or materials that must form or induce a cartilaginous tissue subsequent to implantation. Between these extremes are constructs that are cell-laden tissue, which are not yet fully-functional cartilage. One biological approach is to augment the natural healing process of articular cartilage. Another biological approach is to recapitulate the normal development process of cartilage and, thus, to regenerate articular cartilage.

Several of the current tissue engineering therapies, already used clinically for articular cartilage defects, are described below and provide examples of these approaches. The therapies discussed include osteochondral allografts, osteochondral autografts, and autologous chondrocyte transplantation. There are many other approaches, such as generation *ex vivo* of a preformed cartilaginous tissue from cells seeded onto a meshwork formed from polyglycolic acid (49, 176), transplantation of perichondrial cells in a polylactic acid carrier material (43), transplantation of mesenchymal cells in a collagen carrier material (177). Many of these and other tissue engineering therapies for articular cartilage are progressing rapidly toward or in clinical trials.

For the past several decades, the transplantation of site-matched allogeneic osteochondral fragments has been a reasonably successful treatment for relatively large cartilage defects (29, 42, 57, 62, 64). An osteochondral graft consists of mechanically functional articular cartilage attached to a supporting layer of subchondral bone. Viable chondrocytes within the grafted articular cartilage contribute to the long-term maintenance of the tissue. In 1997, an episode of the popular television series, "ER", introduced the American public to osteochondral allografting for a traumatic defect in the femoral condyle. After implantation and fixation, the subchondral bone of the allograft is ideally resorbed and replaced by host bone. Disadvantages of osteochondral allografts include difficulty in obtaining fresh grafts (with viable chondrocytes) and the possibilities of immune response and disease transmission.

An osteochondral autograft procedure avoids these problems. Recently, the autograft procedure termed "mosaicplasty" was popularized by Hangody (76) and Bobic (21). In this procedure, autogeneic osteochondral fragments, typically in the shape of cylinders, are harvested from relatively "unused" areas of an articulating joint and transplanted into the defect area. However, there are disadvantages with such autografts in comparison to allografts. There is morbidity associated with the donor site. Also, it is unfeasible to create an anatomically-matched graft. Further, there are multiple vertical cartilage interfaces between the multiple tissue fragments. While there is relatively little long-term outcome data, this procedure and various modifications have become popular (20, 89, 137, 152), and a number of companies (including Arthrex, Innovasive Design, Smith and Nephew, Stratec Medical) have marketed surgical tools for the procedures.

An altogether different cartilage tissue engineering approach is the transplantation of autologous chondrocytes, cells derived from cartilage and manipulated *ex vivo*. Unlike procedures that involve only tissue harvest and transplantation, autologous chondrocyte implantation requires the surgical acquisition of non weight-bearing articular cartilage from the patient, *in vitro* isolation of chondrocytes from this tissue, expansion of the number of cells in tissue culture, and then reimplantation (23, 67). In the implant procedure, a piece of periosteum is harvested and sutured over the defect area to form a tissue flap, the cells are injected under the periosteum, and fibrin glue is applied around the margin of the defect to seal the cells in the defect area. In 1995, Genzyme Tissue Repair introduced a service, named Carticel^{TM}, to receive harvested cartilage, process the tissue using a proprietary method, and provide culture-expanded autologous cells for implantation. On August 25, 1997, Genzyme Tissue Repair received a Biologics License from the United States Food and Drug Administration to commercially market Carticel^{TM} for the "repair of clinically significant symptomatic cartilaginous defects of the femoral condyle (medial, lateral, or trochlear) caused by acute or repetitive trauma" (2). Genzyme Tissue Repair has created a registry to track the outcome of Carticel^{TM} procedures (3). As of March, 1999, 2,419 patients have been treated with Carticel^{TM}. Autologous chondrocyte transplantation has attracted widespread clinical interest (11, 65, 99, 110, 124). However, the procedure is still controversial. One issue is how effective the procedure is compared to other surgical procedures, such as microfracture of the subchondral bone, that are currently used to enhance cartilage repair. Another related issue is if the mechanism of repair involves the chondrocytes, the periosteum, or an interaction between these two components.

While many tissue engineering approaches for articular cartilage repair show great promise, long-term experimental studies in animals have shown that a significant proportion of the current procedures result in biological "failures" (22, 43, 122). Dr. Henry Mankin has emphasized that "cartilage does not yield its secrets easily, and... inducing cartilage to heal is not simple... (T)he progression to osteoarthritis is sometimes so slow that we delude ourselves into thinking we are doing better than we are" (114). With clinical application already gaining widespread popularity, there is a pressing need to determine cellular, biochemical, and physical conditions that reliably facilitate cartilage growth, regeneration, repair, and long-term homeostasis.

3.   Principles of Articular Cartilage Tissue Engineering

The repair of a cartilage defect through tissue engineering approaches can be conceptualized as two processes: (1) the integration of the repair tissue with the surrounding host tissue, and (2) the filling of the bulk of the defect with tissue that is characteristic of normal articular cartilage (Fig. 2). In general, tissue repair and regeneration involves cell signaling,

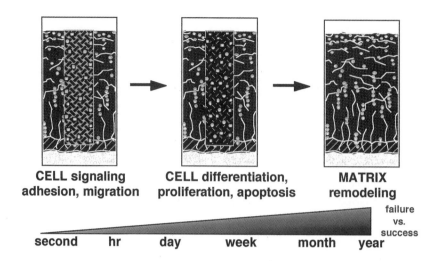

Fig. 2. Cellular and molecular mechanisms, and time course of remodeling processes in cartilage tissue engineering.

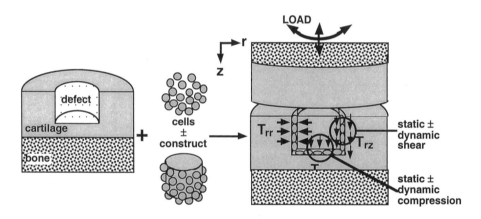

Fig. 3. Biomechanical stimulation during cartilage repair.

adhesion, migration, differentiation, proliferation, apoptosis, and matrix deposition and remodeling (28, 44, 113). Signals to and within cells initiate metabolic processes. These result in regulation of cell adherence (e.g. to local sites where repair is needed), migration to and proliferation at sites needing repair, programmed death of certain cells, and remodeling of tissue-specific matrix components.

Each of these cellular processes may be modulated by regulatory molecules including matrix components, cytokines, and growth factors (Fig. 1) as well as physical forces (Fig. 3). Cells typically respond to regulatory molecules in an age-dependent manner [e.g. (14, 73)]. Cartilage homeostasis reflects a balance between synthesis and degradation of matrix components, and this balance is influenced by a number of growth factors (127), known to be present in the synovial fluid. In addition, physical factors regulate homeostasis (149), and may involve altered transport or chemical, mechanical, or electrical signals that are induced by loading (74, 172).

Each of the repair (Fig. 2) and regulatory (Figs. 1 and 3) processes can be analyzed at different length scales and time scales. Cartilage function and structure can be analyzed at many length scales, ranging from whole organisms, to intact joints, full-thickness regions of cartilage over a joint surface, layers of cartilage tissue at different depths, cellular and extracellular regions of cartilage tissue, molecular constituents within these regions, etc. The nature of the physical environment in and around cell-laden repair constructs (after being placed in a cartilage defect) depends on the loads imposed on the joint as well as the properties of the joint tissues. This environment can be analyzed at the level of a typical joint using "average" cartilage and bone properties, down to the level of individual cells and molecules. Cartilage repair processes can be analyzed at different time scales. For example, signal transduction within cells occurs over seconds-minutes (45). Adhesion, migration, proliferation and apoptosis usually occurs over minutes-days, and matrix remodeling typically occurs over weeks to months (44, 163).

Quantitative models are useful to analyze the extent and nature of various processes over different length and time scales. Models allow extrapolation of tissue engineering strategies from one situation to another. For example, if the controlled release of a repair-promoting chemical substance is needed to achieve a certain concentration range for a certain duration within the joint space, a model of transport may be useful to extrapolate the release method and dose from a small defect in an animal to a larger defect in a human. Another example is if transplanted cells are regulated by certain amplitudes of mechanical compressive and shear stress, biomechanical analysis of the implant may be useful to design a physical milieu that is conducive to tissue repair. In addition, models allow assessment of specific mechanisms. For example, the balance between biosynthetic and degradative processes and the overall accumulation of matrix components can be described simply by a mass balance model (61, 78). Since the overall sequence of biological events in tissue engineering is complex, a major challenge is to integrate models of different processes and test both their ability to describe and predict the evolution of biological processes in time and space.

The development and refinement of tissue engineering methods involves a variety of complementary approaches. Ultimately, a tissue engineering therapy is tested in humans. Ideally, this is in the form of a randomized, blinded prospective clinical trial in which various therapies are compared. Alternatively, a clinical study may be observational, as was the case in the first report of autologous chondrocyte transplantation in humans in Sweden (23). Before human studies, experiments are typically performed in an animal model. An example of this is the repair of a cartilage defect in the medial femoral condyle (178) or patellar groove (67) of the rabbit. *In vivo* studies are often motivated by *in vitro* results. *In vivo*, it is difficult to control or determine the mechanical and biochemical microenvironment of transplanted cells and decipher underlying repair mechanisms. Thus, simplified experimental systems *in vitro* are often used to study such regulation. An example of this is the study of responses of cells in tissue culture to biochemical (127) and physical stimuli (74, 172).

A common theme in tissue engineering is to analyze critical processes, and then to use the results of such analysis to refine tissue engineering methods. In the context of different biological applications, such processes are often tissue specific. The studies, below, provide some specific examples of processes and issues in the tissue engineering of articular cartilage.

4.   Integrative Cartilage Repair: Matrix Metabolism and Biomechanics

At a 1994 workshop on "New Horizons in Osteoarthritis," a panel of basic and clinical scientists concluded, "(m)ethods must be developed to obtain... a firm attachment between (tissue-engineered implants)... and the collagen network and (other extracellular matrix components) of the host cartilage" (25). One general area of study in cartilage tissue engineering is the biomechanics of integrative repair (i.e. implant integration with host articular cartilage) and associated matrix remodeling mechanisms.

4.1.   *Biomechanics of Integrative Cartilage Repair*

From a macroscopic viewpoint, the complete repair of an articular cartilage defect requires the integration of repair tissue with the surrounding host cartilage. Consideration of this process, at least in the early stages, as the formation of an adhesive substance, a material at "the surfaces of two solids (that) can join them together such that they resist separation" (175) suggests several biomechanical approaches for characterizing the properties of the repair tissue (reviewed in (8) and summarized here). Both strength of materials and fracture mechanics approaches for characterizing adhesives have been applied recently to the study of integrative cartilage repair. In the former, the mechanical failure of a material is characterized by the normal or shear strength, defined as the ultimate stress that is sustained just before failure (168). In the latter, the propagation of flaws (intrinsic to a material) to cause failure is dependent on the applied stress and characterized by the fracture toughness of the material (9).

Experimental configurations, such as the single-lap adhesive test (Fig. 4), have been adapted to determine the strength of the biological repair that occurs between sections of cartilage during culture, as well as the strength of bioadhesives that are applied to opposing cartilage surfaces. The repair strength of a cartilage-cartilage interface after *in vitro* culture (see below) is similar to that of glues [e.g. tissue transglutaminase and fibrin sealant (93)]. However, the adhesive strength of photochemical welds (91) can be ∼ 5-fold higher than that of these glues.

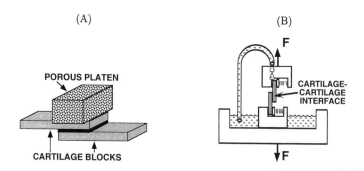

Fig. 4. Integrative repair of bovine cartilage explants. (A) Culture configuration with area of apposing cartilage blackened. (B) Adhesive strength testing using single-lap shear configuration. From (144), with permission.

A variety of different fracture mechanics test procedures, such as the (modified) single edge notch, "T" peel, dynamic shear, and trouser tear tests, have also been used to assess fracture toughness of normal articular cartilage and repair tissue in different failure modes. For bovine cartilage explants that are incubated to allow integration as described above, the fracture toughness (estimated as fracture energy) is 16 J/m^2 (7). This is approximately 10- to 1000-fold lower than the fracture energy of normal articular cartilage (24, 41). Such biomechanical measures of integrative repair may be used not only to compare different repair strategies to each other and to normal cartilage but also to further analyze mechanisms of cartilage repair.

4.2. *An in vitro Model of Integrative Cartilage Repair*

Explant cultures of cartilage have proven useful for examining the mechanisms of biological (127) and physical (74, 172) regulation of cartilage metabolism. Explant cultures involve isolation and incubation of small pieces of tissue, usually with a volume of several mm^3. Such cultures help to maintain the normal structure of the tissue, especially the relationship between cells and extracellular matrix. Incubation of cartilage explants in medium containing fetal bovine serum (FBS) stimulates the chondrocytes to maintain matrix homeostasis (77) and the indwelling cells to retain the cartilaginous phenotype (18). Explant cultures of defined geometry have been used to assess the regulation of chondrocyte metabolism by physical factors that are associated with joint loading (74).

Cartilage explants have been used to study the integrative repair of cartilage. Scully *et al.* (158) examined the *in vitro* healing of experimental lacerations in cartilage explants by histological analysis. Over the first few weeks of incubation in medium including FBS, an acellular matrix filled the defect. Between three and six weeks, cells began to populate the newly formed matrix. While these studies were not designed to allow mechanical characterization of the repair tissue, they suggested that repair tissue could be synthesized and deposited at the interface region between a pair of cartilage explants grown in apposition.

To determine the biomechanical characteristics of integrative repair *in vitro*, cartilage explants have been incubated in a single-lap joint configuration and subsequently tested [Fig. 4 and (144)]. Cartilage blocks were harvested from the patellofemoral groove and patella of adult bovines. Pairs of cartilage blocks were placed in chambers that maintained overlapping areas of tissue (Fig. 4A) during culture. At the end of the incubation period, the samples were tested by applying a uniaxial positive displacement at a constant rate until failure, while measuring the resultant load (Fig. 4B). The adhesive strength of the interface region was calculated as the peak load normalized to the original overlap area.

Incubation of cartilage explant pairs in partial apposition resulted in the development of adhesive strength of the interface between samples under certain culture conditions. After incubation in medium including 20% FBS, the adhesive strength between pairs of bovine cartilage blocks increased at a rate of ~ 11 kPa/week. This repair process appeared dependent on appropriate regulation of the indwelling cells. Either lyophilization of cartilage (to lyse the endogenous chondrocytes) before incubation or inclusion of 100 $\mu g/ml$ cycloheximide [a chemical that inhibits protein synthesis (150)] during incubation completely inhibited the development of adhesive strength. In addition, the repair process was modulated by biochemical factors in the medium. While incubation of sample pairs for 2–3 weeks in medium including 20% FBS resulted in a relatively high adhesive strength (30 kPa), incubation in basal medium (without FBS stimulation of cells) resulted in an adhesive strength that was reduced to ∼ 20% of this level. Supplementation of basal medium with certain growth factors, known to be present during cartilage growth or in a repair scenario, regulated integration. For example, incubation with insulin-like growth factor-I (IGF-I) had a stimulatory effect on integrative repair that was similar to that of FBS; on the other hand, incubation with transforming growth factor beta (TGF-β) did not augment integration (5).

The use of this test geometry allowed assessment of integrative repair between sections of cartilage that were approximately parallel to the articular surface and devoid of the superficial layer of cartilage. Such repair may be particularly relevant to repair of a defect that is restricted to the articular cartilage (e.g. not penetrating subchondral bone [22, 23]), or fissures that extend horizontally in fractured or osteoarthritic articular cartilage. The biological mechanisms involved in such an integrative repair process may depend on whether integrative repair is required to occur across a vertical or horizontal plane within articular cartilage.

4.3. *Matrix Remodeling in Integrative Cartilage Repair*

The cellular and molecular mechanisms that result in integrative repair *in vitro* remained to be elucidated. Identification of the mechanisms underlying the development of adhesive strength may suggest ways to enhance the integration process *in vivo*. Since both IGF-I (78) and TGF-β1 (128) stimulate synthesis and inhibit degradation of proteoglycan, it is unlikely the proteoglycan component of the extracellular matrix was responsible for integrative repair. In addition, while it is believed that proteoglycan contributes primarily to the compressive properties of cartilage, the collagen meshwork is primarily responsible for the tensile and cohesive properties of the tissue (70, 117, 131).

Collagen synthesis and extracellular processing appears to be involved in integrative cartilage repair. The relationship between deposition of newly synthesized collagen and integrative repair was examined in correlative studies (48). Explant pairs were cultured in apposition for 2 weeks in medium supplemented with a radioactive amino acid, [^{3}H]proline, for the first 12 days. Samples were tested biomechanically and also analyzed for incorporated [^{3}H]proline, as an index of collagen synthesis and deposition. Adhesive strength was positively correlated with [^{3}H]proline incorporation ($R = 0.60$). This provided evidence for the role of collagen synthesis and deposition in integrative cartilage repair.

The major collagen in cartilage, type II, is normally crosslinked (55). The stabilization of collagen fibrils in cartilage is dependent on crosslinks, covalent bonds between individual collagen molecules. In articular cartilage, the major collagen crosslinks include the difunctional crosslink, dehydrodihydroxylysinonorleucine (ΔDHLNL), and the trifunctional crosslink, hydroxylysyl pyridinoline (HP) (54). The metabolic pathway for the formation of ΔDHLNL and HP crosslinks is dependent on the enzyme lysyl oxidase (55, 94, 145, 147).

Collagen crosslinking appears to be a critical step in integrative cartilage repair (6). β-aminopropionitrile (BAPN) inhibits lysyl oxidase irreversibly, with 50% inhibition at 3–5 μM (94). Thus, BAPN could be used to block collagen crosslinking and test whether crosslinking was involved in integrative repair. A dose-escalation study was performed on cartilage explants to determine the effect of BAPN on overall cellular metabolism and viability as well as collagen stabilization. With addition of 0 (control) or 0.25 mM BAPN in the presence of FBS, biosynthetic levels of collagen, total protein, and GAG were elevated similarly between day 1 and 14 of culture, and levels of total GAG and DNA were not significantly different between the treatment groups. Concomitantly, treatment with 0.25 mM BAPN did markedly increase the chemical extractability of newly synthesized collagen from cartilage explants compared to control treatment, but did not affect extractability of GAG. Subsequent metabolic studies, using radiolabeling and washout (pulse-chase) protocols with [^{14}C]lysine, directly confirmed that treatment with 0.25 mM BAPN was sufficient to inhibit formation of reducible and mature lysyl oxidase-mediated crosslinks. To assess the functional consequences of collagen crosslink inhibition on integrative cartilage repair, explant pairs were cultured for 2 weeks in apposition with 0 or 0.25 mM BAPN and then tested mechanically. The adhesive strength of control (0 BAPN) cultures was 36 kPa. However, in the presence of 0.25 mM BAPN, functional repair was almost completely inhibited. These results indicated that BAPN treatment coordinately inhibits extracellular processing of collagen and integrative repair, without affecting other indices of tissue metabolism.

Since collagen crosslinking appears to be pivotal to the development of adhesive strength, the kinetics of this process in articular cartilage may be particularly relevant. Biochemical experiments and kinetic modeling have been performed to characterize the rate of formation of the major reducible crosslink, ΔDHLNL, and mature crosslink, HP, in adult bovine articular cartilage explants, incubated in medium including FBS and ascorbate (4). Pulse-chase studies, tracking the conversion of [^{14}C]lysine into collagen metabolites, were performed for up to 39 days of incubation of cartilage explants. Fitting of the experimental data to a kinetic model indicated that the portion of [^{14}C]lysine that was incorporated by chondrocytes and destined for conversion into [^{14}C]hydroxylysine, [^{14}C]DHLNL, and [^{14}C]HP were 35%, 2.1%, and 1.4%, respectively, and the time constants for the formation of the latter two crosslink molecules were 2.1 and 31.6 days, respectively. The relatively short time constant is consistent with the kinetics of chemical stabilization of newly synthesized collagen. The relatively long time constant is generally consistent with the kinetics of crosslink formation *in vivo* (161), where it is difficult to interpret pulse-chase studies in terms of quantitative models.

The exact way in which remodeling of the collagen network leads to integrative cartilage repair still remains to be established. The development of adhesive strength may be related to the formation of collagen crosslink molecules. There may be molecules synthesized by chondrocytes or present in the tissue matrix that inhibit collagen remodeling. Also, tissue remodeling typically involves the coordinated synthesis and degradation of matrix molecules. Nevertheless, even with this information (that the synthesis and extracellular processing of molecules, such as collagen, contributes to functional integrative repair), it can be postulated that strategies to accelerate these processes may be useful for enhancing cartilage repair.

5.   Chondrocyte Transplantation for Cartilage Repair: Biomechanical Regulation

The studies above identified cell-based matrix remodeling, that occurs over a period of days-weeks, as one process involved in integrative cartilage repair. The limited intrinsic repair capacity of articular cartilage may be due, in part, to the low density of chondrocytes within articular cartilage and the modest *in situ* proliferative response of these cells (27, 113). To facilitate cartilage repair, procedures have been developed to transplant isolated chondrocytes or chondrogenic progenitor cells into the site of the defect, either alone (22, 23, 67, 164) or within a natural or synthetic scaffolding material (13, 17, 80, 90, 134, 177). The overriding principle governing these methods is that the introduced cells remain in a position and are sufficient in number to effect repair through the production of new extracellular matrix and the integration of new tissue with the surrounding host tissue. In many of these procedures, a common theme is to isolate cells, grow the cells in culture, resuspend the cells by stripping away the pericellular matrix, and transplant the cultured cells into the defect.

In vivo, transplanted chondrocytes are exposed to a challenging mechanical environment (Fig. 3). *In vitro*, the regulation of biological activity of transplanted chondrocytes by physical stimuli relevant to clinical therapies has been examined. Such studies have identified the extent and mechanisms of adhesion of transplanted chondrocytes to a cartilage surface, the ability of such cells to mediate integrative cartilage repair, and the biosynthetic response of such cells to mechanical stimuli.

5.1.   *An in vitro Model of Chondrocyte Transplantation for Cartilage Repair*

In vitro model systems are often initially characterized by analyzing parameters likely to be important to the outcome being studied. Thus, in a model of chondrocyte transplantation, characterization studies have included determining the efficiency of chondrocyte transplantation onto cartilage explants *in vitro* over a range of densities, characterizing proteoglycan biosynthesis by chondrocytes after transplantation at different seeding densities, and characterizing the phenotype of the transplanted cells (37). Using radioactively tagged chondrocytes for quantitation, the efficiency of transplantation onto a cartilage substrate was found to be > 90% for seeding densities up to 650,000 cells/cm^2 and a seeding duration of one hour followed by gentle rinsing. These findings were confirmed both by visually tracking cells stained with a fluorescent dye and by quantitating total DNA (tissue plus transplanted cells). During the 16 hours duration following seeding onto a cartilage substrate (in which the endogenous cells had been lysed by lyophilization), the transplanted cells synthesized GAG in direct proportion to the number of cells seeded. The transplanted cells retained the chondrocyte phenotype, as judged by a high proportion of newly synthesized macromolecules being in the form of large proteoglycan that was capable of normal aggregation, as well as by positive immunostaining for type II collagen within the transplanted cells. These results indicate that the number of chondrocytes transplanted onto a cut cartilage surface can greatly augment matrix synthesis, an effect which in turn may enhance subsequent repair.

5.2. *Biomechanical Regulation of Integrative Cartilage Repair: Chondrocyte Adhesion*

In the earliest stages of repair by transplanted cells alone, and perhaps also in the case of cells within a carrier matrix, the transplanted cells may need to adhere to the cut surfaces of the surrounding host cartilage. A recent *in vivo* study involving implantation of allogeneic chondrocytes, tagged by adenovirus-mediated transduction with β-galactosidase and carried within a type I collagen matrix, showed a rapid loss of transduced cells from the cartilage defect by 24 hours (13). It has been proposed that articular cartilage provides a relatively antiadhesive surface for cell attachment (88). A variety of studies on chondrocyte adhesion have been performed (50, 59, 81, 109, 123, 133, 142, 146, 167) in order to investigate the adhesion of a particular cell type to isolated and purified forms of extracellular matrix proteins. While information derived from such studies is useful, it does not necessarily indicate the strength with which the cell adheres to the collection of extracellular matrix components that are presented at the tissue surface. Steric and interaction effects, that may exist at the cartilage tissue surface, may not be reproduced in these traditional *in vitro* assays of cell adhesion.

Qualitative studies have shown that chondrocytes adhere to cartilage *in vitro* (12, 37). Quantitative studies indicate the ability of chondrocytes to resist detachment from cartilage when subjected to mechanical perturbation, and the dependence of this on the duration of seeding time (154). The strength of chondrocyte adhesion to articular cartilage sections was determined after seeding durations of up to forty minutes. An existing parallel-plate shear-flow chamber design (174) was modified to incorporate a thin cartilage section as one portion of the lower plate, allowing chondrocytes to be seeded on the cartilage section and subsequently subjected to shear stress induced by fluid flow (Fig. 5A-D). Suspensions of adult bovine articular chondrocytes were prepared from primary, high-density monolayer cultures and infused into a parallel-plate shear-flow chamber where they rapidly settled

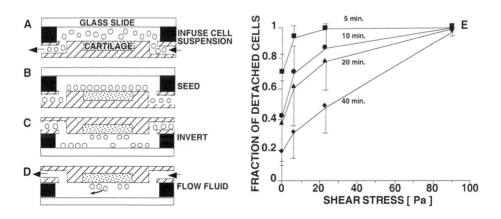

Fig. 5. Schematic of the cell detachment assay. (A) Chondrocyte suspension was infused into parallel-plate shear-flow chamber holding cartilage section. (B) Cells settled to cartilage surface by gravity. (C) Chamber was inverted onto microscope stage and weakly adherent cells detached. (D) Fluid-flow-induced shear stress was applied to cause cell detachment. (E) Cell detachment after various seeding times. From (154), with permission.

onto sections of bovine articular cartilage. The chondrocytes were allowed to attach to the cartilage surface for specific lengths of time of 5-40 minutes (which might be a waiting time that is practical in a clinical situation) in medium including FBS, after which the cells were exposed to fluid flow-induced shear stresses. The fraction of detached cells at each level of shear stress was computed from micrographs. Studies of the kinetics of cell detachment indicated that exposure to shear stress for 30 seconds was sufficient to reach a steady state. With increasing seeding times, chondrocytes became progressively more resistant to detachment from cartilage (Fig. 5E). The increase in resistance to shear stress-induced cell detachment with longer seeding time suggests that it may be beneficial to allow chondrocytes to stabilize in the absence of applied load for some time after chondrocyte transplantation for cartilage repair *in vivo*.

The previous studies examined the adhesiveness of large populations of chondrocytes to cartilage *in vitro*. However, it was unclear if the results were affected by the different depths below the articular surface from which cartilage tissue substrates were harvested, or if a similar time-dependent adhesion would be evident during chondrocyte attachment to a vertical surface. In addition, it has been hypothesized that enzymatic treatment of a cut surface of cartilage with Chondroitinase ABC removes antiadhesive molecules, such as GAG, (87, 88) and enhances adhesion of reparative cells to cartilage, while treatment with proteolytic enzymes enhances repair in a situation in which exogenous repair cells are present (35, 125). Cell adhesion to a cartilage surface may be particularly important to repair since endogenous integration, mediated by cells within cartilage explants *in vitro*, is not affected by enzymatic treatment (106).

Biomechanical studies have demonstrated that the adhesiveness of individual chondrocytes to cartilage can be affected by enzymatic pretreatment of the tissue. Chondrocyte adhesion to a vertical surface of cartilage was examined using micropipette manipulation methods (105). The chondrocyte adhesion force increased with seeding time (15 to 75 min) and Chondroitinase ABC pretreatment but was not affected by the region of articular cartilage (i.e. superficial, middle, deep layer) to which the cells were attached. For normal cartilage, the adhesion force of individual chondrocytes increased from 1.3 mdyne after a 15–30 min seeding period to 5.3 mdyne after 60–75 min. Treatment with Chondroitinase ABC had a marked affect (+144 to +299%) on adhesion during the short (15–30 min) seeding durations, and a lesser (+46%) effect at the longest duration (60–75 min) studied. These results provide direct biomechanical evidence that enzymatic treatment of a cartilage surface can enhance chondrocyte adhesion, and that chondrocyte adhesion to cartilage is similar in different regions of cartilage tissue.

While the above studies provide practical and quantitative information on the adhesiveness of chondrocytes to normal and treated cartilage surfaces, the mechanism by which chondrocytes attach to a cartilaginous tissue substrate was unknown. In articular cartilage, chondrocytes utilize cell surface receptors to attach to matrix components. Chondrocytes, cultured in high-density monolayers in the presence of FBS, are known to express a variety of cell-surface receptors including β1-integrins (33, 50, 53, 108, 142, 151), CD44 (98), and anchorin CII/annexin-V (126). These receptors are also present in cartilage *in situ* (180) and normally mediate interactions between chondrocytes and specific extracellular matrix components (32, 50, 85).

Specific receptors mediate the attachment of transplanted chondrocytes to cartilage. Adhesion studies were performed using the flow chamber technique, described above, and reagents that selectively block these receptors (101). Chondrocytes were released from high density monolayer culture, incubated in the presence or absence of blocking reagents before seeding on normal or Chondroitinase ABC-digested cartilage for 20 min and application of shear stress. As expected, increasing shear stress resulted in increasing cell detachment, and chondrocyte adhesion to cartilage was more marked for tissue pretreated with Chondroitinase ABC. After preincubation with an antibody that blocks $\beta 1$ integrin, chondrocyte adhesion to both normal and Chondroitinase ABC-treated cartilage was markedly inhibited; however, blocking CD44 or anchorin CII did not have a detectable effect. Thus, under these simulated transplantation conditions, $\beta 1$ integrin is a type of surface receptor on chondrocytes that mediates attachment of these cells to cartilage.

5.3. *Biomechanical Regulation of Transplanted Chondrocytes: Cell Proliferation and Matrix Synthesis*

Following the transplantation of chondrocytes into a defect, cell proliferation may modulate local cellularity within the defect. Previous studies have shown that cells can attach to and grow out from explants of articular cartilage (111) and that chondrocytes populate an *in vitro* laceration site after 3–6 weeks of incubation (158). Compression of transplanted cells can occur during cartilage repair as a result of joint loading or "press-fitting", a graft into a cartilage defect (Fig. 3).

A few studies have shown that mechanical loading can modulate the proliferation of chondrocytes in model systems. Some studies involve chondrocytes that are released from extracellular matrix, and then cultured in agarose or collagen gel (46, 51, 181). Other studies have analyzed the proliferative response of chondrocytes after attachment to cartilage and application of static compressive stress between cartilaginous surfaces (107). In the latter case, chondrocytes were isolated from adult bovine cartilage, cultured in high-density monolayer, resuspended, and then transplanted onto the surface of devitalized cartilage at a density of 250,000 cells/cm^2 and maintained in culture medium including FBS and ascorbate. The total DNA content of transplanted cell layers increased 4-fold to a plateau by 5 days. Over this culture period, the level of DNA synthesis ($[^3H]$thymidine incorporation), on a per cell basis, decreased steadily (by $\sim 90\%$ between day 0 and 6). Application of 24 hours of compressive stress in the low physiological range (0.06–0.48 MPa) to the adherent cells at one and four days after transplantation inhibited DNA synthesis by $\sim 80\%$ compared to unloaded controls. After release from load, cell proliferation generally remained at low levels. The marked proliferation of chondrocytes when attached to cartilage without applied load, and the inhibition of this proliferation by relatively low amplitude static compressive stress, may be relevant to the regulation of the cellularity of transplants in cartilage defects, and the occasional overgrowth of tissue that has been noted clinically in some chondrocyte transplantation procedures (2).

Following chondrocyte transplantation into a cartilage defect, synthesis of matrix components by these cells may be a key factor in filling the bulk of the defect as well as mediating integration with the adjacent host tissue. Previous studies have shown that static compression of cartilage results in an inhibition of biosynthesis of matrix components by indwelling chondrocytes [reviewed in (74, 172)]. This effect is common to cartilage explants of various species, ages, and sample geometries, with a 50% inhibition at 0.1–1.5 MPa of applied stress (30, 68, 69, 92, 96, 97, 150, 156, 173).

Matrix synthesis by transplanted chondrocytes is also regulated by mechanical stimuli. (38). Bovine chondrocytes were transplanted onto cartilage disks, allowed to attach, and subjected to compression through overlying devitalized cartilage disks in a confined compression configuration. During a 16-hours radiolabeling period, application of compressive stress of 0.24–0.72 MPa inhibited GAG synthesis by ∼ 50%. The effect of compression on chondrocyte metabolism was reversible; in fact, at 2 days after the release of load, GAG synthesis by the loaded cells was stimulated by 40% compared to transplanted cells that were not subjected to loading. These results suggest that the application of compressive stress to chondrocytes at a cartilage surface affects biosynthesis by these cells and, thus, subsequent integrative cartilage repair.

The above studies indicate that transplanted chondrocytes are strongly regulated by compressive stress at relatively low amplitude. This suggests that the biomechanical environment *in vivo* may have a particularly potent regulatory effect on transplanted chondrocytes, both in terms of proliferation and matrix synthesis. Information on the dose-response relationship between mechanical stimulus and biological response provides biomechanical criteria for designing the tightness of the fit of a cell-laden cartilaginous construct into an articular defect as well as for developing post-operative rehabilitation protocols that would affect the biomechanical environment after a tissue engineering treatment.

6.   Tissue-Scale Biomechanical and Structural Properties of Articular Cartilage

The above studies indicate that biological and biomechanical stimuli are likely to regulate cartilage repair *in vivo*. From such information derived *in vitro* has evolved tissue-engineered implants and procedures that have been tested in a challenging *in vivo* environment. The repair of experimental cartilage defects in joints of animals provides a model for assessing repair mechanisms as well as for developing putative clinical therapies.

In such studies in animals, it is necessary to evaluate the success of the procedure. Such analyses may be noninvasive or invasive, destructive to tissue or nondestructive, and occur before or after termination of the experimental preparation. For example, magnetic resonance imaging and many other radiological techniques are noninvasive and nondestructive. Mechanical probing of the stiffness of cartilage is invasive and, ideally, nondestructive. Removing a biopsy of tissue for histological analysis is invasive and destructive. Whole joint surfaces may be analyzed in detail, post-mortem. Such analysis may include analysis of the repair tissue as well as the surrounding and opposing cartilage. The quality of the repair tissue is typically analyzed biomechanically to assess function, biochemically to assess composition, histologically to assess structure, or metabolically to assess biological activity. Such multidisciplinary analyses is often useful because, although tissue composition, structure, function, and metabolism are somewhat related, the exact relationship between such features of cartilage tissue remains to be fully defined. All of the above analyses are also implicitly or explicitly done at various length scales (i.e. determining a parameter that is spatially-averaged in some sense).

6.1. *Compressive Properties of Articular Cartilage*

The biomechanical material properties and geometry of the articular cartilage and surrounding tissues determine the function of the articulating joint, i.e. its mechanical response to applied load. The biomechanical material properties of cartilage can be analyzed at different length scales, for example, in terms of full-thickness tissue or tissue layers (e.g. ~ 0.1 mm length scale). Theories had been developed to describe depth-varying biomechanical and electromechanical properties [Fig. 6 and (71, 130)]. The value of the biomechanical parameter characterizes each layer in the model. One approach to estimating these parameters is to physically divide full thickness cartilage into tissue sections and to analyze these sections individually (72, 102, 118, 156, 160, 171). Another approach is to leave a fragment of tissue intact through the full thickness, and to use video dimensional analysis (182) adapted to an epifluorescence microscope to view the intra-tissue displacement during mechanical testing (155).

Samples of normal adult articular cartilage have been evaluated in this way (40, 153, 155). Bovine cartilage samples were preincubated with Hoechst 33258, a dye that labels cell nuclei by binding to DNA and increasing fluorescence upon doing so. Such cell nuclei can be used as intrinsic fiducial markers. These nuclei are visible under fluorescence microscopy as objects of diameter of ~ 5 μm and at a position that can be resolved to ~ 0.2 μm (by calculating the centroid). Such labeled cartilage samples were then subjected to compression and allowed to equilibrate to achieve a steady-state mechanical response. The displacement profile was measured (Fig. 7) and axial strain was calculated in sequential 125–250 μm thick cartilage layers. The equilibrium stress-strain data was nonlinear and fit to a finite deformation relationship to allow calculation of the confined compression modulus in each tissue layer (Fig. 6). The compressive modulus varies ~ 25-fold with depth from the articular surface, increasing from 0.08 MPa (superficial, 0–125 μm, layer) to 1.14 MPa (1000–1125 μm)

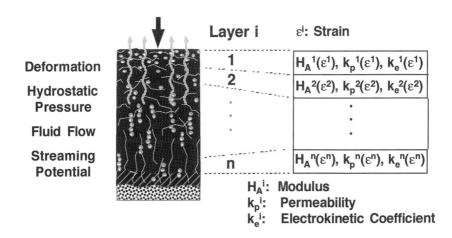

Fig. 6. Depth-dependent compressive properties of articular cartilage. Based on (71).

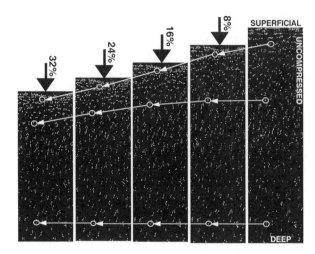

Fig. 7. Determination of depth-dependent compressive properties of articular cartilage. Tracking of Hoechst 33258-labeled chondrocytes (bright spots) in processed images of bovine cartilage subjected to radially-confined compression and viewed by fluorescence microscopy. From (153), with permission.

and 2.10 MPa (250 μm layer adjacent to cartilage-bone interface). Similar trends of increasing modulus with depth from the articular surface have been found to be present in articular cartilage from human femoral head, although the overall compressive modulus was much ($\sim$8-fold) greater (40). The relatively low moduli and compression-induced stiffening of the superficial layers suggest that these regions greatly affect the biomechanical behavior of cartilage during compressive loading.

Dynamic (i.e. time-varying) mechanical testing can also be performed and the resultant measurements can be fit to depth-varying models. Two other parameters in the above model (Fig. 6) that describe the mechanical and electrical behavior of cartilage in a radially-confined compression configuration are hydraulic permeability and electrokinetic coefficient. The hydraulic permeability in such a model is defined as the incremental ratio of the fluid flow through a tissue layer relative to the pressure drop across the tissue layer (117) under the condition of zero current flow (58). Permeability is known to be dependent on strain, with permeability decreasing as tissue is compacted (102). The electrokinetic (streaming potential) coefficient in such a model is defined as the incremental ratio of the electrical potential drop across a tissue layer relative to the pressure drop across the tissue layer (58, 117). The electrokinetic coefficient also is dependent on strain (39, 72). When full-thickness bovine cartilage is described by a layered material, the permeability of various layers is up to 15-fold different in amplitude at zero-strain from that of the permeability value assuming homogeneity and 2-fold different in the strain-dependence parameter, as well as 2-fold different in the amplitude of the electrokinetic coefficient (39).

The magnitude of these differences indicates that the interpretation of the physical behavior of full-thickness cartilage varies greatly when different length scales are considered. The quantitative measures of such material properties are useful for the evaluation of theoretical models of cartilage biomechanical behavior at different length scales in joints, the development of nondestructive probes for analyzing cartilage properties, and the study of the effect of loading on biological responses of chondrocytes in cartilage. Also, by attempting to relate the measured properties to composition and structure, which are known to vary with depth from the articular surface, such measurements help to develop a deeper understanding of the basis for the physical behaviors of cartilage. Such measures of physical properties of normal cartilage may be useful in practice, as design goals for fully functional cartilage after therapeutic tissue engineering procedures.

6.2. *Structure of Articular Cartilage after Joint Injury*

The articular surface is a sensitive indicator of the integrity of an joint. Arthritis is associated with cracking and roughening of the articular cartilage surface. Histological analysis provides detailed (~ 0.1 μm) morphological information on very specific areas through the depth of the tissue, and various characteristics have been used in semi-quantitative grading scales of cartilage degeneration (75, 115, 162) and repair (43, 82, 136, 140). Typically, a histological section for light microscopic analysis of articular cartilage is 6 μm in thickness and through the full depth [0.1–5 mm (165)] of the tissue. However, it is difficult to obtain a complete picture of the joint surface using histological methods.

Alternatively, the structure of the cartilage surface can be analyzed at ~ 0.1 mm resolution after staining of a cartilage surface with India ink. India ink contains carbon black particles, with an individual particle diameter of ~ 40 nm and clusters of particles of ~ 100 nm (112, 170). The relatively large size of the India ink particles prevents them from entering an intact articular cartilage surface with normal proteoglycan-rich matrix (117). In cartilage samples that do stain with India ink, the intensity of ink staining has been related to a reduction in the proteoglycan content through the depth of the specimen (56, 119). India ink particles are not only entrapped by irregularities in the articular surface, but also adhere to fibrillated cartilage (121). Since India ink particles absorb and scatter incident light (112), the reflection of incident light (120) from cartilage depends on the degree of India ink staining. Ink staining has shown that cartilage fibrillation and erosion have a predilection for specific anatomic sites in human osteoarthritis (52, 121). Such patterns of degeneration are likely to be related to the biomechanical environment, which in turn may be modulated by specific types of joint injury or cartilage repair procedures.

In animal models of osteoarthritis, India ink staining has been used to assess the extent and location of cartilage degradation (36). Here, the femora and tibiae were harvested from the operated and contralateral control knees of New Zealand White rabbits that were skeletally mature at the time of anterior cruciate ligament transection (ACLT), nine weeks previously. The knees were positioned to obtain calibrated gray-scale images of the articular cartilage surfaces, painted with India ink, that are opposed with the knee in 90° flexion. Images (Fig. 8) were processed so that areas of normal cartilage gave a relatively high reflectance score, whereas ink-stained fibrillated cartilage and exposed bone gave low scores. Digital image processing allowed registration of the images and averaging of the images (i.e. of the intensity of each $x - y$ pixel position). ACLT led to a 11% decrease in the overall reflectance score. The reflectance score decreased as a traditional morphological grade of degeneration increased. ACLT-induced degeneration had a predilection for the posteromedial aspects of the joint, and to a lesser extent, the anterolateral aspects. In the

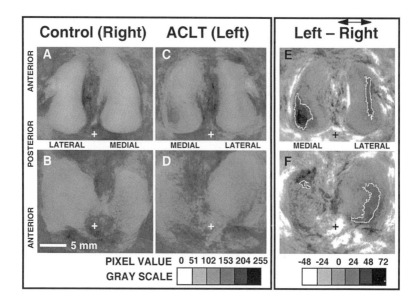

Fig. 8. Digitally-averaged images to highlight areas of cartilage staining with India ink and erosion, and differences between ACL-transected and contralateral control joints. The average images of control femora (A) and tibiae (B), and ACLT femora (C) and tibiae (D), each n=11, are shown. The difference images (E,F) were computed by horizontally flipping the average control (A,B) image, subtracting this from the ACLT (C,D) image, and then contrast-enhancing the result according to the pixel value — gray scale shown. Regions outlined in white (E,F) show the degeneration prone areas. Dark regions correspond to degeneration. From (36), with permission.

tibial plateaus, ACLT caused significant degeneration in the areas of the joint that were covered by the meniscus, but had no detectable effect on the cartilage in the uncovered areas. Image scores of opposing cartilage surfaces were significantly correlated ($R = 0.56$–0.70) in ACLT and control knees.

Such structural analysis of cartilage at different length scales provides insight into the quality of cartilage tissue. Characterization of the surface structure of the cartilage apposing an area of repair may be a sensitive way of assessing the success of a tissue engineering procedure. On the other hand, histological and histochemical analysis of the interface region between an implant and host tissue as well as the implant tissue itself allows determination of features (e.g. cellularity, matrix components, surface smoothness and continuity) that are characteristic of normal articular cartilage.

7.   General Discussion and Outlook

In the 1950's, total joint replacements with nonliving materials were introduced (179) and transformed orthopaedic surgery into a specialty that delivers an improved quality of life for a large number of patients-primarily those who are elderly and suffering from end-stage arthritis. However, many young and active patients still suffer from joint impairment and could benefit from more lasting and effective articular therapies. In the 1990's and the 21st century, tissue engineering promises to deliver another major transformation in the orthopaedic treatment of articular cartilage degeneration.

Acknowledgments

The author thanks Lisa Lottman, Dr. Albert Chen, Melissa Kurtis, and Kelvin Li for their assistance with the figures and critical discussions, and NASA, NIH, and NSF for financial support.

References

1. Tissue Engineering, Keystone Symposium on Molecular and Cellular Biology, Taos, New Mexico, February 20–26, 1994. Abstracts, *J. Cell. Biochem.* **C18**, 265–284 (1994).
2. Carticel™, autologous cultured chondrocytes for implantation, Genzyme Tissue Repair, Cambridge, MA (1997).
3. Cartilage repair registry: Periodic report, Genzyme Tissue Repair, Cambridge, MA (1999).
4. Ahsan, T., Integrative repair and collagen crosslinking in articular cartilage. PhD Thesis, University of California, San Diego (1998).
5. Ahsan, T., Li, K. W., Lau, S. T. and Sah, R. L., Modulation of integrative cartilage repair *in vitro* by IGF-1 and TGF-β1, *Trans. Orthop. Res. Soc.* **20**, 173 (1995).
6. Ahsan, T., Lottman, L. M., Harwood, F. L., Amiel, D. and Sah, R. L., Integrative cartilage repair: Inhibition by β-aminoproprionitrile, *J. Orthop. Res.* **17**, 850–857 (1999).
7. Ahsan, T. and Sah, R. L., Fracture mechanics characterization of integrative cartilage repair using the T-peel test, *Trans. Orthop. Res. Soc.* **21**, 537 (1996).
8. Ahsan, T. and Sah, R. L., Biomechanics of integrative cartilage repair, *Osteoarthritis Cartilage* **7**, 29–40 (1999).
9. Anderson, T. L., Fracture mechanics: Fundamentals and applications, 2nd edition, Boca Raton, F. L., CRC Press (1995), p. 688.
10. Aspenberg, P. and van der Vis, H., Migration, particles, and fluid pressure. A discussion of causes of prosthetic loosening, *Clin. Orthop.* **352**, 75–80 (1998).
11. Bahuaud, J., Maitrot, R. C., Bouvet, R., Kerdiles, N., Tovagliaro, F., Synave, J., Buisson, P., Thierry, J. F., Versier, A., Romanet, J. P., Chauvin, F., Gillet, J. P., Allizard, J. P. and de Belenet, H., Implantation of autologous chondrocytes for cartilagenous lesions in young patients. A study of 24 cases, *Chirurgie* **123**, 568–571 (1998).
12. Baragi, V. M., Renkiewicz, R. R., Jordan, H., Bonadio, J., Hartman, J. W. and Roessler, B. J., Transplantation of transduced chondrocytes protects articular cartilage from interleukin 1-induced extracellular matrix degradation, *J. Clin. Invest.* **96**, 2454–2460 (1995).
13. Baragi, V. M., Renkiewicz, R. R., Qiu, L., Brammer, D., Riley, J. M., Sigler, R. E., Frenkel, S. R., Amin, A., Abramson, S. B. and Roessler, B. J., Transplantation of adenovirally transduced allogeneic chondrocytes into articular cartilage defects *in vivo*, *Osteoarthritis Cartilage* **5**, 275–282 (1997).
14. Barone-Varelas, J., Schnitzer, T. J., Meng, Q., Otten, L. and Thonar, E., Age-related differences in the metabolism of proteoglycans in bovine articular cartilage explants maintained in the presence of insulin-like growth factor-1, *Connect. Tissue Res.* **26**, 101–120 (1991).
15. Basser, P. J., Schneiderman, R., Bank, R., Wachtel, E. and Maroudas, A., Mechanical properties of the collagen network in human articular cartilage as measured by osmotic stress technique, *Arch. Biochem. Biophys.* **351**, 207–219 (1998).
16. Bell, E., ed., Tissue engineering: Current perspectives. Birkäuser, Boston (1993), p. 241.
17. Ben-Yishay, A., Grande, D. A., Schwartz, R. E., Menche, D. and Pitman, M. D., Repair of articular cartilage defects with collagen-chondrocyte allografts, *Tissue Eng.* **1**, 119–133 (1995).

18. Benya, P. D. and Nimni, M. E., The stability of the collagen phenotype during stimulated collagen, glycosaminoglycan, and DNA synthesis by articular cartilage organ cultures, *Arch. Biochem. Biophys.* **192**, 327–335 (1979).

19. Benya, P. D. and Shaffer, J. D., Dedifferentiated chondrocytes reexpress the differentiated collagen phenotype when cultured in agarose gels, *Cell.* **30**, 215–224 (1982).

20. Berlet, G. C., Mascia, A. and Miniaci, A., Treatment of unstable osteochondritis dissecans lesions of the knee using autogenous osteochondral grafts (mosaicplasty), *Arthroscopy* **15**, 312–326 (1999).

21. Bobic, V., Arthroscopic osteochondral autograft transplantation in anterior cruciate ligament reconstruction: A preliminary clinical study, *Knee Surg. Sports Traumatol. Arthrosc.* **3**, 262–264 (1996).

22. Breinan, H. A., Minas, T., Hus, H.-P., Nehrer, S., Sledge, C. B. and Spector, M., Effect of cultured autologous chondrocytes on repair of chondral defects in a canine model, *J. Bone Joint Surg.* **A79**, 1439–1451 (1997).

23. Brittberg, M., Lindahl, A., Nilsson, A., Ohlsson, C., Isaksson, O. and Peterson, L., Treatment of deep cartilage defects in the knee with autologous chondrocyte transplantation, *N. Engl. J. Med.* **331**, 889–895 (1994).

24. Broom, N. D., Oloyede, A., Flachsmann, R. and Hows, M., Dynamic fracture characteristics of the osteochondral junction undergoing shear deformation, *Med. Eng. Phys.* **18**, 396–404 (1996).

25. Buckwalter, J. A., Caterson, B., Howell, D. S. and Rosenberg, L., Molecular versus tissue versus organ repair (functional repair): Future directions. *Osteoarthritic Disorders* (eds. K. E. Kuettner and V. M. Goldberg), American Academy of Orthopaedic Surgeons, Rosemont, IL (1995), pp. 395–400.

26. Buckwalter, J. A. and Mankin, H. J., Articular cartilage. Part I: Tissue design and chondrocyte-matrix interactions, *J. Bone Joint Surg.* **A79**, 600–611 (1997).

27. Buckwalter, J. A. and Mankin, H. J., Articular cartilage. Part II: Degeneration and osteoarthrosis, repair, regeneration, and transplantation, *J. Bone Joint Surg.* **A79**, 612–632 (1997).

28. Buckwalter, J. A. and Mankin, H. J., Articular cartilage repair and transplantation, *Arthritis Rheum.* **41**, 1331–1342 (1998).

29. Bugbee, W. D. and Convery, F. R., Osteochondral allograft transplantation, *Clin. Sports Med.* **18**, 67–75 (1999).

30. Burton-Wurster, N., Vernier-Singer, M., Farquhar, T. and Lust, G., Effect of compressive loading and unloading on the synthesis of total protein, proteoglycan, and fibronectin by canine cartilage explants, *J. Orthop. Res.* **11**, 717–729 (1993).

31. Buschmann, M. D. and Grodzinsky, A. J., A molecular model of proteoglycan-associated electrostatic forces in cartilage mechanics, *J. Biomech. Eng.* **117**, 179–192 (1995).

32. Camper, L., Heinegard, D. and Lundgren-Akerlund, E., Integrin $\alpha2\beta1$ is a receptor for the cartilage matrix protein chondroadherin, *J. Cell. Biol.* **138**, 1159–1167 (1997).

33. Camper, L. and Hellman, U., Lundgren-Akerlund: Isolation, cloning, and sequence analysis of the integrin subunit of $\alpha10$, a $\beta1$-associated collagen binding integrin expressed on chondrocytes, *J. Biol. Chem.* **283**, 20383–20389 (1998).

34. Canale, S. T., ed., Campbell's operative orthopaedics, Mosby, St. Louis (1998), p. 4076.

35. Caplan, A. I., Elyaderani, M., Mochizuki, Y., Wakitani, S. and Goldberg, V. M., Principles of cartilage repair and regeneration, *Clin. Orthop.* **342**, 254–269 (1997).

36. Chang, D. G., Iverson, E. P., Schinagl, R. M., Sonoda, M., Amiel, D., Coutts, R. D. and Sah, R. L., Quantitation and localization of cartilage degeneration following the induction of osteoarthritis in the rabbit knee, *Osteoarthritis Cartilage* **5**, 357–372 (1997).

37. Chen, A. C., Nagrampa, J. P., Schinagl, R. M., Lottman, L. M. and Sah, R. L., Chondrocyte transplantation to articular cartilage explants *in vitro*, *J. Orthop. Res.* **15**, 791–802 (1997).

38. Chen, A. C. and Sah, R. L., The effect of static compression on proteoglycan synthesis by chondrocytes transplanted to articular cartilage *in vitro*, *J. Orthop. Res.* **16**, 542–550 (1998).

39. Chen, A. C., Schinagl, R. M. and Sah, R. L., Inhomogeneous and strain-dependent electromechanical properties of full-thickness articular cartilage, *Trans. Orthop. Res. Soc.* **23**, 225 (1998).

40. Chen, S. S., Falcovitz, Y. H., Schneiderman, R., Maroudas, A. and Sah, R. L., Depth-dependent compressive properties of aged human femoral head articular cartilage, *Trans. Orthop. Res. Soc.* **24**, 643 (1999).

41. Chin-Purcell, M. V. and Lewis, J. L., Fracture of articular cartilage, *J. Biomech. Eng.* **118**, 545–556 (1996).

42. Chu, C. R., Convery, F. R., Akeson, W. H., Meyers, M. and Amiel, D., Articular cartilage transplantation: Clinical results in the knee, *Clin. Orthop.* **360**, 159–168 (1999).

43. Chu, C. R., Dounchis, J. S., Yoshioka, M., Sah, R. L., Coutts, R. D. and Amiel, D., Osteochondral repair using perichondrial cells: A one year study in rabbits, *Clin. Orthop.* **340**, 220–239 (1997).

44. Cotran, R. S., Kumar, V. and Robbins, S. L., Inflammation and repair, *Robbins Pathologic Basis of Disease*, Co., W. B. Saunders, Philadelphia (1989), pp. 39–86.

45. Davies, P. F., Barbee, K. A., Volin, M. V., Robotewskyj, A., Chen, J., Joseph, L., Griem, M. L., Wernick, M. N., Jacobs, E., Polacek, D. C., dePaola, N. and Barakat, A. I., Spatial relationships in early signaling events of flow-mediated endothelial mechanotransduction, *Ann. Rev. Physiol.* **59**, 527–549 (1997).

46. DeWitt, M. T., Handley, C. J., Oakes, B. W. and Lowther, D. A., *In vitro* response of chondrocytes to mechanical loading: The effect of short term mechanical tension, *Connect. Tissue Res.* **12**, 97–109 (1984).

47. Di Cesare, P. E., Surgical management of osteoarthritis, *Clin. Geriatr. Med.* **14**, 613–631 (1998).

48. DiMicco, M. A., Chen, A. C., Lottman, L. M., Copelan, N. B., Silvestre, M. A., Ahsan, T. and Sah, R. L., Integrative cartilage repair: Relationship to biosynthesis by endogenous and transplanted chondrocytes, *Trans. Orthop. Res. Soc.* **24**, 275 (1999).

49. Dunkelman, N. S., Zimber, M. P., LeBaron, R. G., Pavelec, R., Kwan, M. and Purchio, A. F., Cartilage production by rabbit articular chondrocytes on polyglycolic acid scaffolds in a closed bioreactor system, *Biotechnol. Bioeng.* **46**, 299–305 (1995).

50. Durr, J., Goodman, S., Potocnik, A., von der Mark, H. and von der Mark, K., Localization of β1-integrins in human cartilage and their role in chondrocyte adhesion to collagen and fibronectin, *Exp. Cell. Res.* **207**, 235–244 (1993).

51. Elder, S. H., Kimura, J. H., Soslowsky, L. J. and Goldstein, S. A., Effects of compressive loading on chondrocyte differentiation in agarose gel cultures of chick limb bud cells, *Trans. Orthop. Res. Soc.* **23**, 912 (1998).

52. Emery, I. H. and Meachim, G., Surface morphology and topography of patello-femoral cartilage fibrillation in Liverpool necropsies, *J. Anat.* **116**, 103–120 (1973).

53. Enomoto, M., Leboy, P. S., Menko, A. S. and Boettiger, D., β1 integrins mediate chondrocyte interaction with type I collagen, type II collagen, and fibronectin, *Exp. Cell. Res.* **205**, 276–285 (1993).

54. Eyre, D. R., Grypnas, M. D., Shapiro, F. D. and Creasman, C. M., Mature crosslink formation and molecular packing in articular cartilage collagen, *Sem. Arthritis Rheum.* **11**, 46–57 (1981).

55. Eyre, D.R., Paz, M.A. and Gallop, P. M., Cross-linking in collagen and elastin, *Ann. Rev. Biochem.* **53**, 717–748 (1984).

56. Ficat, C. and Maroudas, A., Cartilage of the patella. Topographical variation of glycosamino-glycan content in normal and fibrillated tissue, *Ann. Rheum. Dis.* **34**, 515–529 (1975).

57. Fitzpatrick, P. L. and Morgan, D. A., Fresh osteochondral allografts: A 6-10 year review, *Aust. New Zealand J. Surg.* **68**, 573–589 (1998).

58. Frank, E. H. and Grodzinsky, A. J., Cartilage electromechanics — Part II. A continuum model of cartilage electrokinetics and correlation with experiments, *J. Biomech.* **20**, 629–639 (1987).

59. Frenkel, S. R., Clancy, R. M., Ricci, J. L., DiCesare, P. E., Rediske, J. J. and Abramson, S. B., Effects of nitric oxide on chondrocyte migration, adhesion, and cytoskeletal assembly, *Arthritis Rheum.* **39**, 1905–1912 (1996).

60. Galletti, P. M., Let's do tissue engineering right — Viewpoint, *IEEE Spectrum* **33**, 94 (1996).

61. Garcia, A. M. and Gray, M. L., Dimensional growth and extracellular matrix accumulation by neonatal rat mandibular condyles in long-term culture, *J. Orthop. Res.* **13**, 208–219 (1995).

62. Garrett, J. C., Osteochondral allografts for reconstruction of articular defects of the knee, *Instr. Course Lect.* **47**, 517–522 (1998).

63. Ghadially, F. N., Thomas, I., Oryschak, A. F. and LaLonde, J.-M. A., Long term results of superficial defects in articular cartilage. A scanning electron microscope study, *J. Pathol.* **121**, 213–227 (1977).

64. Ghazavi, M. T., Pritzker, K. P., Davis, A. M. and Gross, A. E., Fresh osteochondral allografts for post-traumatic osteochondral defects of the knee, *J. Bone Joint Surg.* **B79**, 1008–1013 (1997).

65. Gillogly, S. D., Voight, M. and Blackburn, T., Treatment of articular cartilage defects of the knee with autologous chondrocyte implantation, *J. Orthop. Sports Phys. Ther.* **28**, 241–251 (1998).

66. Goodman S. B., Lind, M., Song, Y. and Smith, R. L., *In vitro, in vivo*, and tissue retrieval studies on particulate debris, *Clin. Orthop.* **352**, 25–34 (1998).

67. Grande, D. A., Pitman, M. I., Peterson, L., Menche, D. and Klein, M., The repair of experimentally produced defects in rabbit articular cartilage by autologous chondrocyte transplantation, *J. Orthop. Res.* **7**, 208–218 (1989).

68. Gray, M. L., Pizzanelli, A. M., Grodzinsky, A. J. and Lee, R. C., Mechanical and physicochemical determinants of the chondrocyte biosynthetic response, *J. Orthop. Res.* **6**, 777–792 (1988).

69. Gray, M. L., Pizzanelli, A. M., Lee, R. C., Grodzinsky, A. J. and Swann, D. A., Kinetics of the chondrocyte biosynthetic response to compressive load and release, *Biochim. Biophys. Acta.* **991**, 415–425 (1989).

70. Grodzinsky, A. J., Electromechanical and physicochemical properties of connective tissue, *CRC Crit. Rev. Bioeng.* **9**, 133–199 (1983).

71. Grodzinsky, A. J. and Frank, E. H., Electromechanical and physicochemical regulation of cartilage strength and metabolism, *Connective Tissue Matrix: Topics in Molecular and Structural Biology*, Vol. II (ed. D. W. L. Hukins), CRC Press, Boca Raton (1990), pp. 91–126.

72. Gu, W. Y., Lai, W. M. and Mow, V. C., Transport of fluid and ions through a porous-permeable charged-hydrated tissue, and streaming potential data on normal bovine articular cartilage, *J. Biomech.* **26**, 709–723 (1993).

73. Guerne, P.-A., Blanco, F., Kaelin, A., Desgeorges, A. and Lotz, M., Growth factor responsiveness of human articular chondrocytes in aging and development, *Arthritis Rheum.* **38**, 960–968 (1995).

74. Guilak, F., Sah, R. L. and Setton, L. A., Physical regulation of cartilage metabolism, *Basic Orthopaedic Biomechanics* (eds. V. C. Mow and W. C. Hayes), Raven Press, New York (1997), pp. 179–207.

75. Hacker, S. A., Healey, R. M., Yoshioka, M. and Coutts, R. D., A methodology for the quantitative asessment of articular cartilage histomorphometry, *Osteoarthritis Cartilage* **5**, 343–355 (1997).

76. Hangody, L., Kish, G., Karpati, Z., Szerb, I. and Eberhardt, R., Treatment of osteochondritis dissecans of the talus: Use of the mosaicplasty technique — A preliminary report, *Foot Ankle Int.* **18**, 628–634 (1997).

77. Hascall, V. C., Handley, C. J., McQuillan, D. J., Hascall, G. K., Robinson, H. C. and Lowther, D. A., The effect of serum on biosynthesis of proteoglycans by bovine articular cartilage in culture, *Arch. Biochem. Biophys.* **224**, 206–223 (1983).

78. Hascall, V. C., Luyten, F. P., Plaas, A. H. K. and Sandy, J. D., Steady-state metabolism of proteoglycans in bovine articular cartilage, *Methods in Cartilage Research* (eds. A. Maroudas and K. Kuettner), Academic Press, San Diego (1990), pp. 108–112.

79. Häuselmann, H. J., Fernandes, R. J., Mok, S. S., Schmid, T. M., Block, J. A., Aydelotte, M. B., Kuettner, K. E. and Thonar, E. J., Phenotypic stability of bovine articular chondrocytes after long-term culture in alginate beads, *J. Cell. Sci.* **107**, 17–27 (1994).

80. Hendrickson, D. A., Nixon, A. J., Grande, D. A., Todhunter, R. J., Minor, R. M., Erb, H. and Lust, G., Chondrocyte-fibrin matrix transplants for resurfacing extensive articular cartilage defects, *J. Orthop. Res.* **12**, 485–497 (1994).

81. Hewitt, A. T., Kleinman, H. K., Pennypacker, J. P. and Martin, G. R., Identification of an adhesion factor for chondrocytes, *Proc. Natl. Acad. Sci. USA* **77**, 385–398 (1980).

82. Hjertquist, S. O. and Lemperg, R., Histological, autoradiographic and microchemical studies of spontaneously healing osteochondral articular defects in adult rabbits, *Calcif. Tissue Res.* **8**, 54–72 (1971).

83. Hochberg, M. C., Altman, R. D., Brandt, K. D., Clark, B. M., Dieppe, P. A., Griffin, M. R., Moskowitz, R. W. and Schnitzer, T. J., Guidelines for the medical management of osteoarthritis. Part I: Osteoarthritis of the hip. *Arthritis Rheum.* **38**, 1535–1540 (1995).

84. Hochberg, M. C., Altman, R. D., Brandt, K. D., Clark, B. M., Dieppe, P. A., Griffin, M. R., Moskowitz, R. W. and Schnitzer, T. J., Guidelines for the medical management of osteoarthritis. Part II: Osteoarthritis of the knee, *Arthritis Rheum.* **38**, 1541–1546 (1995).

85. Holmvall, K., Camper, L., Johansson, S., Kimura, J. H. and Lundgren-Akerlund, E., Chondrocyte and chondrosarcoma cell integrins with affinity for collagen type II and their response to mechanical stress, *Exp. Cell. Res.* **221**, 496–503 (1995).

86. Hunter, W., On the structure and diseases of articulating cartilage, *Philos. Trans. Roy. Soc. London* **42**, 514–521 (1743).

87. Hunziker, E. B. and Kapfinger, E., Removal of proteoglycans from the surface of defects in articular cartilage transiently enhances coverage by repair cells, *J. Bone Joint Surg.* **B80**, 144–150 (1998).

88. Hunziker, E. B. and Rosenberg, L. C., Repair of partial-thickness defects in articular cartilage: Cell recruitment from the synovial membrane, *J. Bone Joint Surg.* **A78**, 721–733 (1996).

89. Ishida, O., Ikuta, Y. and Kuroki, H., Ipsilateral osteochondral grafting for finger joint repair, *J. Hand Surg. [Am]* **19**, 372–387 (1994).

90. Itay, S., Abramovici, A. and Nevo, Z., Use of cultured embryonal chick epiphyseal chondrocytes as grafts for defects in chick articular cartilage, *Clin. Orthop.* **220**, 284–303 (1987).

91. Jackson, R. W., Judy, M. M., Matthews, J. L. and Nosir, H., Photochemical tissue welding with 1, 8 naphthalimide dyes: *In vivo* meniscal and cartilage welds, *Trans. Orthop. Res. Soc.* **22**, 650 (1997).

92. Jones, I. L., Klamfeldt, D. D. S. and Sandstrom, T., The effect of continuous mechanical pressure upon the turnover of articular cartilage proteoglycans *in vitro. Clin. Orthop.* **165**, 283–299 (1982).

93. Jürgensen, K., Aeschlimann, D., Cavin, V., Genge, M. and Hunziker, E. B., A new biological glue for cartilage-cartilage interfaces: Tissue transglutaminase, *J. Bone Joint Surg.* **A79**, 185–193 (1997).

94. Kagan, H. M., Characterization and regulation of lysyl oxidase. *Regulation of Matrix Accumulation* (ed. R. P. Mecham), Academic Press, Inc., Orlando (1986), pp. 321–398.

95. Khouri, R. K., Koudsi, B. and Reddi, H., Tissue transformation into bone *in vivo*. A potential practical application, *JAMA* **266**, 1953–1965 (1991).

96. Kim, Y. J., Grodzinsky, A. J. and Plaas, A. H. K., Compression of cartilage results in differential effects on biosynthetic pathways for aggrecan, link protein, and hyaluronan, *Arch. Biochem. Biophys.* **328**, 331–340 (1996).

97. Kim, Y. J., Sah, R. L., Grodzinsky, A. J., Plaas, A. H. K. and Sandy, J. D., Mechanical regulation of cartilage biosynthetic behavior: Physical stimuli, *Arch. Biochem. Biophys.* **311**, 1–12 (1994).

98. Knudson, W., Aguiar, D. J., Hua, Q., Knudson, C. B., CD44-anchored hyaluronan-rich pericellular matrices: An ultrastructural and biochemical analysis, *Exp. Cell. Res.* **228**, 216–228 (1996).

99. Knutsen, G., Solheim, E. and Johansen, O., Treatment of focal cartilage injuries in the knee, *Tidsskr Nor Laegeforen* **118**, 2493–2497 (1998).

100. Kuettner, K. E., Memoli, V. A., Pauli, B. U., Wrobel, N. C., Thonar E. J.-M. A. and Daniel, J. C., Synthesis of cartilage matrix by mammalian chondrocytes *in vitro*. Part II: Maintenance of collagen and proteoglycan phenotype, *J. Cell. Biol.* **93**, 751–757 (1982).

101. Kurtis, M. S., Gaya, O. A., Tu, B. P., Loeser, R. F., Knudson, W., Knudson, C. B. and Sah, R. L., Mechanisms of chondrocyte adhesion to cartilage: Role of β 1 integrins, CD44, and anchorin CII, *Trans. Orthop. Res. Soc.* **24**, 105 (1999).

102. Lai, W. M., Mow, V. C. and Roth, V., Effects of nonlinear strain-dependent permeability and rate of compression on the stress behavior of articular cartilage, *J. Biomech. Eng.* **103**, 61–76 (1981).

103. Langer, R. and Vacanti, J. P., Tissue engineering, *Science* **260**, 920–926 (1993).

104. Lanza, R. P., Langer, R. and Chick, W. L., eds., *Principles of Tissue Engineering*, Academic Press, San Diego (1997), p. 808.

105. Lee, M. C., Kurtis, M. S., Akeson, W. H., Sah, R. L. and Sung K.-L. P., Adhesive force of chondrocytes to cartilage: A micropipette study of the effects of chondroitinase ABC, *Trans. Orthop. Res. Soc.* **24**, 737 (1999).

106. Li, K. W., Ahsan, T. and Sah, R. L., Effect of trypsin treatment on integrative cartilage repair *in vitro. Trans. Orthop. Res. Soc.* **21**, 101 (1996).

107. Li, K. W., Falcovitz, Y. H., Nagrampa, J. P., Chen, A. C., Lottman, L. M., Shyy, Y. J. and Sah, R. L., Effects of compression on proliferation of transplanted chondrocytes, *Trans. Orthop. Res. Soc.* **24**, 627 (1999).

108. Loeser, R. F., Integrin-mediated attachment of articular chondrocytes to extracellular matrix proteins, *Arthritis Rheum.* **36**, 1103–1110 (1993).

109. Loeser, R. F. and Wallin, R., Cell adhesion to matrix Gla protein and its inhibition by an Arg-Gly-Asp-containing peptide, *J. Biol. Chem.* **267**, 9459–9462 (1992).

110. Lohnert, J., Regeneration of hyalin cartilage in the knee joint by treatment with autologous chondrocyte transplants — initial clinical results, *Langenbecks Arch Chir Suppl Kongressbd* **115**, 1205–1217 (1998).

111. Luyten, F. P., Hascall, V. C., Nissley, S. P., Morales, T. I. and Reddi, A. H., Insulin-like growth factors maintain steady-state metabolism of proteoglycans in bovine articular cartilage explants, *Arch. Biochem. Biophys.* **267**, 416–425 (1988).

112. Madsen, S. J., Patterson, M. S. and Wilson, B. C., The use of India ink as an optical absorber in tissue-simulating phantoms, *Phys. Med. Biol.* **37**, 985–993 (1992).

113. Mankin, H. J., The response of articular cartilage to mechanical injury, *J. Bone Joint Surg.* **A64**, 460–476 (1982).

114. Mankin, H. J., Chondrocyte transplantation — one answer to an old question, *N. Engl. J. Med.* **331**, 940–951 (1994).

115. Mankin, H. J., Dorfman, H., Lipiello, L. and Zarins, A., Biochemical and metabolic abnormalities in articular cartilage from osteoarthritic human hips, *J. Bone Joint Surg.* **A53**, 523–537 (1971).

116. Maroudas, A., Balance between swelling pressure and collagen tension in normal and degenerate cartilage, *Nature* **260**, 808–819 (1976).

117. Maroudas, A., Physico-chemical properties of articular cartilage, *Adult Articular Cartilage* (ed. M. A. R. Freeman), Pitman Medical, Tunbridge Wells, England (1979), pp. 215–290.

118. Maroudas, A. and Bullough, P., Permeability of articular cartilage, *Nature* **219**, 1260–1271 (1968).

119. Maroudas, A., Evans, H. and Almeida, L., Cartilage of the hip joint: Topographical variation of glycosaminoglycan content in normal and fibrillated tissue, *Ann. Rheum. Dis.* **32** (1973).

120. McCluney, W. R., *Introduction to Radiometry and Photometry*, Artech House, Boston (1994), p. 402.

121. Meachim, G., Light microscopy of India ink preparations of fibrillated cartilage, *Ann. Rheum. Dis.* **31**, 457–464 (1972).

122. Messner, K. and Gillquist, J., Cartilage repair. A critical review, *Acta. Orthop. Scand.* **67**, 523–529 (1996).

123. Miller, R. R. and McDevitt, C. A., A quantitative microwell assay for chondrocyte cell adhesion, *Anal. Biochem.* **192**, 380–393 (1991).

124. Minas, T. and Peterson, L., Advanced techniques in autologous chondrocyte transplantation, *Clin. Sports Med.* **18**, 13–44, v-vi (1999).

125. Mochizuki, Y., Goldberg, V. M. and Caplan, A. I., Enzymatical digestion for the repair of superficial articular cartilage lesions, *Trans. Orthop. Res. Soc.* **18**, 728 (1993).

126. Mollenhauer, J., Bee, J. A., Lizarbe, M. A., von der Mark, K., Role of anchorin CII, a 31,000-mol-wt membrane protein, in the interaction of chondrocytes with type II collagen, *J. Cell. Biol.* **98**, 1572–1758 (1984).

127. Morales, T. I., Cartilage proteoglycan homeostasis: Role of growth factors, *Cartilage Changes in Osteoarthritis* (ed. K. D. Brandt), Indiana University School of Medicine, Indianapolis, IN (1990), pp. 17–21.

128. Morales, T. I. and Hascall, V. C., Transforming growth factor-$\beta1$ stimulates synthesis of proteoglycan aggregates in calf articular organ cultures, *Arch. Biochem. Biophys.* **286**, 99–106 (1991).

129. Morgan, J. R., Yarmush and M. L., eds., Methods in Molecular Medicine, *Tissue Engineering Methods and Protocols* (ed. J. M. Walker), Humana Press, Totowa **18**, 629 (1999).

130. Mow, V. C., Kuei, S. C. and Lai, W. M., Armstrong CG: Biphasic creep and stress relaxation of articular cartilage in compression: Theory and experiment, *J. Biomech. Eng.* **102**, 73–84 (1980).

131. Mow, V. C. and Ratcliffe, A., Structure and function of articular cartilage and meniscus, *Basic Orthopaedic Biomechanics* (eds. Mow, V. C. and Hayes, W. C.), Raven Press, New York (1997), pp. 113–78.

132. Nerem, R. M. and Sambanis, A., Tissue engineering: From biology to biological substitutes, *Tissue Eng.* **1**, 3–13 (1995).

133. Nevo, Z., Silver, J., Chorev, Y., Riklis, I., Robinson, D. and Yosipovitch, Z., Adhesion characteristics of chondrocytes cultured separately and in co-cultures with synovial fibroblasts, *Cell. Biol. Int. Reports* **17**, 255–273 (1993).

134. Noguchi, T., Oka, M., Fujino, M., Neo, M. and Yamamuro, T., Repair of osteochondral defects with grafts of cultured chondrocytes, *Clin. Orthop.* **302**, 251–258 (1994).

135. O'Driscoll, S. W., The healing and regeneration of articular cartilage, *J. Bone Joint Surg.* **A80**, 1795–813 (1998).

136. O'Driscoll, S. W., Keeley, F. W. and Salter, R. B., The chondrogenic potential of free autogenous periosteal grafts for biological resurfacing of major full-thickness defects in joint surfaces under the influence of continuous passive motion. An experimental investigation in the rabbit, *J. Bone Joint Surg.* **A68**, 1017–1035 (1986).

137. Outerbridge, H. K., Outerbridge, A. R., Outerbridge, R. E. and Smith, D. E., The use of lateral patellar autologous grafts for the repair of large osteochondral defects in the knee, *Acta. Orthop. Belg.* **65**, 129–135 (1999).

138. Palsson, B. and Hubbell, J. A., Tissue engineering: Introduction, *The Biomedical Engineering Handbook*, Bronzino. (ed. J. D.), CRC Press, Boca Raton (1995), pp. 1580–1582.

139. Patrick, C. W., Mikos, A. G. and McIntire, L. V., eds., *Frontiers in Tissue Engineering*, Elsevier (1998).

140. Pineda, S., Pollack, A., Stevenson, S., Goldberg, V. and Caplan, A., A semiquantitative grading scale for histologic grading of articular cartilage repair, *Acta. Anat.* **143**, 335–340 (1992).

141. Praemer, A., Furner, S. and Rice, D. P., *Musculoskeletal Conditions in the United States*, Park Ridge, IL, American Academy of Orthopaedic Surgeons (1992), p. 199.

142. Ramachandrula, A., Tiku, K. and Tiku, M. L., Tripeptide RGD-dependent adhesion of articular chondrocytes to synovial fibroblasts, *J. Cell. Sci.* **101**, 859–871 (1992).

143. Ratner, B. D., ed., *Biomaterials*, Academic Press, San Diego (1996), p. 484.

144. Reindel, E. S., Ayroso, A. M., Chen, A. C., Chun, D. M., Schinagl, R. M. and Sah, R. L., Integrative repair of articular cartilage *in vitro*: Adhesive strength of the interface region, *J. Orthop. Res.* **13**, 751–760 (1995).

145. Reiser, K., McCormick, R. J. and Rucker, R. B., Enzymatic and nonenzymatic cross-linking of collagen and elastin, *FASEB J.* **6**, 2439–2449 (1992).

146. Rich, A. M., Pearlstein, E., Weissmann, G. and Hoffstein, S. T., Cartilage proteoglycans inhibit fibronectin-mediated adhesion, *Nature* **293**, 224–236 (1981).

147. Robins, S. P. and Duncan, A., Cross-linking of collagen: Location of pyridinoline in bovine articular cartilage at two sites of the molecule, *Biochem. J.* **215**, 175–182 (1983).

148. Sah, R. L., Amiel, D. and Coutts, R. D., Tissue engineering of articular cartilage, *Curr. Opin. Orthop.* **6**, 52–60 (1995).

149. Sah, R. L., Grodzinsky, A. J., Plaas, A. H. K and Sandy, J. D., Effects of static and dynamic compression on matrix metabolism in cartilage explants. In: *Articular Cartilage and Osteoarthritis* (eds. K. E. Kuettner, R. Schleyerbach, J. G. Peyron and V. C. Hascall), Raven Press, New York (1992), pp. 373–392.

150. Sah, R. L., Kim, Y. J., Doong, J. H., Grodzinsky, A. J., Plaas, A. H. K., Sandy, J. D., Biosynthetic response of cartilage explants to dynamic compression, *J. Orthop. Res.* **7**, 619–636 (1989).

151. Salter, D. M., Hughes, D. E., Simpson, R. and Gardner, D. L., Integrin expression by human articular chondrocytes, *Br. J. Rheum.* **31**, 231–244 (1992).

152. Sandow, M. J., Proximal scaphoid costo-osteochondral replacement arthroplasty, *J. Hand Surg. [Br]* **23**, 201–218 (1998).

153. Schinagl, R. M., Gurskis, D., Chen, A. C. and Sah, R. L., Depth-dependent confined compression modulus of full-thickness bovine articular cartilage, *J. Orthop. Res.* **15**, 499–506 (1997).

154. Schinagl, R. M., Kurtis, M. S., Ellis, K. D., Chien, S. and Sah, R. L., Strength of chondrocyte adhesion to articular cartilage: Effect of seeding duration, *J. Orthop. Res.* **17**, 121–129 (1999).

155. Schinagl, R. M., Ting, M. K., Price, J. H. and Sah, R. L., Video microscopy to quantitate the inhomogeneous equilibrium strain within articular cartilage during confined compression, *Ann. Biomed. Eng.* **24**, 500–512 (1996).

156. Schneiderman, R., Keret, D. and Maroudas, A., Effects of mechanical and osmotic pressure on the rate of glycosaminoglycan synthesis in the human adult femoral head cartilage: An *in vitro* study, *J. Orthop. Res.* **4**, 393–408 (1986).

157. Scott, D. L., Shipley, M., Dawson, A., Edwards, S., Symmons, D. P. and Woolf, A. D., The clinical management of rheumatoid arthritis and osteoarthritis: Strategies for improving clinical effectiveness, *Br. J. Rheumatol.* **37**, 546–554 (1998).

158. Scully, S. P., Joyce, M. E., Heydeman, A. and Bolander, M. E., Articular cartilage healing *in vitro*: Modulation by bFGF and TGF-β1. *Trans. Orthop. Res. Soc.* **16**, 385 (1991).

159. Sellers, R. S., Peluso, D. and Morris, E. A., The effect of recombinant human bone morphogenetic protein-2 (rhBMP-2) on the healing of full-thickness defects of articular cartilage, *J. Bone Joint Surg.* **A79**, 1452–1463 (1997).

160. Setton, L. A., Zhu, W. and Mow. V. C., The biphasic poroviscoelastic behavior of articular cartilage: Role of the surface zone in governing the compressive behavior, *J. Biomech.* **26**, 581–592 (1993).

161. Shapiro, F., Brickley-Parsons, D. and Glimcher, M. J., Biosynthesis of collagen crosslinks in rabbit articular cartilage *in vivo*, *Arch. Biochem. Biophys.* **198**, 205–211 (1979).

162. Shapiro, F. and Glimcher, M. J., Induction of osteoarthrosis in the rabbit knee joint, Histologic changes following menisectomy and meniscal lesions, *Clin. Orthop.* **147**, 287–295 (1980).

163. Shapiro, F., Koido, S. and Glimcher, M. J., Cell origin and differentiation in the repair of full-thickness defects of articular cartilage, *J. Bone Joint Surg.* **A75**, 532–553 (1993).

164. Shortkroff, S., Barone, L., Hsu, H. P., Wrenn, C., Gagne, T., Chi, T., Breinan, H., Minas, T., Sledge, C. B., Tubo, R. and Spector, M., Healing of chondral and osteochondral defects in a canine model: The role of cultured chondrocytes in regeneration of articular cartilage, *Biomaterials* **17**, 147–154 (1996).

165. Simon, W. H., Scale effects in animal joints, *Arthritis Rheum.* **14**, 493–502 (1971).

166. Skalak, R. and Fox, C. F., eds., Tissue engineering: Proceedings of a workshop held at Granlibakken, Lake Tahoe, California, February 26–29 (1988). UCLA symposia on molecular and cellular biology, Liss, New York **107**, 343 (1988).

167. Sommarin, Y., Larsson, T. and Heinegard, D., Chondrocyte-matrix interactions: Attachment to proteins isolated from cartilage, *Exp. Cell. Res.* **184**, 181–92 (1989).

168. Stevens, K. K., Statics and strength of materials Englewood Cliffs, Prentice-Hall (1987), p. 572.

169. Thomsen, I., High-tech body shop. Doctors are now using cartilage cultured in labs to repair injured knees *Sports, Illustrated* **90**, 18–29 (1999).

170. Torok, A., Physical properties of India ink, Director, Research and Development, Sanford Faber Corporation, Newark, New Jersey (1995).

171. Torzilli, P. A., Measurement of the compressive properties of thin cartilage slices: Evaluating tissue inhomogeneity, *Methods in Cartilage Research* (eds. A. Maroudas and K. Kuettner), Academic Press, London (1990), pp. 304–308.

172. Urban, J. P., The chondrocyte: A cell under pressure, *Br. J. Rheum.* **33**, 901–918 (1994).

173. Urban, J. P. G., Bayliss MT: Regulation of proteoglycan synthesis rate in cartilage *in vitro*: Influence of extracellular ionic composition, *Biochim. Biophys. Acta.* **992**, 59–65 (1989).

174. Usami, S., Chen, H.-H., Zhao, Y., Chien, S. and Skalak, R., Design and construction of a linear shear stress flow chamber, *Ann. Biomed. Eng.* **21**, 77–83 (1993).

175. Vincent, J. F. V., ed., Biomechanics: Materials. The practical approach series (ed. D. Rickwood and B. D. Hames), IRL Press, New York (1992), p. 247.

176. Vunjak-Novakovic, G., Martin, I., Obradovic, B., Treppo, S., Grodzinsky, A. J., Langer, R. and Freed, L. E., Bioreactor cultivation conditions modulate the composition and mechanical properties of tissue-engineered cartilage, *J. Orthop. Res.* **17**, 130–149 (1999).

177. Wakitani, S., Goto, T., Pineda, S. J., Young, R. G., Mansour, J. M., Caplan, A. I. and Goldberg, V. M., Mesenchymal cell-based repair of large, full-thickness defects of articular cartilage, *J. Bone Joint Surg.* **A76**, 579–592 (1994).

178. Wakitani, S., Kimura, T., Hirooka, A., Ochi, T., Yoneda, M., Yasui, N., Owaki. H. and Ono. K., Repair of rabbit articular surfaces with allograft chondrocytes embedded in collagen gel, *J. Bone Joint Surg.* **B71**, 74–80 (1989).

179. Waugh, W., John Charnley: The man and the hip, Springer-Verlag, London (1990), p. 268.

180. Woods, V. L., Schreck, P. J., Gesink, D. S., Pacheco, H. O., Amiel, D., Akeson, W. H. and Lotz, M., Integrin expression by human articular chondrocytes, *Arthritis Rheum.* **37**, 537–544 (1994).

181. Wu, Q. and Chen, Q., Mechanical stretch of extracellular matrix stimulates chondrocyte proliferation/differentiation process in a novel 3-dimensional culture system, *Trans. Orthop. Res. Soc.* **24**, 648 (1999).

182. Yin, F. C., Tompkins, W. R., Peterson, K. L. and Intaglietta, M., A video-dimension analyzer, *IEEE Trans. Biomed. Eng.* **19**, 376–381 (1972).

Journal of Orthopaedic Research
13:751-760 The Journal of Bone and Joint Surgery, Inc.
© 1995 Orthopaedic Research Society

Integrative Repair of Articular Cartilage *In Vitro:* Adhesive Strength of the Interface Region

Eric S. Reindel, Annamarie M. Ayroso, Albert C. Chen, Denise M. Chun, Robert M. Schinagl, and Robert L. Sah

Department of Bioengineering and Institute for Biomedical Engineering, University of California, San Diego, La Jolla, California, U.S.A.

Summary: The objective of this study was to quantify the strength of the repair tissue that forms at the interface between pairs of cartilage explants maintained in apposition in an *in vitro* culture system. Articular cartilage explants were harvested from calves and from adult bovine animals, dissected into uniform blocks, and incubated in pairs within a chamber that maintained a 4×5 mm area of tissue overlap. Following 1-3 weeks of incubation, integrative repair was assessed by testing samples in a tensile single-lap configuration to estimate adhesive strength. After incubation in medium containing 20% fetal bovine serum, the adhesive strength between pairs of calf cartilage blocks and pairs of adult bovine cartilage blocks increased at a rate of 7.0 and 10.5 kPa/week, respectively. This repair process appeared to be dependent on viable cells, since lyophilization of adult bovine cartilage before incubation completely inhibited the development of an interface with a measurable adhesive strength. The repair process was dependent on serum components in the medium. Incubation of sample pairs for 3 weeks in medium supplemented with 20% fetal bovine serum resulted in a relatively high proteoglycan content as well as a relatively high adhesive strength (34 kPa), whereas incubation in basal medium with or without 0.1% bovine serum albumin resulted in a 54-70% lower proteoglycan content and a 65-88% lower adhesive strength. Samples incubated for 3 weeks with serum also had a 20% higher DNA content than samples maintained in basal medium. Histological analysis indicated some cell division at the free surfaces of the explant and also occasional cells within the interface region between explants.

A variety of orthopaedic pathologies and treatments result in mechanical defects in the articular cartilage (7,8,22). The successful repair of these defects typically requires the integration of regions of tissue that are in physical apposition. For example, integrative cartilage repair is required (a) at the cartilage-cartilage interface in intra-articular chondral and osteochondral fractures, (b) at the repair tissue-cartilage interface following subchondral abrasion, (c) at the host tissue-donor tissue interface after implantation of osteochondral allografts or scaffolds composed of natural or synthetic materials, cells (chondrocytes or chondroprogenitor cells), or tissues (perichondrium or periosteum), and (d) in the fissures of osteoarthritic cartilage. The difficulty of achieving complete cartilage repair in such interface regions is evident from *in vivo* studies of cartilage lacerations (23), intra-articular fractures (24), and full-thickness cartilage defects (32). Such integrative repair may be critical to the long-term success of procedures to replace or stimulate the resurfacing of damaged articular cartilage.

The biological processes and physical phenomena that modulate the repair process in cartilage are difficult to assess *in vivo*. Cartilage injuries that penetrate the subchondral bone allow access of blood and marrow components to the defect site, whereas lacerations that are confined to the articular cartilage do not elicit a vascular response. The complex loading patterns in synovial joints and the resultant static and dynamic physical phenomena make it difficult to control the physical milieu and to assess the regulatory role of specific physical forces in the repair process. In addition, the biomechanical properties of fibrocartilaginous tissue at the site of integrative repair are difficult to determine due to the three-dimensional joint geometry and the proximity of subchondral bone.

Explant cultures of cartilage have proven useful for examining the biological and physical regulation of cartilage metabolism *in vitro*. Incubation of cartilage explants in medium containing fetal bovine serum stimulates the chondrocytes to increase synthesis and decrease degradation of matrix proteoglycan (9,16).

Received March 31, 1994; accepted December 1, 1994.
Address correspondence and reprint requests to R. L. Sah at Department of Bioengineering, 9500 Gilman Drive, Mail Code 0412, University of California, San Diego, La Jolla, CA 92093-0412, U.S.A.

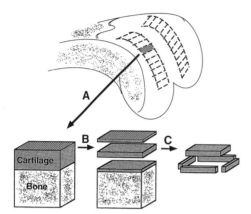

FIG. 1. Explant of bovine cartilage. A: Osteochondral fragments were removed from the distal femur and patella. B: The superficial articular cartilage was removed with use of a microtome and discarded, yielding a flat surface. A cartilage slice 0.5 mm thick was obtained. C: Parallel cuts were made in the slice to form a 9 × 5 × 0.5 mm block of cartilage.

Serum stimulation of adult bovine cartilage results in the maintenance of steady-state levels of matrix proteoglycan (6,16,30), whereas serum stimulation of cartilage from newborn calves results in a net deposition of matrix proteoglycan (29,30). Explant cultures of defined geometry have been used to assess the regulation of matrix metabolism by specific physical factors, such as hydrostatic pressure, cell deformation, fluid flow, and streaming potentials, that are predicted to occur during joint loading (15,20,28).

Few previous studies have utilized cartilage explants *in vitro* to study the integrative repair of cartilage. Scully et al. examined the healing of experimental lacerations in cartilage explants during subsequent culture by histological analysis at the light microscopic level (31). Over the first few weeks of incubation in medium including serum, an acellular matrix filled the defect. At 3-6 weeks, cells began to populate the newly formed matrix. The histological evidence for integrative repair was dependent on the medium being supplemented with serum or a combination of basic fibroblast growth factor and transforming growth factor-β1. These results are consistent with previous studies (21) that have noted the outgrowth of matrix and cells on the surfaces of cartilage explants during incubation in medium with serum. These studies were not designed to allow mechanical characterization of the repair tissue; however, they suggest that cellular repair tissue would form at the interface region between a pair of cartilage explants grown in apposition in the presence of serum.

Biomechanical characterization of a thin layer of repair tissue within a large region of surrounding tissue requires specialized experimental and theoretical methods. Cartilage tissue that is relatively homoge-

neous can be tested and analyzed to assess, for example, compressive or tensile modulus (27). However, it would be difficult to identify and physically isolate the repair tissue forming at cartilage interfaces to perform such tests. Analogous challenges arise in characterizing adhesives that are used for bonding nonbiological load-bearing joints (e.g., cements for wooden or metal sheets) (2). A standard test method for estimating the strength of structural adhesives, based on the theoretical analysis of Goland and Reissner (13), involves the tensile testing of a single-lap joint, in which two overlapping sheets (adherends) are joined together in the overlap area by an adhesive (3).

The objective of the present study was to use a single-lap joint configuration to quantify the strength of the repair tissue that forms at the interface between pairs of cartilage explants in an *in vitro* culture system. The results show that incubation of such samples in partial apposition can result in a measurable increase in the adhesive strength of the interface between explants during culture. This increase appears dependent on viable chondrocytes within the cartilage as well as on serum factors in the medium.

METHODS

Materials

Materials for tissue culture and analytical procedures were obtained as described previously (28,30). In addition, crystalline bovine serum albumin, methylene blue chloride, and sodium cacodylate were from Sigma Chemical (St. Louis, MO, U.S.A.); defined fetal bovine serum was from Hyclone (Logan, UT, U.S.A.); 2-methylbutane was from ICN Biomedical (Irving, CA, U.S.A.); OCT embedding solution was from Miles (Naperville, IL, U.S.A.); specimen embedding molds, Superfrost slides, Hemo-De, and Permount were from Fisher Scientific (Santa Clara, CA, U.S.A.); and electron microscopy grade glutaraldehyde was from EM Science (Gibbstown, NJ, U.S.A.).

Explant and Culture of Cartilage

Cartilage explants were harvested (Fig. 1) from the patellofemoral groove and patella of calves (A. Arena, Hopkinton, MA, U.S.A.) and adult bovine animals (Cuyamaca Meat, National City, CA, U.S.A., or Talone Packing, Escondido, CA, U.S.A.) obtained 12-72 hours after death. The calves were 1-3 weeks old; the adult bovine animals were considered skeletally mature by the criterion of an absent distal femoral growth plate. During all explant procedures, cartilage was kept moist and free of blood by copious irrigation with phosphate buffered saline supplemented with 100 U/ml of penicillin, 100 µg/ml of streptomycin, and 0.25 µg/ml of amphotericin B. Osteochondral fragments were cut from the femoral and patellar surfaces of the joints with a reciprocating saw (Johnson and Johnson, New Brunswick, NJ, U.S.A.). Each osteochondral fragment was clamped into the sample holder of a vibrating microtome (Vibratome 1000; Technical Products International, St. Louis, MO, U.S.A.). The curved articular surface (approximately 0.1-0.5 mm) was cut and discarded to form a planar surface, and one or two 0.5 mm thick plane parallel slices of cartilage were sectioned. From these slices, cartilage blocks that were 9 mm long × 5 mm wide × 0.5 mm thick were formed by making cuts with razor blades that were aligned in parallel in a cutting jig. In some experiments, the cartilage blocks were lyophilized (Lyph-Lock; Labconco, Kansas City, MO, U.S.A.) overnight to lyse the

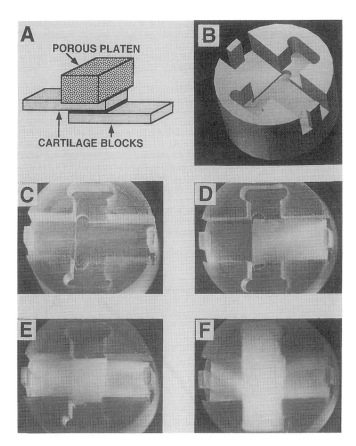

FIG. 2. Culture configuration to evaluate integrative cartilage repair. **A:** Schematic of positioning of cartilage blocks for a 4 × 5 mm area of tissue overlap. A porous polysulfone platen maintained the cartilage blocks in apposition. **B:** CAD/CAM rendition of polysulfone insert for culture plates. **C-F:** Photographs showing insertion of cartilage blocks and overlaying platen into polysulfone insert.

chondrocytes and stop cellular biosynthesis (28) before subsequent incubation. Single knee joints from a total of two calves and eight adult bovine animals were used, and each knee yielded as many as 72 cartilage blocks. As described in the Results section, in each experiment randomly selected cartilage blocks from single knee joints of one to three animals were distributed among the test groups to be compared and the results were pooled.

Pairs of cartilage blocks were placed in culture plate inserts (Fig. 2) that had been fabricated out of polysulfone (Westlake Plastics, Lenni, PA, U.S.A.) with use of a computer-numerically-controlled three-axis milling machine (Bridgeport Machine, Bridgeport, CT, U.S.A., and CNC Software, Tolland, CT, U.S.A.). The inserts were housed within individual wells of 24-well culture plates. Each insert constrained the two cartilage blocks such that 4 × 5 mm areas of tissue were overlapping. The bottom cartilage block protruded 0.1 mm above the support for the top cartilage block, ensuring that the top and bottom blocks were in contact in the overlap region. A porous polysulfone platen with a pore size of 120 μm (Porex, Fairburn, GA, U.S.A.) was applied to the overlap region to provide a small compressive stress (0.1 kPa) that secured the cartilage pairs in apposition. The polysulfone inserts and platens were sterilized by autoclaving before each experiment.

Cartilage pairs were incubated in a humidified 5% CO_2-95% air incubator at 37°C. The basal incubation medium was Dulbecco's modified Eagle medium, 10 mM HEPES, 0.1 mM nonessential amino acids, 0.4 mM proline, 20 μg/ml ascorbate, and antibiotics/antimycotic. In some experiments, the medium was supplemented with 0.1% bovine serum albumin or 20% fetal bovine

serum for the duration of the culture. The medium was changed daily, and the volume of medium per cartilage sample pair was 1 ml to achieve a ratio of tissue to medium (vol/vol) of approximately 1:20. Cartilage samples were maintained for 1, 2, or 3 weeks

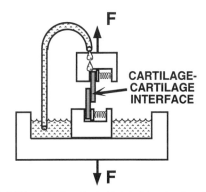

FIG. 3. Schematic of mechanical test to determine the adhesive strength of the interface after integrative cartilage repair. The cartilage sample pair was secured in spring-loaded clamps attached to a mechanical test instrument. A positive displacement was applied and the resultant force, F, was measured. During the test, phosphate buffered saline was recirculated onto the sample pair to prevent dehydration.

TABLE 1. *Structural properties of cartilage sample pairs during testing to failure after 3 weeks of culture (mean ± SD)*

Bovine cartilage source	Culture medium supplement	No.	Displacement at failure (mm)	Structural stiffness (N/mm)
Adult	0.1% BSA	11	0.34 ± 0.21	0.92 ± 0.98
Adult	20% FBS	30	0.99 ± 0.98	1.10 ± 0.89
Calf	20% FBS	7	0.49 ± 0.16	1.40 ± 0.77

BSA = bovine serum albumin and FBS = fetal bovine serum.

before either (a) biomechanical testing followed by biochemical analysis or (b) histological analysis.

Biomechanical Testing

To determine the adhesive strength of the interface region, pairs of cartilage explants were tested using a modification of American Society for Testing and Materials (ASTM) Standard D 3983 (3). Each sample pair was removed carefully from the insert and secured in stainless-steel spring-loaded clamps (Fig. 3) that were attached to the jaws of a uniaxial mechanical spectrometer (Dynastat; Instruments for the Materials and Structural Sciences, Accord, MA, U.S.A.). The jaws were set at an initial clamp-to-clamp separation of 8.0 mm. The clamps were lined with 400 grit wet and dry sandpaper. Compression springs (Lee Spring, Issaguah, WA, U.S.A.) were incorporated into the clamps to provide a normal force of 11 N over the gripped specimen area of 3 × 5 mm (equivalent to a clamping stress of 0.7 MPa).

Samples were tested to failure by application of a uniaxial positive displacement at a rate of 0.5 mm/min, while the resultant load was measured at a sampling rate of 1 Hz (or 0.1 Hz in the earlier studies). Experimental control and data acquisition were implemented by interfacing the Dynastat to a Macintosh IIci computer through a NB-MIO-16XH-42 board controlled by LABVIEW software (both from National Instruments, Austin, TX, U.S.A.). The precision of the applied displacement and measured load was 3 μm and less than 0.01 N (or approximately 0.03 N in the earlier studies), respectively. Throughout the test, the samples were maintained at room temperature and irrigated by recirculation of phosphate buffered saline through a peristaltic pump (WIZ; ISCO, Lincoln, NE, U.S.A.). Since measurements of the length and width of the overlap region at the termination of the 1-3 week experiments indicated that the interface area remained within approximately 10% of the assumed initial area, the adhesive strength of the interface region between explant pairs was calculated as the measured ultimate load divided by the original overlap area (20 mm²). The displacement at which the ultimate load occurred was noted. Since the load-displacement relationship was relatively linear until the ultimate load (Fig. 4), the structural stiffness was calculated as the ultimate load normalized to the corresponding displacement. Samples that slipped out of a clamp rather than failing at the interface region (less than 2% of the sample pairs tested) were reinserted with an approximately 25% greater clamping force and retested. Samples that failed to adhere to each other were assigned an adhesive strength of zero.

Biochemical Analysis

After mechanical testing, the separated halves of the sample pairs were combined and were digested with 0.5 ml of a proteinase K solution (0.5 mg/ml in 0.1 *M* sodium phosphate, 10 m*M* Na₂-EDTA, pH 6.5) at 60°C for 16 hours. The digest was diluted with sterile water, and portions were analyzed for DNA by reaction with Hoechst 33258 (20) by use of a spectrofluorometer (model

F-2000; Hitachi, Weston, MA, U.S.A.) and for sulfated proteoglycan by reaction with dimethylmethylene blue (12) as modified (30) for a spectrophotometric microplate reader fitted with a filter having a center wavelength of 525 nm (EMAX; Molecular Devices, Menlo Park, CA, U.S.A.).

Histological Analysis

Explant pairs from two of the adult animals were incubated for 3 weeks and used for histological assessment (rather than mechanical testing) with use of a modification of a previously described method for the cationic dye, methylene blue (17). Sample pairs were removed from the polysulfone inserts, placed in specimen molds, frozen in OCT embedding solution by immersion in a bath of 2-methylbutane cooled by liquid nitrogen (–159°C), and cryosectioned to 8-10 μm (model 2800 Frigocut E; Cambridge Instruments and Reichert-Jung, Buffalo, NY, U.S.A.). During sectioning, the sample pair was oriented so that the long axis was in the direction of cutting to minimize the compression of one explant onto the other. The cryosection then was adhered to a slide, fixed, stained for 10 minutes with 2% glutaraldehyde, 0.2% methylene blue chloride, 0.115 *M* NaCl, and 0.05 *M* cacodylate at pH 7.4, rinsed with 0.1 *M* sodium acetate at pH 4.0 followed by distilled

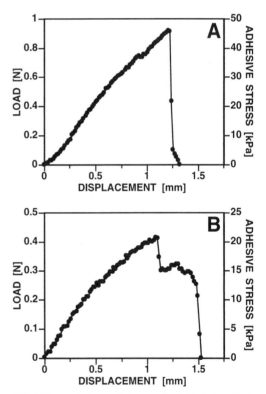

FIG. 4. Representative results from mechanical tests to evaluate integrative repair of adult bovine cartilage sample pairs incubated for 3 weeks in medium supplemented with 20% fetal bovine serum. With use of the test apparatus shown in Fig. 3, the load was measured while a ramp elongation in clamp-to-clamp displacement was applied. The adhesive stress was calculated as the load normalized to the interface area, with the peak taken as the adhesive strength. A and B are examples of two typical patterns of load displacement.

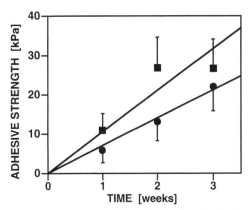

FIG. 5. Integrative repair with increasing duration of incubation in medium supplemented with 20% fetal bovine serum. Calf (●) and adult (■) cartilage pairs were incubated for 1-3 weeks and then tested to determine the adhesive strength of the interface. Linear regression fits through the origin were 7.0 kPa/week (r = 0.40, n = 31, p < 0.001) for calf cartilage pairs and 10.5 kPa/week (r = 0.29, n = 32, p < 0.001) for adult cartilage pairs. Values are given as the mean ± SEM (n = 10-12 at each point).

water, dehydrated through graded ethanols, cleared with Hemo-De, and mounted with Permount. With use of this procedure, the extracellular staining by methylene blue appeared relatively specific for anionic matrix proteoglycan, since control samples that were partially depleted of proteoglycan by trypsin digestion (as determined by biochemical analysis) showed (a) a circumferential region that did not stain and (b) an increase in the area of the nonstained region with increasing digestion and depletion of proteoglycan (data not shown).

Statistical Analysis

Sample quantities are expressed as the mean ± SD, except where indicated. Differences between the means of sample groups were assessed by analysis of variance, and when significant variations were detected, differences between group means were determined by Tukey's *post hoc* test with Systat 5.2 (Systat, Evanston, IL, U.S.A.).

RESULTS

Biomechanical tests were performed to determine the adhesive strength of the interface forming between pairs of cartilage explants maintained *in vitro*. The relationship between clamp-to-clamp displacement and tensile load showed several typical features. Load increased monotonically with displacement until the maximum (ultimate) load. In some cases, this was followed by an abrupt decrease in load to zero as the specimens failed and visibly separated (Fig. 4A). In other cases, this was followed by a drop in load to a nonzero plateau before a subsequent drop in load to zero (Fig. 4B). In all cases, fracture appeared to occur at the interface region between the cartilage explants, and the separated halves of cartilage were grossly similar to the original blocks. In cases such as that in Fig. 4B, fracture appeared to occur initially at one edge of the interface and subsequently propagate

throughout the remainder of the interface. For samples from calves and adults having a measurable adhesive strength after incubation for 3 weeks (Table 1), the average displacement at which the ultimate load occurred was 0.34-0.99 mm, which was 4.3-12.3% of the initial clamp-to-clamp length of 8.0 mm. The corresponding stiffness of the test structures averaged 0.92-1.40 Nmm.

The dependence of the adhesive strength between cartilage samples on the duration of culture in medium supplemented with 20% fetal bovine serum was evaluated in pairs of calf cartilage and pairs of adult bovine cartilage. The results from experiments on samples from two calf joints were pooled, as were the results from two adult bovine joints. Linear regression analysis (Fig. 5) showed that the adhesive strength increased significantly with the duration of culture for both the calf cartilage pairs (7.0 kPa/week, p < 0.001) and the adult cartilage pairs (10.5 kPa/week, p < 0.001).

The relationship between serum components in the medium and the increase in adhesive strength during culture was assessed. The adhesive strength resulting after incubation of cartilage pairs for 3 weeks in (a) basal medium, (b) medium supplemented with 0.1% bovine serum albumin, or (c) medium supplemented with 20% fetal bovine serum were compared. In this experiment, samples from one adult bovine animal were evenly distributed among the three test groups (n = 5 per group), whereas cartilage from two additional adult animals was evenly distributed between groups 1 and 3 (n = 12 per group) and groups 2 and 3

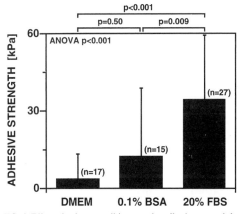

FIG. 6. Effect of culture conditions on the adhesive strength between adult bovine cartilage samples after 3 weeks of culture. Sample pairs (see text for detail) were incubated in basal medium (Dulbecco's modified Eagle medium [DMEM]), medium supplemented with 0.1% bovine serum albumin (BSA), or medium supplemented with 20% fetal bovine serum (FBS). The significance of differences between sample means (±SD) are indicated. ANOVA = analysis of variance.

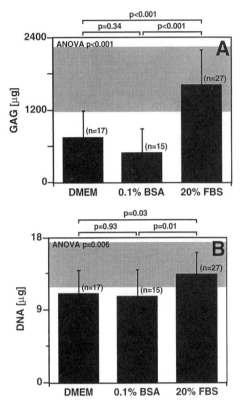

FIG. 7. Effect of culture conditions on (**A**) glycosaminoglycan (GAG) and (**B**) DNA content of adult bovine cartilage sample pairs after 3 weeks of culture. Sample pairs were incubated in basal medium (Dulbecco's modified Eagle medium [DMEM]), medium supplemented with 0.1% bovine serum albumin (BSA), or medium supplemented with 20% fetal bovine serum (FBS). The significance of differences between sample means (±SD) are indicated. The shaded areas show the normal range (mean ± SD) of values from six to eight adult bovine animals. ANOVA = analysis of variance.

(n = 10 per group). The adhesive strength (Fig. 6) was affected by the culture conditions (p < 0.001, analysis of variance). Incubation of sample pairs in medium with 20% fetal bovine serum resulted in a relatively high adhesive strength (34 ± 26 kPa, n = 27), while incubation in basal medium or medium supplemented with 0.1% bovine serum albumin resulted in 88 and 65% lower adhesive strength, respectively (4 ± 10 kPa, n = 17, p < 0.001 and 12 ± 26 kPa, n = 15, p < 0.01). There was no detectable difference in adhesive strength between samples incubated in basal medium or medium with 0.1% bovine serum albumin (p = 0.50). The adhesive strength resulting from incubation with 20% fetal bovine serum in these experiments was consistent with the rate of increase in adhesive strength determined above (Fig. 5).

The modulation of integrative repair by serum components in the medium during the 3 weeks of incu-

bation was accompanied by marked effects on proteoglycan metabolism (Fig. 7A) and slight changes in cell proliferation (Fig. 7B). Sample pairs that were incubated in medium with 20% fetal bovine serum for 3 weeks maintained tissue proteoglycan and DNA at levels comparable with that of adult bovine cartilage terminated within 1 day after explant and normalized to the same volume as the sample pairs (i.e., 93% of 1,724 ± 540 μg proteoglycan and 92% of 14.6 ± 2.8 μg DNA; both, mean ± SD of the average values from eight and six animals, respectively). The proteoglycan content of cartilage explant pairs incubated with basal medium or medium with 0.1% bovine serum albumin was decreased by 54 or 70%, respectively (both, p < 0.001), compared with that of cartilage incubated in medium with 20% fetal bovine serum. The DNA content of the explants incubated with basal medium or medium with 0.1% bovine serum albumin was decreased by 19% (p = 0.03) or 21% (p = 0.01), respectively, compared with that of cartilage incubated in medium with 20% fetal bovine serum.

To distinguish between an integrative repair mechanism that was dependent on active cellular metabolism and one that may be an independent chemical or physical reaction in the matrix, two groups of samples of adult bovine cartilage were lyophilized to lyse the chondrocytes prior to culture. After 3 weeks of subsequent incubation in medium supplemented with 0.1% bovine serum albumin or 20% fetal bovine serum, none of the samples showed any functional integrative repair, with an adhesive strength of zero in all cases (n = 6 and n = 5, respectively). Biochemical analysis of these samples indicated a drop in the DNA content to 20% that of cartilage samples that were terminated immediately after explant (data not shown), consistent with lyophilization-induced cell death and autolysis.

Histological sections of adult bovine cartilage pairs incubated for 2-3 weeks in medium with 20% fetal bovine serum showed nucleated cells throughout the tissue and methylene blue staining throughout the matrix (Fig. 8). A matrix staining lightly positive for methylene blue was present at the boundary between cartilage explants (Fig. 8A and B). In sections of sample pairs from one of the two adult animals analyzed histologically, cells were discernible in this interface region (Fig. 8B). At the free cartilage surfaces, cells appeared to erupt and form small protruding nodules (Fig. 8C).

DISCUSSION

The results described here demonstrate that (a) a process of integrative repair occurs at the interface between two cartilage explants incubated *in vitro*, and (b) the repair process is dependent on active cellular metabolism. Incubation of pairs of calf cartilage

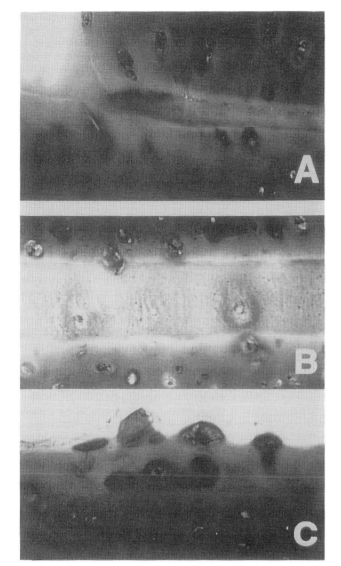

FIG. 8. Histological sections of adult bovine cartilage sample pairs after incubation for 3 weeks in partial apposition in medium with 20% fetal bovine serum. Samples were processed and stained with the cationic dye, methylene blue, as described in the text. (**A**) Edge and (**B**) central area of the interface region of integrative repair between cartilage sample pairs. (**C**) Free cartilage surface.

blocks and pairs of adult bovine cartilage blocks in medium supplemented with 20% serum resulted in an increase in the adhesive strength between sample pairs over a 3-week period (Fig. 5). This integrative repair process appeared dependent on viable cells, since the increase in adhesive strength was nullified by lyophilization-induced lysis of chondrocytes. In addition, the increase in adhesive strength was accompanied by the ability of serum factors to stimulate the chondrocytes to synthesize and deposit proteoglycan (Figs. 6 and 7).

There are a number of possible approaches, each with advantages and disadvantages, for characterizing

the biomechanical properties of the repair tissue that forms in the region between cartilage explants during *in vitro* incubation. The approach taken here was to assess the strength as the maximum stress that is borne by the repair region. The configuration chosen for this study, involving the formation of a single-lap joint by a pair of cartilage explants, allowed the estimation of adhesive strength as the maximum force normalized to the overlap area. This is the most common configuration for testing adhesives (2) and was practical for implementation with articular cartilage blocks as the adherends. However, as described in detail below, the exact state of stress in the interface region is difficult to ascertain for this geometry. Thus, it is difficult to compare our estimates of the strength of the repair tissue directly with that of normal cartilage. Nevertheless, the method of culture and mechanical testing used here may be particularly useful for early stages of integrative cartilage repair where the interface region is relatively weak.

An alternative configuration to assess adhesive strength involves the application of torsional forces to adherent cylindrical specimens that form a butt joint or annular ("napkin-ring") specimens (4). The distribution of stress in idealized geometries is simple; however, nonuniformities in the properties or geometry of the repair tissue would complicate the actual stress distribution (2). In addition, the gripping of specimens in order to apply the necessary torsional forces may require axial compression of the specimens (36) and, thus, compression of the interface region.

Another approach is to assess the energy dissipated at a fracture tip during propagation (34). Peel (11) and tear (10) tests provide an estimate of this fracture energy; the latter tests have been applied to normal cartilage. Such techniques may allow characterization of tissue during all the stages of a repair process that progresses to normal cartilage and thus permit direct comparison of repair tissue and normal cartilage. However, analysis of such experiments requires differentiation of dissipation of energy at the fracture tip from the storage and dissipation of energy elsewhere in the test structure. The amount of elastic energy stored and energy dissipated through solid-fluid interactions that would be expected to occur within the bulk of the tissue matrix away from the fracture tip (14,26) have yet to be estimated in these tests. Indeed, the energy required for crack propagation appears to be inversely related to the proteoglycan content of the tissue (10).

In the single-lap test, the degree to which fluid pressure gradients arise, induce flow relative to the solid matrix, and contribute to the total tissue stress can be estimated from the geometry of the test specimen and the poroelastic (biphasic) model of cartilage mechanics (26). In the test configuration, the characteristic length, a, is half of the thickness of the test structure; i.e., a = 0.25-0.50 mm in the nonoverlap and overlap regions. The displacement rate of 0.5 mm/min, relative to the sample length (jaw-to-jaw distance) of 8 mm, corresponds to a strain rate $d\varepsilon/dt = 0.001$ s^{-1}. The poroelastic behavior of cartilage in the tensile lap test is analogous to that in the unconfined compression test, in which the contribution of interstitial fluid pressurization to load support is related to the dimensionless quantity, $\tau \cdot d\varepsilon/dt$, where $\tau = a^2/(H_A \cdot k_p)$ is the gel diffusion time, H_A is the aggregate modulus of the elastic solid matrix, and k_p is the hydraulic permeability (5). With use of typical cartilage properties (27) of $H_A = 0.5$-5 MPa (in compression and tension), $k_p = 2 \cdot 10^{-15}$ m^2/(Pa·s) and the previously given parameters, $\tau \cdot d\varepsilon/dt = 0.006$-0.3. Since this quantity is somewhat smaller than unity, the strain rate is slow enough to ensure that equilibration of fluid essentially is achieved throughout the test and that the stress distribution within the cartilage is governed by the elastic properties of the solid matrix (5,26).

If cartilage is assumed to behave in this elastic limit, the stress distribution in the single-lap joint test configuration could be estimated analytically or computationally with additional assumptions, according to methods employed for other adherends and adhesives. Linear elastic analysis of the test configuration has revealed that tensile, compressive, and shear stresses would exist within the cartilage sample pair as well as the repair tissue at the interface (2). A common analytical result, if a square edge of repair tissue is assumed (Fig. 2A), is that the normal and shear stresses vary across the length of the interface, with maxima occurring near the edges of the repair tissue along the width of the specimen. However, since the forces applied to the two cartilage adherends are not colinear, both an in-plane tension and a bending moment are applied to the joint (13). This moment complicates the estimate of strength, since the shear and normal stress are not simply proportional to the applied load. In addition, the actual geometry of the edge along the width of the specimen (Fig. 8B) may not be square. Such a geometry may significantly affect the distribution of stress, precisely in the region where a maximum in stress is likely to occur (2). Furthermore, three-dimensional finite element analysis and Moiré interferometric experiments (1,33) indicate that variations in stress exist along the width of the adhesive, with peak shear stresses at the corners and peak tensile stresses in the central region of the edges. Indeed, the pattern of stress concentration near the edges of the repair tissue would be consistent with the pattern of fracture in some samples (Fig. 4B) where the fracture occurred initially at one edge along the width of the interface and subsequently propagated throughout the interface.

Despite these complexities in defining the exact stress distribution, the measured adhesive strength averaged less than 40 kPa (Fig. 6) under the most favorable conditions and was several orders of magnitude less than the tensile strength (4-40 MPa) of normal cartilage (35). Thus, the quality of the repair tissue at the interface is likely to be the predominant factor influencing the measured adhesive strength. Nevertheless, the reported values for adhesive strength, calculated as the force normalized to the original overlap area, should be taken only as an estimate of the actual strength of the interface region.

The marked variability in the measured adhesive

strength within an individual experimental group may have arisen for a variety of reasons. The precise geometry of the repair tissue could affect the stress distribution, as described above, and thus the measured adhesive strength. Also, variability in biological or physical properties that affect repair could be substantial, since the cartilage blocks were relatively large (e.g., compared with the 3 mm diameter disks we used in previous explant studies [30]) and were selected randomly from locations in the patellofemoral groove. In addition, the experimental results that were pooled from three animals (Fig. 6) did not involve equal sampling for all test conditions. However, analysis of the data (adhesive strength and proteoglycan and DNA) contents in which basal medium was compared with medium containing 20% fetal bovine serum and medium containing 0.1% bovine serum albumin was compared with medium with 20% fetal bovine serum in individual animals yielded similar statistical trends by unpaired t test (data not shown).

The use of this test geometry allowed assessment of integrative repair between sections of cartilage that were approximately parallel to the articular surface and devoid of the superficial layer of cartilage. Such repair may be more relevant to fissures that extend horizontally in fractured or osteoarthritic articular cartilage than to fissures that extend vertically, since the orientation of the collagen fibers adjacent to the fissure may modulate the repair process. Nevertheless, the cellular and molecular mechanisms involved in such an integrative repair process may be similar whether or not integrative repair is required to occur across a vertical or horizontal plane within articular cartilage.

The measured increase in adhesive strength between cartilage explants during incubation in medium with 20% serum is consistent with several mechanisms of tissue repair. In general, tissue repair requires cell adhesion, migration, proliferation, and differentiation as well as cell-mediated deposition and remodeling of extracellular matrix (7,8,22). The requirements for each of these events and their coordinated occurrence may be critical for the formation of normally functioning cartilage with its characteristic biomechanical properties. The time course of matrix deposition in a laceration gap between cartilage surface that has been observed previously (31) is consistent with our histological analysis (Fig. 8) and the increased adhesive strength over time in culture (Fig. 5). The migration or proliferation of cells on surfaces of, or in gaps within, the cartilage blocks also may have contributed to the repair process. Our biochemical analysis (Fig. 7B) indicated only a slight proliferation of chondrocytes, even when samples were maintained in medium supplemented with serum; however, histological analysis (Fig. 8C) did reveal that some chondro-

cytes were present in the interface region. Indeed, other studies have shown that cells can attach to and grow out from explants of articular cartilage (21) and that chondrocytes populate an *in vitro* laceration site after 3-6 weeks of incubation (31). However, the magnitude of cellular response may depend on specific interactions between the cells and extracellular surfaces. The absence of a more marked cellular proliferative response in the present study may be due to the inability of cells to attach to the contacting polysulfone surfaces during incubation or to differences in components of the medium that may mediate chondrocyte attachment.

The cellular and molecular mechanisms that result in integrative repair remain to be elucidated. It is interesting that serum not only stimulates chondrocytes to synthesize proteoglycan (16) but also induces an integrative repair process. Whether the regulation of proteoglycan molecules and the regulation of structural or enzymatic molecules involved in the repair process are coordinated remains to be determined. For example, insulin-like growth factor-1 is the component within serum that is predominantly responsible for stimulating chondrocytes to synthesize proteoglycan (21). However, whether growth factors that regulate chondrocyte metabolism of proteoglycan also regulate integrative repair processes is not yet known. It was somewhat surprising that the adhesive strength of the interface forming between pairs of newborn calf cartilage blocks and pairs of adult bovine cartilage blocks were not markedly different (Fig. 5). Although experiments were performed only on cartilage samples from two calves and definitive conclusions on the relative repair rates in calf compared with adult bovine cartilage would require additional population sampling, the results suggest that even the high density of cells within newborn calf cartilage may not be sufficient to accelerate a repair response. Indeed, preliminary reports suggest that specific matrix molecules may inhibit cell adhesion and migration on cartilage surfaces (18,25).

Acknowledgment: This work was supported by a postdoctoral fellowship and a Hulda Irene Duggan Investigator Award from the Arthritis Foundation (R.L.S.), National Institutes of Health Grant 5-T32-HL07089 (R.M.S.), a National Science Foundation Young Investigator Award (R.L.S.), the UCSD Charles Lee Powell Foundation (R.L.S.), the UCSD Academic Senate (R.L.S.), the UCSD Undergraduate Research Experience and Undergraduate Scholastics Grants Programs (A.M.A.), and the UCSD Department of Applied Mechanics and Engineering Sciences. We thank Dr. A. J. Grodzinsky and Dr. S. B. Trippel for critical discussions, Steve Porter and Gary Foreman for crafting the chamber inserts, and Linda Kitabayashi for assistance in the histological analysis.

REFERENCES

1. Adams RD, Peppiatt NA: Effects of Poisson's ratio strains in adherend on stresses of an idealized lap joint. *J Strain Analysis* 8:134-139, 1973

2. Adams RD, Wake WC: *Structural Adhesive Joints in Engineering.* New York, Elsevier Applied Science, 1984
3. American Society for Testing and Materials: Standard test method for measuring strength and shear modulus of nonrigid adhesives by the thick-adherend tensile-lap specimen. In: *Annual Book of ASTM Standards,* pp 312-321. Philadelphia, American Technical Publishers, 1992
4. American Society for Testing and Materials: Standard test method for shear strength and shear modulus of structural adhesives. In: *Annual Book of ASTM Standards,* pp 441-445. Philadelphia, American Technical Publishers, 1992
5. Armstrong CG, Lai WM, Mow VC: An analysis of the unconfined compression of articular cartilage. *J Biomech Eng* 106:165-173, 1984
6. Barone-Varelas J, Schnitzer TJ, Meng Q, Otten L, Thonar E-J: Age-related differences in the metabolism of proteoglycans in bovine articular cartilage explants maintained in the presence of insulin-like growth factor I. *Connect Tissue Res* 26:101-120, 1991
7. Buckwalter JA, Rosenberg LC, Coutts R, Hunziker E, Reddi AH, Mow V: Articular cartilage: injury and repair. In: *The American Academy of Orthopaedic Surgeons: Symposium on Injury and Repair of the Musculoskeletal Soft Tissues,* pp 465-482. Ed by SL-Y Woo and JA Buckwalter. Park Ridge, Illinois, American Academy of Orthopaedic Surgeons, 1988
8. Buckwalter JA, Mow VC: Cartilage repair in osteoarthritis. In: *Osteoarthritis: Diagnosis and Medical/Surgical Management,* 2nd ed, pp 71-107. Ed by RW Moskowitz, DS Howell, VM Goldberg, and HJ Mankin. Philadelphia, W.B. Saunders, 1992
9. Campbell MA, Handley CJ, Hascall VC, Campbell RA, Lowther DA: Turnover of proteoglycans in cultures of bovine articular cartilage. *Arch Biochem Biophys* 234:275-289, 1984
10. Chin-Purcell MV, Lewis JL, Oegema TR, Thompson RC Jr: Effect of collagen and proteoglycan content on fracture toughness of articular cartilage. *Trans Orthop Res Soc* 17:130, 1992
11. Dong C, Mead E, Skalak R, Fung YC, Debes JC, Zapata-Sirvent RL, Andree C, Greenleaf G, Cooper M, Hansbrough JF: Development of a device for measuring adherence of skin grafts to the wound surface. *Ann Biomed Eng* 21:51-55, 1993
12. Farndale RW, Buttle DJ, Barrett AJ: Improved quantitation and discrimination of sulphated glycosaminoglycans by use of dimethylmethylene blue. *Biochim Biophys Acta* 883:173-177, 1986
13. Goland M, Reissner E: The stresses in cemented joints. *J Appl Mech* 11:A17-A27, 1944
14. Grodzinsky AJ: Electromechanical and physicochemical properties of connective tissue. *CRC Crit Rev Bioeng* 9:133-199, 1983
15. Hall AC, Urban JPG, Gehl KA: The effects of hydrostatic pressure on matrix synthesis in articular cartilage. *J Orthop Res* 9:1-10, 1991
16. Hascall VC, Handley CJ, McQuillan DJ, Hascall GK, Robinson HC, Lowther DA: The effect of serum on biosynthesis of proteoglycans by bovine articular cartilage in culture. *Arch Biochem Biophys* 224:206-223, 1983
17. Hunziker EB, Ludi A, Hermann W: Preservation of cartilage matrix proteoglycans using cationic dyes chemically related to ruthenium hexaammine trichloride. *J Histochem Cytochem* 40:909-917, 1992

18. Hunziker EB, Rosenberg LC: Biological basis for repair of superficial articular cartilage lesions. *Trans Orthop Res Soc* 17:231, 1992
19. Kim YJ, Sah RLY, Doong JYH, Grodzinsky AJ: Fluorometric assay of DNA in cartilage explants using Hoechst 33258. *Anal Biochem* 174:168-176, 1988
20. Kim YJ, Sah RL, Grodzinsky AJ, Plaas AHK, Sandy JD: Mechanical regulation of cartilage biosynthetic behavior: physical stimuli. *Arch Biochem Biophys* 311:1-12, 1994
21. Luyten FP, Hascall VC, Nissley SP, Morales TI, Reddi AH: Insulin-like growth factors maintain steady-state metabolism of proteoglycans in bovine articular cartilage explants. *Arch Biochem Biophys* 267:416-425, 1988
22. Mankin HJ: Current concepts review. The response of articular cartilage to mechanical injury *J Bone Joint Surg [Am]* 64:460-466, 1982
23. Meachim G, Roberts C: Repair of the joint surface from subarticular tissue in the rabbit knee. *J Anat* 109:317-327, 1971
24. Mitchell N, Shepard N: Healing of articular cartilage in intra-articular fractures in rabbits. *J Bone Joint Surg [Am]* 62:628-634, 1980
25. Mochizuki Y, Goldberg VM, Caplan AI: Enzymatical digestion for the repair of superficial articular cartilage lesions. *Trans Orthop Res Soc* 18:728, 1993
26. Mow VC, Holmes MH, Lai WM: Fluid transport and mechanical properties of articular cartilage: a review. *J Biomech* 17:377-394, 1984
27. Mow VC, Zhu W, Ratcliffe A: Structure and function of articular cartilage and meniscus. In: *Basic Orthopaedic Biomechanics,* pp 143-198. Ed by VC Mow and WC Hayes. New York, Raven Press, 1991
28. Sah RL-Y, Kim Y-J, Doong J-YH, Grodzinsky AJ, Plaas AHK, Sandy JD: Biosynthetic response of cartilage explants to dynamic compression. *J Orthop Res* 7:619-636, 1989
29. Sah RL, Doong JYH, Grodzinsky AJ, Plaas AHK, Sandy JD: Effects of compression on the loss of newly synthesized proteoglycans and proteins from cartilage explants. *Arch Biochem Biophys* 286:20-29, 1991
30. Sah RL, Chen AC, Grodzinsky AJ, Trippel SB: Differential effects of bFGF and IGF-I on matrix metabolism in calf and adult bovine cartilage explants. *Arch Biochem Biophys* 308:137-147, 1994
31. Scully SP, Joyce ME, Heydeman A, Bolander ME: Articular cartilage healing in vitro: modulation by bFGF and TGF-β1. *Trans Orthop Res Soc* 16:385, 1991
32. Shapiro F, Koide S, Glimcher MJ: Cell origin and differentiation in the repair of full-thickness defects of articular cartilage. *J Bone Joint Surg [Am]* 75:532-553, 1993
33. Tsai MY, Morton J: Three-dimensional deformations in a single-lap joint. *J Strain Analysis* 29:137-145, 1994
34. Williams JG: *Fracture Mechanics of Polymers.* New York, Wiley, 1984
35. Woo SL-Y, Akeson WH, Jemmott GF: Measurements of nonhomogeneous, directional mechanical properties of articular cartilage in tension. *J Biomech* 9:785-791, 1976
36. Zhu W, Mow VC, Koob TJ, Eyre DR: Viscoelastic shear properties of articular cartilage and the effects of glycosidase treatments. *J Orthop Res* 11:771-781, 1993

CHAPTER 8

CELL ACTIVATION IN THE CIRCULATION

GEERT W. SCHMID-SCHÖNBEIN

Department of Bioengineering and Whitaker Institute for Biomedical Engineering,
University of California, San Diego

1. Introduction

As a young and broad discipline, bioengineering offers a rich variety of opportunities to develop and improve techniques to treat disease. In this chapter, I want to draw attention to a related but somewhat different question. What could be the actual cause and mechanisms that may lead to disease in the first place? Fewer bioengineers focus on this question, yet there exists an immense opportunity to contribute to one of the most important issues in life.

Our aspiration is that it may be possible in the 21st century to examine a detailed analysis of the failure of human tissues, perhaps with mathematical precision. This analysis should predict in a quantitative fashion the actual progression of the events that start with early organ dysfunction and eventually lead to catastrophic failure, such as stroke or heart attack. At the moment we do not have such an analysis on hand. Yet the formulation of such an analysis will change the way we think about disease. It will give us a multitude of new suggestions how to reduce the risks and how to anticipate complications and perhaps give us new ideas how to treat disease.

At the moment this possibility still looks far in the future. But this book is about the future, your future. We should keep in mind that one does not understand a problem, to the degree that one can control it in detail, until one has carried out an engineering analysis. In spite of the fact that the sinking of the oceanliner Titanic was observed by thousands of witnesses, only a recent stress analysis revealed the actual cause of the sinking of the ship after it rammed an iceberg. The medical literature is rich in detailed descriptions of the events that surround organ failure in patients, from modern imaging of blood flow and cell metabolism to detection of individual genes, yet we have only a limited possibility to identify the origin of the problem or to predict the outcome in any tissue, and then only in general terms.

Today's computers are powerful, and we are in a position to solve complex sets of equations. Since the turn of the millennium virtually all genes and proteins in man and other species have become known in their exact sequence. Thus, you may argue, we should be in a position to develop a quantitative theory that will *predict* how living tissues fail. Bioengineering is the natural home for such an effort, but it will require many generations of students to develop models, to test them and improve them, so that they become truly useful. No model will ever be perfect.

What are the issues we need to focus our attention on and what are the elements that need to enter into such an analysis of organ failure and disease? There are many entry points. In fact we need many experimental observations and measurements on living tissues in order to start the analysis. The greatest challenge lies in the question how to formulate the problem.

In this chapter, we will look at cardiovascular diseases, which include such conditions as stroke, myocardial infarction (MI), and physiological shock. These are truly important problems since more than 50% of all people die of them. On the surface, stroke, MI or shock look different, but they may have some common features. One such feature is a deficiency, if not complete failure, of the microcirculation. The microcirculation represents the region of the circulation that is made up of billions of capillaries designed to supply nutrients to the tissue cells and remove their metabolites (Zweifach and Lipowsky, 1984). Thus in this chapter we will discuss the flow in microscopic vessels and examine mechanisms for its failure. If a blood vessel does not have flow, we wish to identify the cause. This discussion serves only as an introduction, in later classes of the bioengineering curriculum you will learn more details. Today you may just start to think about an engineering analysis of stroke, MI or shock; any new insight you will gain in the process will be rewarding.

2.   The Microcirculation

All organs have a microcirculation. Interruption of the blood flow in the microcirculation leads to rapid cell death. Thus we will focus on the microcirculation. All the cells that make up a microcirculation have now been identified. They are the circulating red cells (erythrocytes) (Fig. 1), platelets (thrombocytes) (Fig. 2), and the white cells (leukocytes) (Fig. 1), as well as the cells in the wall of the blood vessels which include the endothelial cells (Fig. 3), smooth muscle cells, pericytes and a few others, which for simplicity we will not mention further. But they may have to enter the analysis in the future.

The red cells are carriers of hemoglobin, which serves to transport oxygen, nitric oxide and other gases, and the platelets serve to control bleeding out of blood vessels in case the wall of the blood vessels has been damaged. There are five classes of white cells, which in an

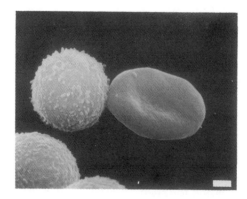

Fig. 1. Scanning electron micrograph of a typical human red cell (right) and leukocyte (left). Note that such cells have to be fixed and dehydrated to be examined by scanning electron microscopy. Thus the cells are shrunk (about 30–40%) but their shapes are preserved. Note that the white cell has many membrane folds, there are none on the red cell. The length of the crossbar is 1 μm.

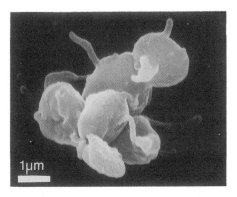

Fig. 2. Scanning electron micrograph of typical human platelets. Note they are much smaller than red cells or white cells.

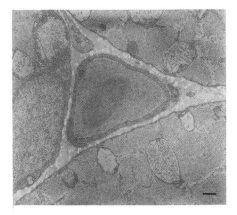

Fig. 3. Transmission electron micrograph of a capillary in rat skeletal muscle. The capillary endothelium is surrounded by three skeletal muscle cells. In transmission electron microscopy you are looking at the surface of a thin (about 0.1 μm) slice through the tissue. The length of the crossbar is 1 μm.

adult are produced in the bone marrow: the B and T lymphocytes (which make antibodies or also can kill tumor cells), the neutrophils and monocytes (which can clean up tissue debris by phagocytosis), the eosinophils and basophils which have more specific activities, like the release of physiological mediators (Fig. 4). These circulating cells are suspended in plasma which is an aqueous medium made up of water salts, proteins, lipids, sugars, i.e. it is a complex mixture of biological molecules, all at relatively low concentrations.

The walls of the blood vessels are lined by the endothelium. Some investigators refer to the endothelium as the *container* for the blood, because all blood vessels are lined by endothelium. The endothelium is made up of individual cells which serve many purposes: they prevent the blood from clotting, they retain the plasma and its molecules inside the blood vessels. The endothelial cells permit some fluid to leak in small amounts out of the blood vessels into the tissue (the leakage is organ dependent), they permit circulating cells to adhere to their membrane, and facilitate leukocytes to crawl across into the tissue (Ohashi *et al.*, 1996). Endothelial cells metabolize hormones, synthesize and release cellular

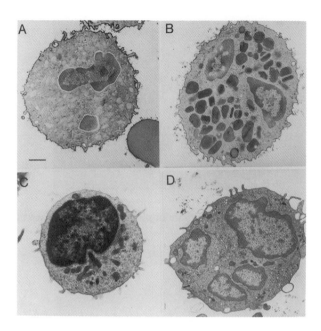

Fig. 4. Transmission electron micrograph of typical passive human white cells. Four types are shown: (A) Neutrophil, (B) eosinophil, (C) lymphocyte, (D) monocyte. The length of the crossbar is 1 μm.

mediators into the blood stream or into adjacent smooth muscle cells, and they divide to form new blood vessels. Remarkably, the endothelial cells sense the magnitude of the fluid stresses on their membrane due to the blood flow (Levesque and Nerem, 1985) and they seem to adjust according to the blood stream. Keep in mind, for none of these cells do we know all functions, the list is still quite incomplete.

The endothelial cells in the arterioles of the microcirculation are surrounded by a specialized muscle cell, the vascular smooth muscle (Rhodin, 1967). Vascular smooth muscle cells form coils around the arterioles, and upon contraction are able to reduce the diameter of the lumen of the arterioles. Capillaries, our smallest blood vessels (their diameter is about 5 to 10 μm depending on the tissue) are also covered by a specialized cell, the pericyte, which is also present in venules (Shepro and Morel, 1993).

The blood flow in the microcirculation depends on the pressure generated by the contraction of the heart muscle. The remarkable feature of the capillaries in many organs is that they are narrower than the dimensions of the red cells and the leukocytes. The only way by which circulating cells can pass from the arteries and arterioles to the venules and back into the large veins is by passing through the capillaries. In capillaries, most red and white cells have to deform in order to fit into the lumen. Thus the resistance to flow in the capillaries depends on the deformability of the blood cells and the endothelium (Skalak and Öskaya, 1987). This problem has been studied by a number of investigators and you can find excellent summaries (Chien, 1987; Skalak and Öskaya, 1987). The red cell is a remarkably deformable particle, with a low viscous cytoplasm (the hemoglobin solution is only about 7 times more viscous than water, although it is saturated with hemoglobin) and a flexible membrane that can be sheared, bent, twisted, but resists an increase of its area.

In contrast to the red cells, the white cells and the endothelial cells have a nucleus, and their cytoplasm consists of a fiber network composed predominantly of actin filame's (Satcher *et al.*, 1997) that are held together into a meshwork by "actin binding proteins" and "actin associated proteins" (Stossel and Hartwig, 1976; Mullins *et al.*, 1998). This actin is similar to the actin fibers found in muscle cells, but is used by the cell to form a meshwork that serves to build the actual cytoplasmic volume, just like the poles in a circus tent. Our cells and organs are made up to a large extent by actin.

The cell cytoplasm is viscoelastic, i.e. it exhibits an elastic response and also creeping deformation (Sato *et al.*, 1985; Sato *et al.*, 1987). Compared to the red cells, the white cells are much stiffer and have a larger volume than red cells. This leads to a situation that the resistance imposed by a single white cell in capillaries is larger than that of a single red cells (Schmid-Schönbein *et al.*, 1980). Fortunately, we have much fewer white cells than red cells in the circulation. Platelets are smaller particles which usually have less influence on capillary blood flow. But platelets may aggregate into large clusters, in which case they may obstruct even larger blood vessels, especially if the endothelium has been damaged significantly (Palabrica *et al.*, 1992; Garcia *et al.*, 1994).

3. Cardiovascular Cell Activation

All cells can be activated, but in the following we will focus only on the cells in the microcirculation. While the majority of endothelial cells or leukocytes in a healthy microcirculation are in a relatively low state of activation, there is a minority of cells on which several signs of activation can be detected. The activation can be identified in the form of pseudopod formation (Fig. 5), which is a slow deformation of the cell cytoplasm due to actin polymerization (Zhelev *et al.*, 1996). Pseudopods are cytoplasmic regions which are mobile due to a dynamic actin polymerization (Fig. 6), but are actually stiffer than the remaining regions of the cytoplasm away from the pseudopods (Schmid-Schönbein, 1990). Endothelial cells can also project pseudopods, which in capillaries will block the blood flow to a significant degree (Fig. 7).

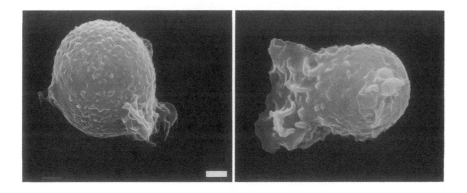

Fig. 5. Two examples of a spontaneously activated human white cell with characteristic pseudopods as seen by scanning electron microscopy. Compare this shape with passive leukocytes (Fig. 1) without such pseudopods. The length of the crossbar is 1 μm.

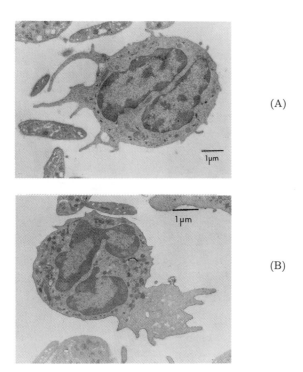

(A)

(B)

Fig. 6. Transmission electron micrograph of human (A) monocyte and (B) neutrophil with pseudopod formation. The pseudopods contain actin and fewer cell organelles.

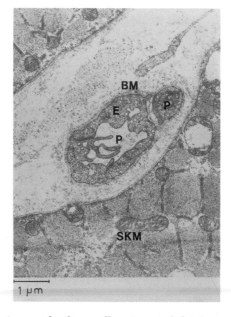

Fig. 7. Transmission electron micrograph of a capillary in rat skeletal muscle after stimulation [for details see (Lee and Schmid-Schönbein, 1995)].

The activation may also be detected in the form of cellular activities, such as the formation of oxygen free radicals, which are oxygen molecules with one or more free electrons on their outer shell. Oxygen free radicals are comparatively reactive and interact with many different biological molecules. Radicals can peroxidize lipid membranes or break DNA (McCord, 1987). In fact, neutrophils or monocytes utilize this process during phagocytosis to peroxidize the membrane of bacteria as the first step of their destruction.

Another form of activation is manifest in the form of the ability of the cell membrane to establish adhesive contact. Adhesion in biological cells is a regulated process mediated by a specific set of glycoproteins located in the plasma membrane (Albelda *et al.*, 1994). In leukocytes L-selectin and several isoforms of integrins have been identified. Blockade of the selectins or integrins with monoclonal antibodies eliminates the ability of leukocytes to adhere to biological surfaces. There is also a form of selectin on endothelium, P-selectin, as well as several members of the immunoglobulin superfamily, such as the intercellular adhesion molecules (ICAM-1, ICAM-1) and the vascular adhesion molecules (VCAM-1, VCAM-2). P-selectin is stored in the cytoplasm of the endothelial cell in membrane bound vesicles, which upon stimulation of the endothelium leads to rapid discharge of the vesicles and incorporation of P-selectin into the endothelial membrane. In this way, the endothelial cell becomes in a short time period an adhesive surface for white cells, but less for red cells. Platelets may also adhere to endothelium, using their own adhesion molecules, and so may red cells, but only in selected diseases such as sickle cell anemia.

There are other forms of cell activation. White cells may release cytoplasmic granules which contain a spectrum of proteolytic enzymes and may lead to tissue destruction (Weiss, 1989). Endothelial cells, which under normal conditions express specific genes that serve to scavenge oxygen free radicals (superoxide dismutase, catalase, nitric oxide synthase, cyclooxygenase) (Topper *et al.*, 1996), may start to express new genes, such as cellular growth factor (e.g. platelet derived growth factor, vascular endothelial derived growth factor), cytokines (e.g. tumor necrosis factor, interleukin 1), which in turn can activate white cells, as well as genes which promote coagulation of the blood (Gimbrone *et al.*, 1997).

The activation of cells in the microcirculation has many consequences, most of which cause a disturbance of the normal microcirculation, and thus may constitute the beginning of a disease process. Instead of normal passage of passive white cells through the capillary network, activation of white cells leads to an elevation of the hemodynamic resistance (Sutton and Schmid-Schönbein, 1992; Helmke *et al.*, 1997) and entrapment of white cells in the capillaries (Warnke and Skalak, 1992; Harris and Skalak, 1993; Ritter *et al.*, 1995) with attachment to the endothelial surface (Fig. 8). In capillaries, the slower moving white cells disturb the otherwise well aligned motion of the red cells and raise the hemodynamic resistance even without attachment to the endothelium (Helmke *et al.*, 1997). Activated white cells become trapped in capillaries and may occlude them (Bagge *et al.*, 1980; Engler *et al.*, 1983; Del Zoppo *et al.*, 1991). The activated leukocytes may also release humoral mediators which signal the arterioles to contract (Mehta *et al.*, 1991). Activation of leukocytes by means of mediators that are released during oxygen free radical formation leads to production of oxygen free radicals by both leukocytes and the endothelium (Suematsu *et al.*, 1992). Interestingly, activation of the microcirculation with inflammatory substances, like platelet activating factor (which in spite of its name is also an effective activator for endothelial

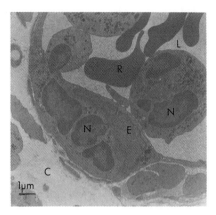

Fig. 8. Transmission electron micrograph of white cells crawling across the endothelium (E) in a small venule. Only part of the venular wall is shown with collagen (C) fiber in the tissue and red cells (R) in the lumen (L) of the venule. Note that one of the neutrophils has projected a major part of its cytoplasm *underneath* the endothelium.

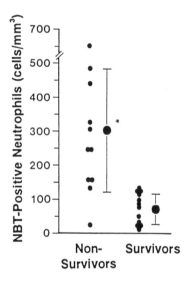

Fig. 9. The number of superoxide forming neutrophils (as determined by nitroblue-tetrazolium reduction) in the rat circulation *before* hemorrhagic shock by reduction of the blood pressure (to a mean arterial pressure of 40 mmHg for 90 minutes). The survival was determined by observation over a period of 24 hours after the hemorrhagic shock. Note the separate levels of cell activation between the two groups. There are animals which initially have low levels of cell activation (left), but in the course of hemorrhagic shock develop higher levels and die.

cells or white cells), leads to migration of leukocytes across the wall of postcapillary venules into the tissue. But in a healthy tissue leukocyte migration leads to parenchymal cell injury mostly in the presence of two stimulators and less in the presence of a single stimulator (Tung *et al.*, 1997). Thus, a single stimulus is less effective in mediating cell activation and tissue cell injury compared with a stimulation by a combination of stimulators. The exposure to multiple stimulations (e.g. from smoking, infections, dietary risk factors, arterial hypertension) may be an important issue also in human cardiovascular disease.

Perhaps one of the most direct ways to illustrate the importance of cell activation can be demonstrated in hemorrhagic shock. To induce hemorrhagic shock, the central blood pressure is reduced for a selected period of time by withdrawal of blood from a central artery or vein. Trauma victims with major blood loss may be subject to hemorrhagic shock. Such a reduction of the blood pressure may lead to significant organ dysfunction and lack of perfusion of the microcirculation in many organs (Bagge *et al.*, 1980; Barroso-Aranda *et al.*, 1988). Depletion of white cells in the circulation serves to enhance the probability of survival following hemorrhagic shock. But in any group of animals, some survive and some will succumb to this type of challenge of the circulation by temporary pressure reduction. What could be the reason? While there are a number of differences between survivors and nonsurvivors, like heart rate, white cell counts and others, these parameters are only different as an average between the two groups, individual animals show large deviations from the average. But the level of cell activation prior to the pressure reduction exhibits a difference between the two groups: initially and during the course of the pressure reduction survivors have a low level of white cell activation, nonsurvivors have elevated levels (Barroso-Aranda and Schmid-Schönbein, 1989; Barroso-Aranda *et al.*, 1991) (Fig. 9).

4.   Cell Activation in Patients

There is evidence that patients with myocardial ischemia and stroke have elevated levels of cell activation in the circulation. The activation can be detected in the form of pseudopod formation on white cells (Chang *et al.*, 1992), oxygen free radical production (Ott *et al.*, 1996), adhesion (Grau *et al.*, 1992), reduced cell deformability (Bauersachs *et al.*, 1997), or gene expression (Marx *et al.*, 1997). Patients which have a known risk factor for these conditions also exhibit signs of activation in their white blood cells, such as diabetics (Pécsvarády *et al.*, 1994), patients with arterial hypertension (Lacy *et al.*, 1998), as well as smokers (Pitzer *et al.*, 1996). As we have seen above, activated cells may be trapped in capillaries. Thus the population of cells that we collect from the venous or in selected cases also arterial blood of patients may contain fewer of the activated cells. The activated cells may be trapped in the microcirculation of those organs that are subject to complications. Unfortunately there are currently no noninvasive techniques to study endothelial cell activation in patients. Thus, there is this need that alternative techniques be developed to study the activation in patients. One such approach was developed by Pitzer *et al.*, who collected fresh plasma and incubated it with *naive* donor cells from an individual without symptoms (typically a young college student who does not smoke or show other signs of activation). Incubation of the plasma of smokers with such *naive* leukocytes leads to significantly elevated levels of activation (Pitzer *et al.*, 1996). This experiment illustrates that the plasma of smokers may contain a factor that serves to activate the white cells. Similar tests in hemorrhagic shock reveal that activated cells may not be detectable in blood samples from central blood vessels (Shen *et al.*, 1995). If the plasma contains an activator which can reach every organ and every capillary of the microcirculation, even innocent bystander organs, then we may be dealing with a form of activation which is less desirable and may lead to complications in microvascular perfusion and organ function.

5.   Mechanisms of Cell Activation

The fundamental question which we are confronted with in several cardiovascular diseases is: What are the mechanisms for cell activation? There are four general categories (Table 1) that could lead to cell activation.

(1) The most frequently proposed mechanism is via a positive feedback involving a cell activator, i.e. a molecule or set of molecules which directly stimulate leukocytes, endothelial cells or platelets. The list of molecules in this category is large and may span from bacterial (like endotoxin) or virus derived products to direct activators such as complement fragments, lipid membrane derived products (e.g. platelet activating factor, lipoxygenase derived products), small peptides (cytokines, lymphokines) and fragments derived from extracellular proteins, thrombotic products (e.g. thrombin, fibrin fragments), to peroxidized products (e.g. oxidized cholesterol) (Mazzoni and Schmid-Schönbein, 1996).

(2) There are natural mediators that serve to downregulate cells *in vivo*, such as glucocorticoid derived from the adrenal gland, adenosine (a metabolite), nitric oxide, and albumin (one of the main plasma proteins). Depletion of these mediators may serve to upregulate the cells. This constitutes a negative feedback mechanism for cell activation.

(3) There is evidence that adhesion molecules, like integrins, may serve not only as molecules for membrane attachment, but also as signalling molecules across the membrane to the cell cytoplasm. Thus it is possible that passive white cells may become activated upon attachment to activated endothelium, and vice versa activated white cells may stimulate passive endothelial cells upon membrane contact. This has been referred to as junxtacrine activation (Zimmerman *et al.*, 1996).

(4) Fluid shear stress serves to control the state of activation of cardiovascular cells. Perhaps this is the most surprising mechanism that controls cell activation.

In spite of the fact that fluid shear stress (the force tangential to the cell surface per unit area, typically about 10 dyn/cm^2) in the microcirculation is much lower than the fluid normal stress (the negative of the fluid pressure and about 10^3 to 10^5 dyn/cm^2), it has a powerful effect on endothelial cells (Dewey *et al.*, 1981; Davies, 1995; Resnick and Gimbrone, 1995), red cells (Johnson, 1994), and platelets (Konstantopoulos *et al.*, 1995). The mechanics in the microcirculation controls biology. Endothelial cells respond to shear stress by ion exchange, induction of membrane signalling pathways, by phosphorylation of G-proteins, rearrangement of the cytoplasmic actin, expression of genes with shear stress specific promotors (Resnick *et al.*, 1993; Kuchan *et al.*, 1994; Shyy *et al.*, 1995). Furthermore, endothelial cells are quite sensitive with respect to the details of the shear field; unsteady stresses give a different response than steady fluid shear stress (Davies *et al.*, 1986; DePaola *et al.*, 1992).

The application of steady laminar shear stress serves to induce a specific set of genes in endothelial cells which protect against the detrimental action of the oxygen free radicals (cyclooxygenase, endothelial nitric oxide synthase, superoxide dismutase, and catalase) (Topper *et al.*, 1996) and may constitute the best of all situations in an adult circulation. Unsteady flow or the application of cytokines serves to disturb this unique pattern of gene expression (Frangos *et al.*, 1996; Topper *et al.*, 1997) and leads to disturbance of the microcirculation. White cells also respond instantaneously to shear stress (Moazzam *et al.*, 1997). Physiological levels of fluid shear stress serve to downregulate white cells. Cessation of flow in the microcirculation at reduced fluid shear stress upregulates the cells. Thus, just the reduction of fluid shear stress may serve to activate cells in the circulation.

As we saw above, in physiological shock high levels of cell activation are detected. Recently, we traced their origin to a surprising source, the pancreas. Besides the production of insulin, this organ synthesizes and releases a set of enzymes that serve the digestion in the intestine. The enzymes can break down proteins, lipids, carbohydrates and nucleotides. In the ischemic intestine these enzymes are not restricted to the inner lumen of the intestine, but penetrate through the intestinal mucosal barrier (the brush border cell layer) into the wall of the intestine where new activators are produced. The article following this Chapter by Mitsuoka *et al.* (2000) provides details and a method to prevent the formation of the activators. Note, in the absence of activator formation, many of the detrimental and lethal events in shock are prevented.

6. Synopsis

Which of the mechanisms is responsible for the disturbance of the microcirculation in myocardial ischemia, stroke or hemorrhagic shock? As we have seen, there is evidence for both a change of fluid shear stress and for an activator in the plasma. Which of these mechanisms constitutes the trigger mechanism? We do not know for certain. Are there other pathways to disrupt the microcirculation? Possibly. Whatever mechanisms you favor, they are likely to differ from case to case. But at the moment we do not know any case with certainty. The study of cell activation may serve as a key to examine the problem and to develop a systematic model of vascular disease. This is the message of this chapter.

Acknowledgments

I am deeply indebted to the colleagues and students in the Department of Bioengineering, who have inspired many of the ideas presented here. Drs. Jorge Barroso-Aranda, Brian Helmke, Michelle Mazzoni, Fariborz Moazzam, Scott Simon, Thomas C. Skalak, Allan Swei, Don Sutton, David Tung, the students Jennifer J. Costa, Jo-Ellen Pitzer, Kevin L. Ohashi, Drs. Ricardo Chavez-Chavez, Shunichi Fukuda, Armin Grau, Anthony Harris, Fred Lacy, Jye Lee, Shou-Yan Lee, Phillip Pfeiffer, Kai Shen, Makoto Suematsu, Hidekazu Suzuki, Shinya Takase, my long term associate Frank A. Delano, and our friend and founder of Bioengineering at UCSD, the late Benjamin W. Zweifach. Special thanks to Drs. Robert L. Engler and Daniel T. O'Connor in the Department of Medicine at UCSD, Dr. John J. Bergan in the Department of Surgery, and Dr. Gregory J. Del Zoppo at Scripp Research Institute in La Jolla for their clinical perspective and enthusiastic support. All of this work was only possible by support that makes it possible to study microcirculation and cardiovascular disease, recently by National Science Foundation Grant IBN-9512778, National Institute of Health grants HL 10881 and HL 43024.

References

1. Albelda, S. M., Smith, C. W. and Ward, P. A., Adhesion molecules and inflammatory injury, *FASEB J.* **8**, 504–512 (1994).
2. Bagge, U., Amundson, B. and Lauritsen, C., White blood cell deformability and plugging of skeletal muscle capillaries in hemorrhagic shock, *Acta. Physiol. Scand.* **108**, 159–163 (1980).
3. Barroso-Aranda, J., Chavez-Chavez, R. and Schmid-Schönbein, G. W., Spontaneous neutrophil activation and the outcome of hemorrhagic shock in rabbits, *Circ. Shock* **36**, 185–190 (1991).
4. Barroso-Aranda, J. and Schmid-Schönbein, G. W., Transformation of neutrophils as indicator of irreversibility in hemorrhagic shock, *Am. J. Physiol.* **257**, H846–H852 (1989).
5. Barroso-Aranda, J., Schmid-Schönbein, G. W., Zweifach, B. W. and Engler, R. L., Granulocytes and no-reflow phenomenon in irreversible hemorrhagic shock, *Circ. Res.* **63**, 437–447 (1988).
6. Bauersachs, R. M., Moessmer, G., Koch, C., Neumann, F. J., Meiselman, H. J. and Pfafferott, C., Flow resistance of individual neutrophils in coronary artery disease: Decreased pore transit times in acute myocardial infarction, *Heart* **77**, 18–23 (1997).
7. Chang, R. R. K., Chien, N. T. Y., Chen, C.-H., Jan, K.-M., Schmid-Schönbein, G. W. and Chien, S., Spontaneous activation of circulating granulocytes in patients with acute myocardial and cerebral diseases, *Biorheology* **29**, 549–561 (1992).
8. Chien, S., Red cell deformability and its relevance to blood flow, *Ann. Rev. Physiol.* **49**, 177–192 (1987).
9. Davies, P., Flow-mediated endothelial mechanotransduction, *Physiol. Rev.* **75**, 519–560 (1995).
10. Davies, P. F., Remuzzi, A., Gordon, E. J., Dewey, C. F. and Gimbrone, M. A., Turbulent fluid stress induces vascular endothelial cell turnover *in vitro*, *Proc. Natl. Acad. Sci.* **83**, 2114–2117 (1986).
11. Del Zoppo, G. J., Schmid-Schönbein, G. W., Mori, E., Copeland, B. R. and Chang, C.-M., Polymorphonuclear leukocytes occlude capillaries following middle cerebral artery occlusion and reperfusion, *Stroke* **22**, 1276–1283 (1991).
12. DePaola, N., Gimbrone, M. A., Davis, P. F. and Dewey, C. F. J., Vascular endothelium responds to fluid shear stress gradients, *Arterioscl. Throm.* **12**, 1254–1257 (1992).
13. Dewey, C. F., Bussolari, S. R., Gimbrone, M. A. J. and Davies, P. F., The dynamic response of vascular endothelium to fluid shear stress, *J. Biomech. Eng.* **103**, 177–185 (1981).
14. Engler, R. L., Schmid-Schönbein, G. W. and Pavelec, R. S., Leukocyte capillary plugging in myocardial ischemia and reperfusion in the dog, *Am. J. Pathol.* **111**, 98–111 (1983).
15. Frangos, J. A., Huang, T. Y. and Clark, C. B., Steady shear and step changes in shear stimulate endothelium via independent mechanisms — Superposition of transient and sustained nitric oxide production, *Biochem. Biophys. Res. Com.* **224**, 660–665 (1996).
16. Garcia, J. H., Liu, K. F., Yishida, Y., Lian, J., Chen, S. and Del Zoppo, G. J., The influx of leukocytes and platelets in an evolving brain infarct (Wistar rat), *Am. J. Pathol.* **144**, 188–199 (1994).
17. Gimbrone, M. A. J., Nagel, T. and Topper, J. N., Biomechanical activation: An emerging paradigm in endothelial adhesion biology, *J. Clin. Invest.* **100**, S61–S65 (1997).
18. Grau, A. J., Berger, E., Sung, K.-L. P. and Schmid-Schönbein, G. W., Granulocyte adhesion, deformability, and superoxide formation in acute stroke, *Stroke* **22**, 33–39 (1992).
19. Harris, A. G. and Skalak, T. C., Leukocyte skeletal structure determines capillary plugging and network resistance, *Am. J. Physiol.* **265**, H1670–H1675 (1993).
20. Helmke, B. P., Bremner, S. N., Zweifach, B. W., Skalak, R. and Schmid-Schönbein, G. W., Mechanisms for increased blood flow resistance due to leukocytes, *Am. J. Physiol.* **273**, H2884–H2890 (1997).

21. Johnson, R. M., Membrane stress increases cation permeability in red cells, *Biophys. J.* **67**, 1876–1881 (1994).

22. Konstantopoulos, K., Wu, K. K., Uden, M. M., Banez, E. I., Shattil, S. J. and Hellums, J. D., Flow cytometric studies of platelet response to shear stress in whole blood, *Biorheology* **32**, 73–93 (1995).

23. Kuchan, M. J., Hanjoong, J. and Frangos, J. A., Role of *G* proteins in shear stress-mediated nitric oxide production by endothelial cells, *Am. J. Physiol.* **267**, C753–C758 (1994).

24. Lacy, F., O'Connor, D. T. and Schmid-Schönbein, G. W., Plasma hydrogen peroxide production in hypertensives and normotensives subjects at genetic risk for hypertension, *J. Hypertension* **16**, 291–303 (1998).

25. Lee, J. and Schmid-Schönbein, G. W., Biomechanics of skeletal muscle capillaries: Hemodynamic resistance, endothelial distensibility, and pseudopod formation, *Ann. Biomed. Eng.* **23**, 226–246 (1995).

26. Levesque, M. J. and Nerem, R. M., The elongation and orientation of cultured endothelial cells in response to shear stress, *J. Biomech. Eng.* **107**, 341–347 (1985).

27. Marx, N., Neumann, F. J., Ott, I., Gawaz, M., Koch, W., Pinkau, T. and Schömig, A., Induction of cytokine expression in leukocytes in acute myocardial infarction, *J. Am. Coll. Cardiol.* **30**, 165–70 (1997).

28. Mazzoni, M. C. and Schmid-Schönbein, G. W., Mechanisms and consequences of cell activation in the microcirculation, *Cardiovasc. Res.* **32**, 709–719 (1996).

29. McCord, J. M., Oxygen-derived radicals: A link between reperfusion injury and inflammation, *Fed. Proc.* **46**, 2402–2406 (1987).

30. Mehta, J. L., Lawson, D. L., Nicolini, F. A., Ross, M. H. and Player, D. W., Effects of activated polymorphonuclear leukocytes on vascular smooth muscle tone, *Am. J. Physiol.* **261**, H327–H334 (1991).

31. Moazzam, F., DeLano, F. A., Zweifach, B. W. and Schmid-Schönbein, G. W., The leukocyte response to fluid stress, *Proc. Natl. Acad. Sci. USA* **94**, 5338–5343 (1997).

32. Mullins, R. D., Heuser, J. A. and Pollard, T. D., The interaction of Arp2/3 complex with actin: Nucleation, high affinity pointed end capping, and formation of branching networks of filaments, *Proc. Natl. Acad. Sci. USA* **95**, 6181–6186 (1998).

33. Ohashi, K. L., Tung, D. K.-L., Wilson, J. M. and Schmid-Schönbein, G. W., Transvascular and interstitial migration of neutrophils in rat mesentery, *Microcirculation* **3**, 199–210 (1996).

34. Ott, I., Neumann, F. J., Gawaz, M., Schmitt, M. and Schömig, A., Increased neutrophil-platelet adhesion in patients with unstable angina [see comments], *Circulation* **94**, 1239–1246 (1996).

35. Palabrica, T., Lobb, R., Furie, B. C., Aronovitz, M., Benjamin, C., Hsu, Y. M., Sajer, S. A. and Furie, B., Leukocyte accumulation promoting fibrin deposition is mediated *in vivo* by *P*-selectin on adherent platelets, *Nature* **359**, 848–851 (1992).

36. Pécsvarády, Z., Fisher, T. C., Darwin, C. H., Fabók, A., Maqueda, T. S., Saad, M. F. and Meiselman, H. J., Decreased polymorphonuclear leukocyte deformability in NIDDM, *Diabetes Care* **17**, 57–63 (1994).

37. Pitzer, J. E., Del Zoppo, G. J. and Schmid-Schönbein, G. W., Neutrophil activation in smokers, *Biorheology* **33**, 45–58 (1996).

38. Resnick, N., Collins, T., Atkinson, W., Bonthron, D. T., Dewey, C. F. J. and Gimbrone, M. A., Platelet-derived growth factor *B* chain promotor contains a cis-acting fluid shear-stress-responsive element, *Proc. Natl. Acad. Sci. USA* **90**, 4591–4595 (1993).

39. Resnick, N. and Gimbrone, M. A., Jr, Hemodynamic forces are complex regulators of endothelial gene expression, *FASEB J.* **9**, 874–882 (1995).

40. Rhodin, J. A. G., The ultrastructure of mammalian arterioles and precappillary sphincters, *J. Ultrastruct. Res.* **18**, 181–223 (1967).

41. Ritter, L. S., Wilson, D. S., Williams, S. K., Copeland, J. G. and McDonagh, P. F., Early in reperfusion following myocardial ischemia, leukocyte activation is necessary for venular adhesion but not capillary retention, *Microcirculation* **2** (1995).

42. Satcher, R. L. J., Dewey, C. F. J. and Hartwig, J. H., Mechanical remodeling of the endothelial surface and actin cytoskeleton induced by fluid flow, *Microcirculation* **4**, 439–453 (1997).

43. Sato, M., Leimbach, G., Schwarz, W. H. and Pollard, T. D., Mechanical properties of actin, *J. Biol. Chem.* **260**, 8585–8592 (1985).

44. Sato, M., Levesque, M. J. and Nerem, R. M., An application of the micropipette technique to the measurement of the mechanical properties of cultured bovine aortic endothelial cells, *J. Biomech. Eng.* **109**, 27–34 (1987).

45. Schmid-Schönbein, G. W., Leukocyte biophysics, *Cell Biophys.* **17**, 107–135 (1990).

46. Schmid-Schönbein, G. W., Usami, S., Skalak, R. and Chien, S., The interaction of leukocytes and erthrocytes in capillary and postcapillary, *Microvasc. Res.* **19**, 45–70 (1980).

47. Shen, K., Chavez-Chavez, R., Loo, A. K. L., Zweifach, B. W., Barroso-Aranda, J. and Schmid-Schönbein, G. W., Interpretation of leukocyte activation measurements from systemic blood vessels, *Cerebrovascular Diseases*, 19th Princeton Stroke Conference" (eds. M. A. Moskowitz and L. R. Caplan), Butterworth-Heinemann, Boston, Mass. (1995), pp. 59–73.

48. Shepro, D. and Morel, N. M. L., Pericyte physiology, *FASEB J.* **7**, 1031–1038 (1993).

49. Shyy, Y.-J., Lin, M.-C., Han, J., Lu, Y., Petrime, M. and Chien, S., The *cis*-acting phorbolester "12-O-tetradecanoylphorbol 13-acetate" — Responsive element is involved in shear stress-induced monocyte chemotactic protein 1 gene expression, *Proc. Natl. Acad. Sci.* **92**, 8069–8073 (1995).

50. Skalak, R. and Öskaya, N., Models of erythrocyte and leukocyte flow in capillaries, *Physiological Fluid Dynamics II* (eds. L. S. Srinath and M. Singh), Tata McGraw Hill, New Delhi (1987), pp. 1–10.

51. Stossel, T. P. and Hartwig, J. H., Interactions of actin, myosin, and a new actin-binding protein of rabbit pulmonary macrophages, *J. Cell. Biol.* **68**, 602–619 (1976).

52. Suematsu, M., Schmid-Schönbein, G. W., Chavez-Chavez, R. H., Yee, T. T., Tamatami, T., Miyasaka, M., DeLano, F. A. and Zweifach, B. W., *In vivo* visualization of oxidative changes in microvessels during neutrophil activation, *Am. J. Physiol.* **264**, H881–H891 (1992).

53. Sutton, D. W. and Schmid-Schönbein, G. W., Elevation of organ resistance due to leukocyte perfusion, *Am. J. Physiol.* **262**, H1646–H1650 (1992).

54. Topper, J. N., Cai, J., Falb, D. and Gimbrone, M. A. J., Identification of vascular endothelial genes differentially responsive to fluid mechanical stimuli: Cyclooxygenase-2, manganese superoxide dismutase, and endothelial cell nitric oxide synthase are selectively up-regulated by steady laminar shear stress, *Proc. Natl. Acad. Sci. USA* **93**, 10417–10422 (1996).

55. Topper, J. N., Wasserman, S. M., Anderson, K. R., Cai, J., Falb, D. and Gimbrone, M. A., Jr, Expression of the bumetanide-sensitive Na-K-Cl cotransporter BSC2 is differentially regulated by fluid mechanical and inflammatory cytokine stimuli in vascular endothelium, *J. Clin. Invest.* **99**, 2941–2949 (1997).

56. Tung, D. K.-L., Bjursten, L. M., Zweifach, B. W. and Schmid-Schönbein, G. W., Leukocyte contribution to parenchymal cell death in an experimental model of inflammation, *J. Leukocyte. Biol.* **62**, 163–175 (1997).

57. Warnke, K. C. and Skalak, T. C., Leukocyte plugging *in vivo* in skeletal muscle arteriolar trees, *Am. J. Physiol.* **262**, H1149–H1155 (1992).

58. Weiss, S. J., Tissue destruction by neutrophils, *N. Engl. J. Med.* **320**, 365–376 (1989).

59. Zhelev, D. V., Alteraifi, A. M. and Hochmuth, R. M., F-actin network formation in tethers and in pseudopods stimulated by chemoattractant, *Cell Motility and the Cytoskeleton* **35**, 331–344 (1996).

60. Zimmerman, G. A., McIntyre, T. M. and Prescott, S. M., Adhesion and signaling in vascular cell–cell interactions, *J. Clin. Invest.* **98**, 1699–1702 (1996).

61. Zweifach, B. W. and Lipowsky, H. H., Pressure-flow relations in blood and lymph microcirculation, *Handbook of Physiology, Section 2: The Cardiovascular System*, eds. E. M. Renkin and C. C. Michel, American Physiological Society, Bethesda, M.D. (1984), pp. 251–307.

ELSEVIER

Cardiovascular Research 32 (1996) 709–719

Cardiovascular
Research

Review

Mechanisms and consequences of cell activation in the microcirculation

Michelle C. Mazzoni, Geert W. Schmid-Schönbein *

Department of Bioengineering, University of California, San Diego, 9500 Gilman Drive, La Jolla, CA 92093-0412, USA

Received 12 December 1995; accepted 13 June 1996

Abstract

Cells undergo activation in response to a wide range of stimuli. In vascular cells (leukocytes, endothelial cells, and platelets), the different forms of activation include degranulation, oxygen free radical formation, expression of membrane adhesion proteins, and biophysical changes such as pseudopod formation and increased cytoplasmic viscosity. Cell activation and low flow are common features of many cardiovascular diseases. There is evidence that plasma from patients contains an activating factor for neutrophils as well as other vascular cells. Activated neutrophils have the ability to impair microcirculatory transit by elevation of endothelial permeability, leukocyte adhesion to the endothelium, leukocyte capillary plugging, release of vasoactive products, and capillary deformation and compression due to oxygen-radical-mediated interstitial edema and cell dysfunction. In addition to reduced organ perfusion, cell activation can also cause cell dysfunction via release of cytotoxic mediators. A lower degree of neutrophil activation prior to acute circulatory challenge (i.e., low preactivation) correlates with improved survival rates after challenge and suggests that elevated levels of in vivo cell preactivation is a risk factor for cell injury and organ failure. Under conditions of low in-vivo cell preactivation (e.g., as is the case in endotoxin-tolerant animals), there is reduced tissue injury and lower mortality after challenge. We hypothesize that in-vivo cell preactivation due to everyday activity (infection, diet, smoking) may be a mechanism for microvascular low blood flow with leukocyte accumulation and may represent a risk factor for various cardiovascular diseases.

Keywords: Neutrophils; Neutrophilic activation; Endothelium; Platelets; Free radicals; Infection; Endotoxins

1. Introduction

The cells in the microcirculation can be encountered in a relatively quiescent state and in various stages of activation. Cellular activation is a normal physiological response which is essential for survival from infection through a regulated series of functions to kill bacteria and pathogens. Recent evidence, however, also suggests that in cardiovascular complications such as myocardial infarction, cerebral stroke, venous ulceration, and ischaemia/reperfusion (I/R) injury may be associated with an activation of cells in the circulation. The reasons for cell activation in these conditions is less well-defined, with as yet no identification of the underlying stimulus in most cases.

An important question can be raised: Are activated cells just a phenomenon secondary to an evolving cardiovascular disease, or can cell activation contribute to the progression of the disease process? While the literature provides data for an association between cell activation and microvascular low-flow disorders, there is currently a lack of convincing evidence that cardiovascular disease may manifest from chronic cell activation. To this end, this review will present a new hypothesis to consider cardiovascular disease by assembling pertinent information available in the literature.

The potential importance of plasma as the incipient site for global cell activation, which in turn leads to local flow disturbances in the microcirculation, is one of the key objectives of this review. Emerging in the literature are the first accounts of the ability of plasma, derived from the blood of patients with a variety of different cardiovascular risk factors and diseases, to activate naive leukocytes from a healthy donor. We will review these reports in man along with animal studies in order to examine the concept that plasma-derived activating factors may circulate and may

---

* Corresponding author. Tel.: (+1-619) 534-3852; fax: (+-1-619) 534-5722.

Time for primary review 42 days.

PII S0008-6363(96)00146-0

reach higher levels in circulatory disorders. Furthermore, we will discuss the endotoxin-tolerant animal, an experimental model in which plasma after acute circulatory challenges induces minimal cell activation. The tolerance model may provide additional insight into the relationship between in vivo cell activation and microvascular pathologies.

2. Quiescent and activated cells in the microcirculation

With focus on the factors affecting local blood perfusion of the tissue, the cells of interest are the red blood cells (RBCs), leukocytes, platelets, and the endothelial cells lining the microvessels. Unlike the other vascular cells, RBCs exhibit little evidence for an 'activated' state as defined by a rapid oxygen-dependent increase in metabolic activity. The deformation behavior of a suspension of flowing RBCs is a key determinant of microcirculatory blood flow (e.g., see article by Pries in this issue). Normally at low flow rates, RBCs form reversible aggregates (rouleaux) due to membrane cross-linking by large-molecular-weight proteins, especially fibrinogen [1]. This condition may be exacerbated in clinical disorders with decreased flow and/or increases in plasma levels of fibrinogen or globulin. Moreover, enhanced aggregation in a microvessel increases the local apparent blood viscosity and can contribute to flow impairment up to the point of complete stasis.

Vascular cells exist in different metabolic states ranging from a passive state with virtually undetectable cell activity to a highly dynamic state of activation. Under steady-state conditions, cells use primarily glycolytic pathways to maintain basal activity. But when stimulated, the cells switch on oxidative pathways to promote release of preformed intracellular products and synthesis of selected mediators. A component of the respiratory burst can be generation of highly reactive oxygen radicals. Due to their availability, leukocytes have been the most extensively studied cell group in terms of activation although there has been extensive research addressing the implications of activated endothelial cells and platelets in low-flow states.

2.1. The impact of the leukocyte count on microvascular flow

After leaving the bone marrow, leukocytes travel in the blood stream at a ratio of about one leukocyte to every 1000 RBCs. Despite their relatively low numbers in the circulation, epidemiological studies in man have demonstrated a correlation between the total leukocyte count and the risks for myocardial infarction [2,3] and cerebral stroke [4]. It is not evident whether the elevated counts serve as a consequence of inflammation, and/or whether leukocytes participate in the evolving vascular injury. The mere physical presence of even passive leukocytes can affect microcirculatory blood flow because the cells are large with high internal cytoplasmic stiffness as compared to RBCs, and have diameters greater than those of capillaries in most tissues [5]. Thus, there is a natural propensity for passive leukocytes to slow blood flow as they deform at the entrance of and then transit through the capillary [6]. The phenomenon is exaggerated by an increased number of leukocytes entering into the capillaries, a decreased perfusion pressure, or an activation-induced stiffening of the leukocyte [6–9].

2.2. Forms of activation in leukocytes

In vitro studies using isolated leukocytes have provided most of our understanding of the different forms of cellular activation, while in vivo models have been instrumental in demonstrating the microvascular implications of cell activation. The neutrophil, one of the chief cells involved in the inflammatory response, is the type of leukocyte focussed on in this discussion.

Neutrophils express their state of activation by physical parameters, notably pseudopod formation [10] and the upregulation or downregulation of adhesion receptors [11], and by biochemical changes including degranulation with release of proteolytic enzymes (e.g., elastase, myeloperoxidase) [12] and the generation of the oxygen free radicals, superoxide anion (O_2^-) and hydrogen peroxide (H_2O_2) [13]. Activated neutrophils also secrete eicosanoids such as leukotrienes and thromboxane A_2 [14]. The projection of pseudopods reflects cytoplasmic actin polymerization and is a requirement for transendothelial migration and phagocytosis. Pseudopods are stiffer than the main cell body, and therefore circulating activated neutrophils have greater difficulty in passing through capillaries [15]. Membrane adhesion mediated by receptors is also necessary for many cell functions including emigration from the hematopoietic pool, migration, and phagocytosis. Specifically for attachment of neutrophils to post-capillary endothelium, adhesive interaction involves several families of adhesion molecules (integrins, selectins, immunoglobulins, sialyl-Lewis x) (see article by Ley in this issue).

Activation of neutrophils within the circulation even before migration is evident from various soluble markers in the plasma (e.g., granular products, L-selectin). Using neutrophil elastase as an indirect marker of neutrophil activation, studies in man have found a positive correlation between increased levels of elastase and patients with myocardial infarction, diabetes, or hypertension [16], and acute cerebral stroke [17]. Further, it is possible to detect elevated elastase values in a defined group of controls who were without symptoms at the time of phlebotomy, but who also had one or more of the established risk factors for stroke [17]. The liberation of proteases and oxygen free radicals from activated neutrophils may lead to degeneration of endothelial cell membrane functions.

Neutrophils from patients with cerebral injury have significantly greater adhesion to laminin and fibronectin (components of the endothelial extracellular matrix), but no increase in cell deformability or O_2^- production as compared to an age-matched control group [18]. One explanation for this result could be that the most activated cells do not circulate (see below), and there may have been a partially activated subset of cells which display only limited forms and degrees of activation. Furthermore, the control group itself had elevated levels of activation, given that more than half in this group had stroke risk factors; conditions like hypertension [19], diabetes [20], and smoking [21] are associated with leukocyte activation. This in vivo 'preactivation' of circulating neutrophils may reflect global cell activation due to activating factors in the plasma.

2.3. Forms of endothelial cell activation

Endothelium has been identified to be a dynamic tissue cell capable of secreting or metabolizing substances with vasoactive, inflammatory, and coagulant activities [22]. Among the important vasodilator species produced by endothelial cells are adenosine and two platelet inhibitors, prostacyclin (PGI_2) and nitric oxide (NO). Endothelin is an endothelial-derived vasoconstrictor which also stimulates release of PGI_2 and NO, so there exists an interplay of factors regulating vascular tone which can be upset by pathophysiologic stimuli [22].

When activated, endothelium produces pro-inflammatory substances such as platelet-activating factor (PAF), oxygen free radicals, and expression of surface adhesion molecules. In particular, PAF release is coincident with rapid upregulation of endothelial P-selectin to induce spontaneous rolling and activation of leukocytes [23]. The constituitively expressed endothelial immunoglobulin receptor ICAM-1 can then bind with an upregulated expression of CD11/CD18 on the neutrophil surface. Cytokine stimulation of endothelium causes *de novo* synthesis and delayed expression of E-selectin which binds to sialyl-Lewis *x* ligand on the neutrophil [24]. Production of O_2^- and H_2O_2 by the endothelium has been documented during post-ischemic reperfusion when oxygen readmitted to the tissue combines with endothelial xanthine oxidase, an enzyme enriched during the ischemic period [25,26].

2.4. Forms of activation in platelets

Platelets prevent bleeding when the integrity of the blood vessel wall is disrupted. Upon activation, these small cells degranulate, project pseudopods, express P-selectin, and produce vasoactive, pro-inflammatory products and endothelial growth factors. Activated platelets release both vasoconstrictors (e.g., thromboxane A_2 and serotonin) and vasodilators (e.g., adenosine diphosphate). The receptors P-selectin and integrin gpIIb/IIIa expressed on the acti-

vated cell surface bind neutrophils and potentially create platelet aggregates which can circulate as a thromboembolism or plug at sites of damaged endothelium. Upon activation, there is also release of mediators capable of amplifying the inflammatory response such as platelet factor 4, thromboxane A_2, PAF, and also oxygen free radicals. An increase in the number of circulating platelets is associated with deleterious cardiovascular effects [27], and elevated levels of circulating activated platelets have been shown to be involved in a broad range of conditions including inflammatory bowel disease [28], coronary artery disease [22], ischemic heart disease [29], and diabetes [30].

2.5. Leukocyte assessment from central blood samples

There is as yet only indirect evidence to link activated leukocytes in the circulation with the nature and severity of human cardiovascular disease. One of the underlying problems in the investigation of cell activation is the fundamental limitation in collecting activated leukocytes from central vessels. By virtue of the hemodynamic mechanisms described above, activated leukocytes tend to become trapped in the microcirculation and cease to circulate. But even a slight increase in the proportion of activated cells in the circulating pool could imply significant local accumulation in the tissue, with the full magnitude of activation actually hidden from view when the analysis is based on central blood samples.

Circulating leukocytes are continually replenished from the hematopoietic pool, so there could be a gradual and progressive buildup of activated leukocytes in the microcirculation due to capillary entrapment and venular accumulation if stimuli are persistently present as new cells

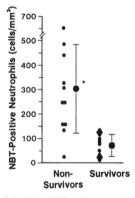

Fig. 1. Number of circulating NBT-positive neutrophils in whole blood taken immediately before acute hemorrhagic shock in rats. Superoxide anion formation in activated neutrophils reduces NBT solution to black formazan crystals which are detected microscopically in the cytoplasm of single cells. Within a wide range of values, animals which survived the challenge (24 h) had significantly lower levels of circulating activated cells. Large circles represent mean and bars represent standard deviation. $^*P < 0.005$, t-test. Modified from Barroso-Aranda et al. [32].

enter the circulation. Thus, even relatively low levels of systemic leukocyte activation could indicate compromised organ function at the microcirculatory level when faced with a circulatory challenge such as hemorrhagic shock. Support for this premise comes from acute experiments in animals which have different levels of activated neutrophil counts in the blood. Low levels of central activation prior to shock result in significantly improved survival rates after shock, whereas high levels of activation are associated with a high risk of death [31,32] (see Fig. 1). This result suggests that the degree of activation of leukocytes is potentially a risk factor for a favorable outcome after circulatory challenge.

3. Mechanisms of cell activation

The expanding body of evidence implicating activated cells in the pathogenesis of microvascular disorders makes identification of the source(s) of activation a research topic of key importance. Supposing that a small, but significant, subpopulation of vascular cells normally exists in a preactivated state, it is interesting to ask which stimuli could be responsible, and whether chronic preactivation predisposes an individual to vascular and parenchymal cell dysfunction. In this section, a brief review of the major chemical and physical mechanisms of activation for leukocytes, endothelial cells, and platelets will be given with emphasis on activating factors discovered in plasma from patients with cardiovascular disorders characterized by chronic low blood flow.

3.1. Neutrophils

There are numerous substances known to activate vascular cells, with stimuli for one cell type often the release product of another cell type. For neutrophils, the list includes cytokines (e.g., TNFα, IL-1, IL-8, NAP-2), leukotrienes (e.g., LTB_4), PAF, opsonized bacteria, complement products (e.g., C3a, C5a), endotoxins, oxidized low-density lipoprotein, and fMLP (an exogenous peptide). There are also compounds that attenuate neutrophil activation such as adenosine [33], glucocorticoids [34], and PGI_2 [35]. NO has been reported to have an effect on neutrophil adhesion in vivo [36], but it was found to act indirectly via induction of an oxidative stress after NO synthesis inhibition [37]. The beneficial function of these endogenous inhibitors may, in part, serve to modulate the inflammatory process. For example, PGI_2 is released from endothelium following a period of activation [38], presumably to prevent further neutrophil activation. Estrogen has also been found to reduce the responsiveness of neutrophils to activators in vitro [39], which may contribute partially to the fact that women have a lower incidence of cardiovascular disease. Furthermore and interestingly, the activator IL-8 has been shown to reduce neutrophil recruitment into dermal inflammatory sites in an in vivo rabbit model, suggesting it may influence multiple signal transduction pathways [40].

Gram-negative bacterial endotoxin (lipopolysaccharide, LPS) is a prime mediator of septic shock. Patients with acute bacterial infection have a subpopulation of neutrophils which are 'primed', indicating that these cells exhibit a greatly enhanced activation response to stimuli [41]. The proportion of primed cells as assessed by H_2O_2 production was on average 40% in patients (range 0 to 80%) as compared to 0% in healthy controls. Primed cells theoretically would improve host defense to infection, but a relatively large number of neutrophils activating in response to low concentrations of stimuli could create transient episodes of local tissue damage. Evidence to support the premise that infection compromises the circulation comes from an epidemiological study of the elderly suggesting that a 30% increase in death from cardiovascular disease in winter, as compared to the remainder of the year, in these subjects is due to an increased incidence of respiratory infection [42]. Recently, infection has been shown to be a risk factor for cerebral stroke in persons of all ages [43]. Further investigation is needed to define the role of elevated levels of activated and primed neutrophils in the development of cardiovascular disorders under circumstances with viral and bacterial infection.

Neutrophil preactivation detected in healthy humans and animals in vivo could be due to the presence of a low-grade infection (without apparent clinical symptoms) or due to other environmental factors such as normal food consumption. Blood samples taken from laboratory rats of comparable age which were born throughout the year show a variable subpopulation of activated neutrophils (measured on fresh unseparated blood samples by O_2^- reduction of nitroblue tetrazolium, NBT), with higher levels of activation in the fall and winter than in spring and summer (see Fig. 2). In an in vitro study by Chadwick et al. [12], blood samples were taken from human subjects either after a 12 h fast or 30–60 min after a standard breakfast. Autologous neutrophils from individuals were hyperresponsive to fMLP stimulation after eating as compared to after fasting, based on the detection of a secreted binding protein. In a study on individuals without symptoms performed in our laboratory, O_2^- production in neutrophils (NBT test) was detected in a larger percentage of cells after as compared to before breakfast (data not published). Thus, nutrition may emerge as an important factor in the activation status of circulating neutrophils. A prime example has recently been demonstrated in the form of oxidized cholesterol as a mechanism to induce leukocyte adhesion to arterial as well as microvascular endothelium (see article by Lehr and Messmer in this issue) [44].

3.2. Endothelial cells

Endothelial cells are rapidly activated by exposure to thrombin, histamine, PAF, or H_2O_2. The significance of

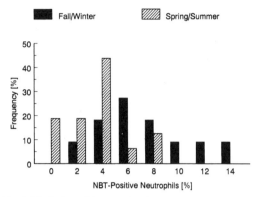

Fig. 2. Distributions of fraction of NBT-positive neutrophils in whole blood from the same rat species at various times throughout the year ($n = 27$ rats). The level of preactivated cells ranged from 2 to 16%, with a significantly higher mean percentage during fall and winter as compared to spring and summer season (8 ± 4 vs. $5 \pm 2\%$, $P = 0.0049$, t-test).

H_2O_2 production is that it is frequently enhanced after adhesive endothelial–leukocyte interaction. Cytokines (TNFα, IL-1), LTB$_4$, bradykinin, and endotoxin are also endothelial agonists. Endothelial cell function is influenced at the level of gene expression by the shear stress of flowing blood on the luminal surface including the regulation of PGI$_2$ synthesis [45]. Further, ICAM-1 and E-selectin expression in cultured endothelium are downregulated after prolonged exposure to flow of varying shear rates [46]. Thus, endothelial cell activation may be exacerbated in chronic low-flow conditions by an upregulation of key adhesion molecules and lack of inhibition by PGI$_2$.

3.3. Platelets

The primary agonists for platelets include thrombin which is formed after activated endothelium converts prothrombinase complex [47], and the protease cathepsin G released from neutrophil granules after activation [48]. Activated platelets have been shown to interact with activated neutrophils in a synergistic manner [49], and in a dual capacity to either prime or inhibit neutrophil O$_2^-$ generation depending on concentration and time course of platelet release products [50]. Moreover, enhanced release of serotonin and thromboxane A$_2$ by activated platelets in hypercholesterolemic humans may contribute to abnormal vascular responses in atherosclerosis [51]. PGI$_2$ is a potent inhibitor of platelet aggregation [52].

3.4. Examples of plasma activating factors

Chang and co-workers [10] launched one of the first investigations into the presence of activating factors for neutrophils in the plasma of patients with cardiovascular diseases. By applying patient plasma to naive neutrophils from sedimented donor blood, they found a more rapid rate

of neutrophil pseudopod formation in patients with acute myocardial infarction (AMI). It could be demonstrated that a plasma-derived factor was responsible for the activation. Recently, Siminiak et al. [53] also found that the plasma from AMI patients contains soluble stimuli capable of inducing in neutrophils the upregulation of adhesion molecules, L-selectin shedding, and oxygen free radical production.

Patients undergoing arterial or myocardial revascularization are subject to local tissue ischemia followed by reperfusion with re-establishment of the blood supply. In a study by Paterson et al. [54], plasma taken 15 min after reperfusion was introduced into chambers overlying dermabrasions on the skin of anesthetized rabbits. Significantly more neutrophils accumulated in the chamber exudate 3 h after exposure to patient plasma as compared to saline or control plasma (general surgery patients, no ischemia). The cells were activated as indicated by pronounced H$_2$O$_2$ generation. The increased microvascular permeability to protein in the chamber was attributed to the elevated concentrations of neutrophil activation release products LTB$_4$ and thromboxane A$_2$. The authors concluded that plasma taken during reperfusion contained a soluble chemotactic and activating factor for neutrophils, but no identification was made.

Recently, Petrasek et al. [55] used an animal model to simulate this clinical paradigm by subjecting rabbits to 5 h of hindlimb ischemia followed by 48 h reperfusion. Plasma was obtained at regular time intervals (comparisons were made to pre-ischemia plasma) and applied to neutrophils isolated from donor rabbit blood to test its ability to alter CD18 expression and related functions. The key result was that plasma collected late in reperfusion, but not early reperfusion or end ischemia, produced significant increases in CD18 expression, and also in vitro adhesion to coated plastic and oxidant production. Both human and animal I/R studies provide evidence for a potent neutrophil activator, but with apparently different time courses, although the patient plasma was sampled only at one time point early in reperfusion. To directly compare the studies, patient plasma would need to be analyzed later in reperfusion.

Elgebaly and co-workers have described a cardiac-derived neutrophil chemotactic factor obtained from the coronary sinus effluent after cardioplegia arrest of the heart in humans and animals [56–58]. The human factor appears to be a novel 3-kDa peptide which is not a cleaved product of a larger protein [59]. In addition to inducing significant neutrophil chemotactic activity, the factor was also found to stimulate release of IL-8, but not TNFα or IL-1, and expression of ICAM-1 and E-selectin in cultured endothelial cells. [60] Senoh and co-workers [61] have also identified a neutrophil chemotactic factor in the coronary effluent and they approached its characterization by eliminating known activating substances. They concluded that it was not complement, adenosine, LTB$_4$, 5-lipoxygenase, throm-

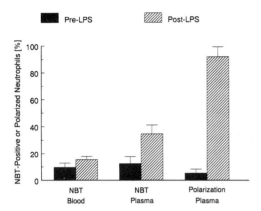

Fig. 3. Spontaneous neutrophil activation during endotoxic shock as measured by NBT and polarization (pseudopod formation) assays with cells in whole blood or isolated donor cells from human volunteers. The significant increase in autologous neutrophil activation (post-LPS vs. pre-LPS, $P < 0.05$) is amplified when plasma from the same conditions is analyzed using naive cells. In particular, the extent of pseudopod formation in neutrophils after 40 s of plasma exposure indicates the rapid and sensitive nature of the polarization assay. Results ($n = 5$–6 rats in each case) indicate that there is an activating factor in plasma which is enhanced during the 5 h shock period.

boxane, or PAF, but that it was a PAF-related substance since PAF antagonists abolished the activity. Thus, there is evidence for one or more newly described factors which activate neutrophils, endothelium, and possibly other cell types after I/R.

Barroso-Aranda et al. [62] and Shen et al. [63] recently published the time course of activation of naive neutrophils in donor rat blood by plasma taken from rats before and during a 3 h hypotensive (40 mmHg) shock period. There is evidently development over time of an activator in the plasma as detected by an increased percentage of neutrophils which produce O_2^- (NBT test). It is interesting to note that the strength of the activator was greater in those animals which did not survive for 24 h after the shock as compared to those which did survive. This result reaffirms the observation that the level of neutrophil activation is critical to the outcome after a circulatory challenge (Fig. 1).

Endotoxic shock is another example in which a neutrophil-activating factor has been detected in the plasma of a compromised circulation [64]. Fig. 3 compares three different indices of spontaneous neutrophil activation in shock after high-dose LPS using data recently obtained in our laboratory. The findings clearly demonstrate that use of shock (post-LPS) plasma with donor neutrophils instead of autologous neutrophils provides a clearer distinction between pre- and post-LPS cell activation. Gel chromatography separates out the most active fraction in plasma after shock at about 100-kDa size, which has been characterized further as a protein of slight negative charge [65]. The significance of this neutrophil-activating factor is evident

in endotoxin-tolerant rats which do not express it during shock, and possibly, as such, they almost all survive a shock protocol lethal to most normal rats (see below).

Smoking is a documented risk factor for cardiovascular disease due presumably, in part, to the elevated leukocyte count in smokers [66]. However, there is recent evidence showing that smoking directly activates vascular cells. A study by Lehr et al. [67] found increased postcapillary leukocyte–endothelium interaction in animals exposed to cigarette smoke, which suggests a selectin upregulation on endothelial cells. In the blood of human smokers, Pitzer and co-workers [21] detected elevated levels of neutrophil activation as compared to non-smokers in terms of O_2^- production (NBT test). Moreover, in comparison, smoker plasma caused in donor neutrophils a shedding of L-selectin and enhanced O_2^- production and pseudopod formation compared with control plasma from non-smokers. This is yet another example of a soluble and transferable activating factor for neutrophils in the plasma of persons at risk for or already affected by circulatory disorders.

4. Cell activation and vascular resistance

Organ blood flow critically depends on the hydraulic resistance of its microvascular network comprised of arterioles, venules, and capillaries. Vascular resistance in each region is a function of blood viscosity and vessel geometry including the number of vessels available for perfusion and the diameter and length of each vessel. The primary effect of cell activation on viscosity is to increase it via formation of cell aggregates which impede laminar blood flow in large blood vessels and potentially block blood flow in the smaller vessels of the microcirculation.

4.1. Arteriolar resistance

The functional requirements of an organ depend largely on an adequate blood supply and patent capillaries to distribute the oxygen and nutrients to each cell. Tissue damage can result after an extended period of either decreased arterial flow or obstructed capillary flow. Vascular resistance resides in most organs, primarily in the arterioles, and production of vasoactive substances from endothelial cells and platelets may produce diameter and flow changes. For example, O_2^- has been shown to annihilate NO in vitro [68]. Prolonged activation of the endothelium may induce arteriolar constriction. Further, the aggregation of activated platelets in atherosclerotic coronary arteries has been shown to be a factor contributing to acute myocardial ischemia [22].

4.2. Capillary resistance

Leukocytes may elevate microvascular resistance despite their relatively small numbers. For skeletal muscle, the amount of this resistance increase has been estimated in the range from a low value of about 2% [5] up to 20% at physiological leukocyte counts [9]. After leukocyte activation, the contribution to resistance further increases to 15–50% of the whole organ resistance [5,6,9] due to a concomitant increase in cytoplasmic viscosity [69], pseudopod formation, and possibly cell swelling [70]. Narrowed capillaries with decreased luminal diameters due to endothelial cell swelling further hinder leukocyte passage [8], but whether or not activation of the endothelium in vivo produces volume changes is unknown.

Under conditions of low blood flow, activated leukocytes have even greater difficulty in traversing capillaries and may become trapped in the capillary lumen and obstruct flow (i.e., capillary no-reflow). Leukocyte trapping has been observed in hemorrhagic shock [71], endotoxic shock [64], and ischemic organs such as heart [72], brain [73], and skeletal muscle [74,75]. Assessment of leukocyte trapping at low perfusion pressures by direct microscopy suggests that it is largely a pressure-dependent effect and may occur even without the adhesive contribution of CD11/CD18 [6], especially if the circulating leukocytes are activated [76].

4.3. Venular resistance

The strength of the membrane adhesion between leukocytes and endothelium in postcapillary venules is increased by activation of one or both cells and their interaction is facilitated by reduced shear rates. In the extreme case of a complete paving of activated leukocytes on activated endothelium, the decrease in the effective venular diameter has been estimated to be about 5 μm [77]. For venules with diameters in the range between 20 and 40 μm, this magnitude of diameter decrease is expected to produce a significant increase in local hemodynamic resistance [8].

5. Cell activation and tissue injury

Many early forms of tissue injury which lead to cardiovascular disease are manifest in the microcirculation. Specific conditions include maldistribution of capillary blood flow, elevated microvascular permeability, accumulation of activated leukocytes in capillaries and postcapillary venules, as well as activation of endothelium and possibly platelets. Demonstration of these circumstances in man is limited at the present time. However, patients with chronic venous insufficiency have ulcers on the skin which facilitate tissue and exudate sampling. Leukocyte entrapment is prevalent and is enhanced when venous pressure is raised by lowering the legs [78]. Further, immunohistochemistry indicated T-lymphocyte and macrophage infiltration into the perivascular space. Parenchymal cell injury occurs when proteolytic enzymes appear, such as elastase which is known to break down endothelial matrix and connective tissue components [79].

The association between cell activation and tissue injury can be documented in animal experiments with acute circulatory challenges. Two commonly used models are I/R and hemorrhagic shock. Animals treated with a monoclonal antibody against CD18 prior to hemorrhagic shock were found to have mitigated cellular injury and improved survival on reperfusion [80]. Reflow in skeletal muscle capillaries was also found to improve with antibody CD18 pretreatment following I/R [81]. Removal of leukocytes from the circulation prior to I/R has lead to conflicting results—e.g., no improvement in flow recovery in brain [82], but in skeletal muscle prevention of otherwise increased vascular permeability and resistance [74]. In addition to the inherent difficulties in introducing leukopenia, a possible explanation for these discrepancies is that the degree of activation in the circulating leukocytes was not controlled although this activation status may be important to determine baseline flow and resistance values.

One consequence of activated cells in the circulation has been investigated with the use of agents that inhibit the formation of oxygen-free radicals or scavenge the oxygen free radicals at the time of production. Reactive oxygen metabolites have been implicated in the pathogenesis of I/R injury [83]. Inhibition or scavenging oxygen free radicals serves to significantly attenuate vascular permeability and total resistance increases in skeletal muscle I/R [84]. There is also a reduction in the number of capillaries with no-reflow [85], suggesting that a loss of endothelial integrity in I/R may induce interstitial edema which in turn causes narrowing of capillaries and increased resistance.

The source of oxygen free radicals may be from individual vascular cells or the result of a synergistic interaction between neutrophils, endothelial cells, and platelets. The beneficial results derived from monoclonal antibodies against CD18 may be due to attenuation of neutrophil adherence and enhanced oxygen free radical release. Suematsu et al. [26] have studied the involvement of endothelial-derived oxygen free radicals in initial cellular injury on reperfusion in ischemic skeletal muscle. The major source of oxygen free radical in this model was derived from xanthine oxidase in endothelium, causing initial cell death shortly after reoxygenation. Later periods during reperfusion were accompanied by adherence and trapping of leukocytes and further enhancement of cell death in the tissue.

Animal models have thus provided clear evidence that activation of vascular cells causes tissue damage and contributes to the low-flow state which, if sustained, may lead to organ failure. An interesting point is that microvascular injury has been found in organs remote from the initial insult such as intestinal perforation [86], intestinal I/R [87], and lower torso ischemia [88]. In particular, lung injury that occurs as a consequence of reperfusion of

M.C. Mazzoni, G.W. Schmid-Schönbein / Cardiovascular Research 32 (1996) 709–719

ischemic lower extremities [89] or intestine [90] was found to be mediated in part by neutrophil CD18. Circulating plasma factors may be responsible for this phenomenon of remote organ injury and, further, may provide the stimulus for multiple organ failure which is a typical sequelae in hemorrhagic shock.

6. Model of low cell activation

There is clear evidence that the outcome following a circulatory challenge is improved if the level of circulating activated leukocytes is maintained low. A case in point are animals tolerant to endotoxin. Tolerance is achieved by daily administration of sublethal graded doses of endotoxin (LPS) for several days and is characterized by the reduced effect of subsequent challenge with a high-dose LPS. In endotoxin-tolerant rats, the proportion of circulating neu-trophils that are activated is kept low as compared to control rats during hypotensive periods in hemorrhagic [91] and endotoxic shock [64]. Corresponding survival rates were significantly improved in tolerant rats. The improved survival after a challenge to the cardiovascular system in the tolerant rats is likely due to the low level of activated circulating leukocytes. The relevance for study-ing an endotoxin-tolerant model is that it possesses desir-able features, which, if characterized, could possibly be applied to the design of new therapeutics.

In endotoxic shock, no-reflow in about 40% of the myocardial capillaries of control rats is strongly correlated with trapped leukocytes [64]. In contrast, the myocardium of tolerant rats has significantly fewer obstructed capillar-ies, due presumably to the lower level of neutrophil activa-tion. Recent evidence from a myocardial infarction model in rats shows greatly reduced infarct size in tolerant as compared to control rats [92]. Furthermore, intravital mi-croscopy indicates a mitigated leukocyte adherence to the venular endothelium in mesentery of tolerant rats during hemorrhagic shock and reperfusion, which suggests that the endothelium in tolerant animals is only minimally activated and the expression of adherence receptors is attenuated [91]. In addition, tolerant animals may have lower mortality after shock since their systemic blood pressure is better regulated (although they are still hy-potensive), due in part to a reduction of inducible NO and NO-mediated vascular hyporeactivity [93].

As indicated, the plasma of control rats after endotoxic shock contains an activating factor for neutrophils which is not present in the plasma prior to shock (see Figs. 3 and 4). Interestingly, the plasma of tolerant rats after shock only minimally activates donor neutrophils as compared to the plasma before shock using the NBT [64] and the polarization assays (see Fig. 4). The activating factor is either not produced during shock or is diminished in potency by inhibition with plasma components developed in tolerance. The plasma endotoxin levels are nearly 10

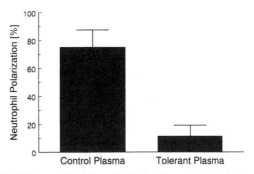

Fig. 4. Plasma activity during endotoxic shock for control and endotoxin-tolerant rats ($n = 5$ in each group) according to the cell polarization assay (due to pseudopod projection) using donor human neutrophils. The strength of the activating factor present in control plasma is much greater than that in tolerant plasma. The low neutrophil activation and microcir-culatory entrapment in tolerant animals during acute circulatory challenge closely correlates with their improved survival rates.

times less in tolerant than in control rats, suggesting that the stimulus for production of an activating factor is greatly reduced in the tolerant state (unpublished data). Further, tolerant plasma contains a component(s) induced by LPS which inhibits activating factor(s) in post-LPS control plasma [94]. It is known that tolerance is accompa-nied by the production of acute-phase proteins (e.g., α_2-macroglobulin) which bind proteases. This distinction, along with enhanced blood clearance of endotoxin [95], could serve as mechanisms for the superior response of tolerant animals to acute circulatory challenges. Tolerant animals typically have lower levels of circulating preacti-vated neutrophils. For example, analysis of plasma applied to the same donor neutrophils showed 15% NBT-positive cells in control rat plasma as compared to 5% NBT-posi-tive cells in tolerant rat plasma.

7. Summary and future directions

A hypothesis has been advanced in this review that activation of blood cells and endothelium in the general circulation may initiate the early stages of microvascular low-flow disorders and thereby sow the seed for cardio-vascular diseases. The limited ability to retrieve activated cells from the circulation for examination requires the analysis of plasma using naive control cells. The presence of an activator for neutrophils in plasma has been shown for all circulatory disorders explored thus far. The bio-chemical nature and source of this in vivo activating factor(s) in each case is currently unexplored. A systematic analysis of the ability of patient plasma to activate primary vascular cells could be carried out with current techniques. Once human activating factors are isolated and character-ized, patients could be screened for levels of activating factor and thus a predisposition for future cardiovascular diseases may be identified. Further, animal models could be developed to facilitate identification of the source of the activating factors in vivo.

We propose that chronic microcirculatory flow impairment and tissue injury associated with cardiovascular disease may originate from persistent release of cytotoxic mediators into the general circulation, possibly due to viral, bacterial, or environmental sources. The goal of such a research frontier is to suggest lifestyle changes or therapeutic intervention for the early prevention of the progression of cardiovascular diseases, especially in high-risk patients, as opposed to merely improving modalities to treat existing disease.

Acknowledgements

Supported by USPHS Program Project grant HL 43026.

References

[1] Somer T, Meiselman HJ. Disorders of blood viscosity. Ann Med 1993;25:31–39.

[2] Ensrud K, Grimm RH. The white blood cell count and risk for coronary heart disease. Am Heart J 1992;124:207–213.

[3] Friedman GD, Klatsky AL, Siegelaub AB. The leukocyte count as a predictor of myocardial infarction. N Engl J Med 1974;290:1275–1278.

[4] Prentice RL, Szatrowski TP, Kato H, Mason MW. Leukocyte counts and cerebrovascular disease. J Chronic Dis 1982;35:703–714.

[5] Warnke KC, Skalak TC. The effects of leukocytes on blood flow in a model skeletal muscle capillary network. Microvasc Res 1990;40:118–136.

[6] Harris AG, Skalak TC. Effects of leukocyte activation on capillary hemodynamics in skeletal muscle. Am J Physiol 1993;264:H909–H916.

[7] Hansell P, Borgström P, Arfors KE. Pressure-related capillary leukostasis following ischemia-reperfusion and hemorrhagic shock. Am J Physiol 1993;265:H381–H388.

[8] Mazzoni MC, Warnke KC, Arfors KE, Skalak TC. Capillary hemodynamics in hemorrhagic shock and reperfusion: in vivo and model analysis. Am J Physiol 1994;267:H1928–H1935.

[9] Sutton DW, Schmid-Schönbein GW. Elevation of organ resistance due to leukocyte perfusion. Am J Physiol 1992;262:H1646–H1650.

[10] Chang RR.K, Chien NT.Y, Chen CH, Jan KM, Schmid-Schönbein GW, Chien S. Spontaneous activation of circulating granulocytes in patients with acute myocardial and cerebral diseases. Biorheology 1992;29:549–561.

[11] Von Andrian UH, Hansell P, Chambers JD, et al. L-selectin function is required for beta 2-integrin-mediated neutrophil adhesion at physiological shear rates in vivo. Am J Physiol 1992;263:H1034–H1044.

[12] Chadwick VS, Ferry DM, Butt TJ. Assessment of neutrophil leukocyte secretory response to fMLP in whole blood in vitro. J Leukocyte Biol 1992;52:143–150.

[13] Babior BM. Oxygen-dependent microbial killing by phagocytes. N Engl J Med 1978;298:659–668.

[14] Malmsten CL. Prostaglandins, thromboxanes, and leukotrienes in inflammation. Am J Med 1986;80:11–17.

[15] Schmid-Schönbein GW. Leukocyte kinetics in the microcirculation. Biorheology 1987;24:139–151.

[16] Jackson MH, Collier A, Nicoll JJ, et al. Neutrophil count and activation in vascular disease. Scot Med J 1992;37:41–43.

[17] Grau AJ, Seitz R, Immel A, Steichen-Wiehn C, Hacke W. Increased levels of leukocyte elastase in ischemic stroke and in subject with vascular risk factors. Cerebrovasc Dis 1995;5:50–54.

[18] Grau AJ, Berger E, Sung KL.P, Schmid-Schönbein GW. Granulocyte adhesion, deformability, and superoxide formation in acute stroke. Stroke 1992;23:33–39.

[19] Schmid-Schönbein GW, Seiffge D, Delano FA, Shen K, Zweifach BW. Leukocyte counts and activation in spontaneously hypertensive and normotensive rats. Hypertension 1991;17:323–330.

[20] Schröder S, Palinski W, Schmid-Schönbein GW. Activated monocytes and granulocytes, capillary nonperfusion, and neovascularization in diabetic retinopathy. Am J Pathol 1991;139:81–100.

[21] Pitzer JE, Del Zoppo GJ, Schmid-Schönbein GW. Neutrophil activation in smokers. Biorheology 1996;33:45–58.

[22] Mehta JL, Nicolini FA, Donnelly WH, Nichols WW. Platelet-leukocyte–endothelial interactions in coronary artery disease. Am J Cardiol 1992;69:8B–13B.

[23] McIntyre TM, Patel KD, Zimmerman GA, Prescott SM. Oxygen radical-mediated leukocyte adherence. In: Granger, DN, Schmid-Schönbein, GW, eds. Physiology and Pathophysiology of Leukocyte Adhesion. New York: Oxford, 1995:261–277.

[24] Bevilacqua MP, Pober JS, Mendrick DL, Cotran RS, Gimbrone MA. Identification of an inducible endothelial-leukocyte adhesion molecule. Proc Natl Acad Sci USA 1987;84:9238–9242.

[25] Roy RS, McCord JM. Superoxide and ischemia: conversion of xanthine dehydrogenase to xanthine oxidase. In: Greenwald, RA, Cohen, G, eds. Oxy Radicals and Their Scavenger Systems II. New York: Elsevier, 1983:145–153.

[26] Suematsu M, DeLano FA, Poole D, et al. Spatial and temporal correlation between leukocyte behavior and cell injury in postischemic rat skeletal muscle microcirculation. Lab Invest 1994;70:684–695.

[27] Flores NA, Sheridan DJ. The pathophysiological role of platelets during myocardial ischaemia. Cardiovasc Res 1994;28:295–302.

[28] Collins CE, Cahill MR, Newland AC, Rampton DS. Platelets circulate in an activated state in inflammatory bowel disease. Gastroenterology 1994;106:840–845.

[29] Elwood PC, Renaud S, Sharp DS, Beswick AD, O'Brien JR, Yarnell JW. Ischemic heart disease and platelet aggregation;the Caerphilly collaborative heart disease study. Circulation 1991;83:38–44.

[30] Colwell JA, Winocour PD, Halushka PV. Do platelets have anything to do with diabetic microvascular disease? Diabetes 1983;32:14–19.

[31] Barroso-Aranda J, Chavez-Chavez RH, Schmid-Schönbein GW. Spontaneous neutrophil activation and the outcome of hemorrhagic shock in rabbits. Circ Shock 1992;36:185–190.

[32] Barroso-Aranda J, Schmid-Schönbein GW. Transformation of neutrophils as indicator of irreversibility in hemorrhagic shock. Am J Physiol 1989;257:H846–H852.

[33] Nolte D, Lehr HA, Messmer K. Adenosine inhibits postischemic leukocyte–endothelium interaction in postcapillary venules of the hamster. Am J Physiol 1991;261:H651–H655.

[34] Cronstein BN, Kimmel SC, Levin RI, Martiniuk F, Weissmann G. A mechanism for the antiinflammatory effects of corticosteroids: the glucocorticoid receptor regulates leukocyte adhesion to endothelial cells and expression of endothelial-leukocyte adhesion molecule-1 and intercellular adhesion molecule-1. Proc Natl Acad Sci USA 1992;89:9991–9995.

[35] Asako H, Kubes P, Wallace J, Gaginella T, Wolf RE, Granger DN. Indomethacin-induced leukocyte adhesion in mesenteric venules: role of lipoxygenase products. Am J Physiol 1992;262:G903–G908.

[36] Kubes P, Suzuki M, Granger DN. Nitric oxide: an endogenous modulator of leukocyte adhesion. Proc Natl Acad Sci USA 1991;88:4651–4655.

[37] Suematsu M, Tamatani T, Delano FA, et al. Microvascular oxidative stress preceding leukocyte activation elicited by in vivo nitric oxide suppression. Am J Physiol 1994;266:H2410–H2415.

[38] Baeziger NL, Force LE, Bechner PR. Histamine stimulates prostacyclin synthesis in cultured human umbilical vein endothelial cells. Biochem Biophys Res Commun 1980;92:1435–1440.

[39] Buyon JP, Korchak HM, Rutherford LE, Ganguly M, Weissmann G. Female hormones reduce neutrophil responsiveness in vitro. Arthritis Rheum 1984;27:623–630.

[40] Ley K, Baker JB, Cybulsky MI, Gimbrone MA, Luscinskas FW. Intravenous interleukin-8 inhibits granulocyte emigration from rabbit mesenteric venules without altering L-selectin expression or leukocyte rolling. J Immunol 1993;151:6347–6357.

[41] Bass DA, Olbrantz P, Szejda P, Seeds MC, McCall CE. Subpopulations of neutrophils with increased oxidative product formation in blood of patients with infection. J Immunol 1986;136:860–866.

[42] Woodhouse PR, Khaw KT, Plummer M, Foley A, Meade TW. Seasonal variations of plasma fibrinogen and factor VII activity in the elderly: winter infections and death from cardiovascular disease. Lancet 1994;343:435–439.

[43] Grau AJ, Buggle F, Heindl S, et al. Recent infection as a risk factor for cerebrovascular ischemia. Stroke 1995;26:373–379.

[44] Lehr HA, Becker M, Marklund SL, et al. Superoxide-dependent stimulation of leukocyte adhesion by oxidatively modified LDL in vivo. Arterioscler Thromb 1992;12:824–829.

[45] Frangos JA, Eskin SG, McIntire LV, Ives CL. Flow effects on prostacyclin production by cultured human endothelial cells. Science 1985;227:1477–1479.

[46] Sampath R, Kukielka GL, Smith CW, Eskin SG, McIntire LV. Shear stress-mediated changes in the expression of leukocyte adhesion receptors on human umbilical vein endothelial cells in vitro. Ann Biomed Eng 1995;23:247–256.

[47] Bevilacqua MP, Pober JS, Majeau GR, Cotran RS, Gimbrone MA. Interleukin-1 (IL-1) induces biosynthesis and cell surface expression of procoagulant activity in human vascular endothelial cells. J Exp Med 1984;160:618–623.

[48] LaRosa CA, Rohrer MJ, Benoit SE, Rodino LJ, Barnard MR, Michelson AD. Human neutrophil cathepsin G is a potent platelet activator. J Vasc Surg 1994;19:306–319.

[49] Coëffier E, Delautier D, LeCouedic JP, Chignard M, Denizot Y, Benveniste J. Cooperation between platelets and neutrophils for paf-acether (platelet-activating factor) formation. J Leukocyte Biol 1990;47:234–243.

[50] Naum CC, Kaplan SS, Basford RE. Platelets and ATP prime neutrophils for enhanced O_2^- generation at low concentrations but inhibit O_2^- generation at high concentration. J Leukocyte Biol 1991;49:83–89.

[51] Aviram M, Brook JG. Platelet activation by plasma lipoproteins. Prog Cardiovasc Dis 1987;30:61–72.

[52] Moncada S, Palmer RM.J, Higgs EA. Prostacyclin and endothelium derived relaxing factor: biological interactions and significance. Thromb Haemost 1987;25:597–618.

[53] Siminiak T, Egdell RM, O'Gorman DJ, Dye JF, Sheridan DJ. Plasma-mediated neutrophil activation during acute myocardial infarction: role of platelet-activating factor. Clin Sci 1995;89:171–176.

[54] Paterson IS, Smith FC.T, Tsang GM.K, Hamer JD, Shearman CP. Reperfusion plasma contains a neutrophil activator. Ann Vasc Surg 1993;7:68–75.

[55] Petrasek PF, Lindsay TF, Romaschin AD, Walker PM. Plasma activation of neutrophil CD18 after skeletal muscle ischemia—a potential mechanism for late systemic injury. Am J Physiol 1996;in press.

[56] Elgebaly SA, Hashmi FH, Houser SL, Allam ME, Doyle K. Cardiac-derived neutrophil chemotactic factors: detection in coronary sinus effluents of patients undergoing myocardial revascularization. J Thorac Cardiovasc Surg 1992;103:952–959.

[57] Elgebaly SA, Houser SL, Elkerm AF, Doyle K, Gillies C, Dalecki

[58] K. Evidence of cardiac inflammation after open heart operations. Ann Thorac Surg 1994;57:391–396.

[58] Elgebaly SA, Masetti P, Allam M, Forouhar F. Cardiac derived neutrophil chemotactic factors; preliminary biochemical characterization. J Mol Cell Cardiol 1989;21:585–593.

[59] Elgebaly SA, Tyles E, Houser SL, Elkerm AF, Mauri F. partial purification of a novel cardiac-derived neutrophil chemotactic factor: Nourin-1. Circulation 1993;88:I-240(Abstract).

[60] Tyles E, Houser SL, Shams NK, Consoli KA, Elgebaly SA. Coronary sinus effluents from cardiopulmonary bypass patients stimulate the secretion of cytokines and adhesion molecules by vascular endothelial cells. Circulation 1994;100:I-465(Abstract).

[61] Senoh M, Aosaki N, Ohsuzu F, et al. Early release of neutophil chemotactic factor from isolated rat heart subjected to regional ischaemia followed by reperfusion. Cardiovasc Res 1993;27:2194–2199.

[62] Barroso-Aranda J, Zweifach BW, Mathison JC, Schmid-Schönbein GW. Neutrophil activation, tumor necrosis factor, and survival after endotoxic and hemorrhagic shock. J Cardiovasc Pharmacol 1995;25:S23–S29.

[63] Shen K, Chavez-Chavez RH, Loo AK.C, Zweifach BW, Barroso-Aranda J, Schmid-Schönbein GW. Interpretation of leukocyte activation measurements from systemic blood vessels. In: Moskowitz, MA, ed. The Proceedings of the 19th Princeton Conference. Boston, MA: Butterworth Heineman, 1996:in press.

[64] Barroso-Aranda J, Schmid-Schönbein GW, Zweifach BW, Mathison JC. Polymorphonuclear neutrophil contribution to induced tolerance to bacterial lipopolysaccharide. Circ Res 1991;69:1196–1206.

[65] Pfeifer PH, Mazzoni MC, Hugli TE, Schmid-Schönbein GW. A neutrophil activating factor is enhanced in plasma from rats after endotoxic shock. FASEB J 1995;9:A888(Abstract).

[66] Grimm RH, Neaton JD, Ludwig W. Prognostic importance of the white blood cell count for coronary, cancer, and all-cause mortality. J Am Med Assoc 1985;254:1932–1937.

[67] Lehr HA, Kress E, Menger MD. Involvement of 5-lipoxygenase products in cigarette smoke-induced leukocyte/endothelium interaction in hamsters. Int J Microcirc Clin Exp 1993;12:61–73.

[68] Gryglewski RJ, Palmer RM.J, Moncada S. Superoxide anion is involved in the breakdown of endothelium-derived vascular relaxing factor. Nature 1986;320:454–456.

[69] Lipowsky HH, Riedel D, Shi GS. In vivo mechanical properties of leukocytes during adhesion to venular endothelium. Biorheology 1991;28:53–64.

[70] Grinstein S, Furuya W, Cragoe EJ. Volume changes in activated human neutrophils: the role of Na^+/H^+ exchange. J Cell Physiol 1986;128:33–40.

[71] Barroso-Aranda J, Schmid-Schönbein GW, Zweifach BW, Engler RL. Granulocytes and no-reflow phenomenon in irreversible hemorrhagic shock. Circ Res 1988;63:437–447.

[72] Engler RL, Schmid-Schönbein GW, Pavelec RS. Leukocyte capillary plugging in myocardial ischemia and reperfusion in the dog. Am J Pathol 1983;111:98–111.

[73] Del Zoppo GJ, Schmid-Schönbein GW, Mori E, Copeland BR, Chang CM. Polymorphonuclear leukocytes occlude capillaries following middle cerebral artery occlusion and reperfusion. Stroke 1991;22:1276–1283.

[74] Korthuis RJ, Grisham MB, Granger DN. Leukocyte depletion attenuates vascular injury in postischemic skeletal muscle. Am J Physiol 1988;254:H823–H827.

[75] Bagge U, Amundson B, Lauritzen C. White blood cell deformability and plugging of skeletal muscle capillaries in hemorrhagic shock. Acta Physiol Scand 1980;180:159–163.

[76] Ritter LS, Wilson DS, Williams SK, Copeland JG, McDonagh PF. Early in reperfusion following myocardial ischemia, leukocyte activation is necessary for venular adhesion but not capillary retention. Microcirculation 1995;2:315–327.

[77] House SD, Lipowsky HH. Leukocyte–endothelium adhesion: micro-

hemodynamics in mesentery of the cat. Microvasc Res 1987;34:363–379.

[78] Edwards J, McMullin GM, Scott HJ, Wilkinson L. White blood cell distribution in chronic venous insufficiency. In: Coleridge Smith PD, ed. Microcirculation in Venous Disease. Austin: R.G. Landes Company, 1994:113–128.

[79] Shields D. White cell activation. In: Coleridge Smith PD, ed. Microcirculation in Venous Disease. Austin: R.G. Landes Company, 1994:113–128.

[80] Vedder NB, Winn RK, Rice CL, Chi EY, Arfors KE, Harlan JM. A monoclonal antibody to the adherence-promoting leukocyte glycoprotein, CD18, reduces organ injury and improves survival from hemorrhagic shock and resuscitation in rabbits. J Clin Invest 1988;81:939–944.

[81] Jerome SN, Smith CW, Korthuis RJ. CD18-dependent adherence reactions play an important role in the development of the no-reflow phenomenon. Am J Physiol 1993;264:H479–H483.

[82] Bednar MM, Raymond S, McAuliffe T, Lodge PA, Gross CE. The role of neutrophils and platelets in a rabbit model of thromboembolic stroke. Stroke 1991;22:44–50.

[83] Korthuis RJ, Granger DN. Reactive oxygen metabolites, neutrophils, and the pathogenesis of ischemic-tissue/reperfusion. Clin Cardiol 1993;16:I19–I26.

[84] Korthuis RJ, Granger DN, Townsley MI, Taylor AE. The role of oxygen-derived free radicals in ischemia-induced increases in canine skeletal muscle vascular permeability. Circ Res 1985;57:599–609.

[85] Menger MD, Steiner D, Messmer K. Microvascular ischemia-reperfusion injury in striated muscle: significance of 'no-reflow'. Am J Physiol 1992;263:H1892–H1900.

[86] Donais K, Gill RS, Macris N, Ellis CG. Remote microvascular injury in an acute septic model. Microcirculation 1995;2:69(Abstract).

[87] Koike K, Moore EE, Moore FA, Read RA, Carl VS, Banerjee A. Gut ischemia/reperfusion produces lung injury independent of endotoxin. Crit Care Med 1994;22:1438–1444.

[88] Fantini GA, Conte MS. Pulmonary failure following lower torso ischemia: clinical evidence for a remote effect of reperfusion injury. Am Surg 1995;61:316–319.

[89] Welbourn R, Goldman G, Kobzik L, et al. Role of neutrophil adherence receptors (CD18) in lung permeability following lower torso ischemia. Circ Res 1992;71:82–86.

[90] Hill J, Lindsay T, Valeri CR, Shepro D, Hechtman HB. A CD18 antibody prevents lung injury but not hypotension after intestinal ischemia-reperfusion. J Appl Physiol 1993;74:659–664.

[91] Barroso-Aranda J, Chavez-Chavez RH, Mathison JC, Suematsu M, Schmid-Schönbein GW. Circulating neutrophil kinetics during tolerance in hemorrhagic shock using bacterial lipopolysaccharide. Am J Physiol 1994;266:H415–H421.

[92] Eising GP, Mao L, Schmid-Schönbein GW, Engler RL, Ross J. Effects of induced tolerance to bacterial lipopolysaccharide on myocardial infarct size in rats. Cardiovasc Res 1996;31:73–81.

[93] Szabó C, Thiemermann C, Wu CC, Perretti M, Vane JR. Attenuation of the induction of nitric oxide synthase by endogenous glucocorticoids accounts for endotoxin tolerance in vivo. Proc Natl Acad Sci USA 1994;91:271–275.

[94] Mazzoni MC, Pfeifer PH, Hugli TE, Schmid-Schönbein GW. Plasma of endotoxin-tolerant rats has inhibitory properties for endogenous neutrophil activation during endotoxic shock. Microcirculation 1995;2:105(Abstract).

[95] Greisman SE, DuBuy B. Mechanisms of endotoxin tolerance. IX. Effect of exchange transfusion. Proc Soc Exp Biol Med 1975;148:675–678.

Generation of *in vivo* activating factors in the ischemic intestine by pancreatic enzymes

Hiroshi Mitsuoka, Erik B. Kistler, and Geert W. Schmid-Schönbein*

Department of Bioengineering and The Whitaker Institute for Biomedical Engineering, University of California at San Diego, La Jolla, CA 92093-0412

Communicated by Yuan-Cheng B. Fung, University of California at San Diego, La Jolla, CA, December 13, 1999 (received for review November 1, 1999)

One of the early events in physiological shock is the generation of activators for leukocytes, endothelial cells, and other cells in the cardiovascular system. The mechanism by which these activators are produced has remained unresolved. We examine here the hypothesis that pancreatic digestive enzymes in the ischemic intestine may be involved in the generation of activators during intestinal ischemia. The lumen of the small intestine of rats was continuously perfused with saline containing a broadly acting pancreatic enzyme inhibitor (6-amidino-2-naphthyl *p*-guanidino-benzoate dimethanesulfate, 0.37 mM) before and during ischemia of the small intestine by splanchnic artery occlusion. This procedure inhibited activation of circulating leukocytes during occlusion and reperfusion. It also prevented the appearance of activators in portal venous and systemic artery plasma and attenuated initiating symptoms of multiple organ injury in shock. Intestinal tissue produces only low levels of activators in the absence of pancreatic enzymes, whereas in the presence of enzymes, activators are produced in a concentration- and time-dependent fashion. The results indicate that pancreatic digestive enzymes in the ischemic intestine serve as an important source for cell activation and inflammation, as well as multiple organ failure.

rat | splanchnic arterial occlusion | shock | multiple organ failure | microcirculation

S hock is a life-threatening cardiovascular complication (1). Cellular activation in the circulation is a relatively early event in shock that can be detected by leukocyte or endothelial superoxide production, pseudopod projection, expression of membrane adhesion molecules, and many other cell functions (2, 3). Cell activation fundamentally alters the biomechanics of microvascular blood flow by a shift in rheological, adhesive, and cytotoxic cell properties. The interaction between activated leukocytes and endothelial cells is followed by cell and organ failure (4). The level of activation correlates with survival after the shock (5), but the mechanism for production and the source of activating factors has remained an unresolved problem. Several candidates have been proposed, including endotoxins, lipid-derived products, and cytokines (6–9).

Recently, we have demonstrated that the supernatant of the homogenized pancreas, but less so homogenates of other organs, mediates a powerful activation of cardiovascular cells (10). Incubation of homogenates from nonactivating organs, such as the liver or intestine, with low concentration of serine proteases also increases the ability to activate leukocytes. An enzyme inhibitor could block the activation, suggesting that pancreatic enzymes may play a central role in the production of activating factors.

Pancreatic enzymes are discharged via the pancreatic duct into the duodenum and intestine as a requirement for digestion. We hypothesize that pancreatic enzymes in the ischemic intestine may be involved in the production of activating factors for circulatory cells in shock. Ischemia in the intestine may cause an increased epithelial and endothelial permeability (11) and allow such activators to enter the systemic circulation via the portal vein and/or intestinal lymphatics (12, 13). Thus, after entering the systemic circulation, digestive enzymes and their products

may play a central role in activation of leukocytes or endothelial cells in shock. The pathophysiological role of pancreatic enzymes in intestinal ischemia has not been elucidated.

The objective of this study was to investigate the influence of digestive enzymes on activation of circulating leukocytes during shock produced by splanchnic arterial occlusion. A broad-acting enzyme inhibitor placed into the intestinal lumen served to prevent the formation of central activation and initiation of multiple organ failure in shock.

Materials and Methods

Splanchnic Arterial Occlusion After Pancreatic Ligation. Male Wistar rats (290–340 g; Charles River Breeding Laboratories) were maintained on a standard rat chow and water ad libitum. All experiments were reviewed and approved by the University of California San Diego Animal Use and Care Committee. After general anesthesia (sodium pentobarbital, 25 mg/kg, i.p.), the left femoral artery was cannulated with polyethylene tubing (PE-50, Clay Adams) to record arterial pressure. The left femoral vein was cannulated (PE-50, Clay Adams) for administration of anesthetic agent. The celiac and superior mesenteric artery were isolated through an abdominal midline incision and looped with silk thread (3-0) at their aortic origin. The threads were guided out of the abdomen through polyethylene tubing (PE-50), and arteries were occluded by application of tension to the threads.

In the first experimental sequences, three groups were formed: I, a nonischemic sham shock ($n = 5$ rats); II, a splanchnic artery occlusion shock ($n = 10$); and III, a splanchnic artery occlusion shock group with pancreatic vessel ligation ($n = 10$). In group III, all pancreatic arteries and draining veins were ligated before splanchnic ischemia. Considering that the rat pancreas consists of three main parts, the splenic, duodenal, and gastric gland, the pancreatic ligation was achieved as follows. The communication between splenic and duodenal gland was dissected to expose the origin of the splenic artery or vein from the celiac artery or the superior mesenteric vein. The splenic artery and vein were ligated at their origin. Short gastric arteries and veins and other small communicators from the stomach and spleen were also ligated (silk 4-0) to isolate the splenic gland from the systemic circulation. To isolate the blood supply to the duodenal gland, the vessels connecting along the superior mesenteric vein and also the duodenum were ligated, because there are small anastomoses that originate from the marginal artery and vein of the duodenum. The duodenum with its marginal artery was ligated at the junction to the stomach and the jejunum. Isolation of blood supply to the pancreas was confirmed by central injection of heparinized and filtered India ink (30% in saline) and detailed examination of a color changes in the surrounding tissue but not in the pancreas *per se*.

Abbreviations: IF, intestinal fluid; SAL, saline; ANGD, 6-amidino-2-naphthyl *p*-guanidino-benzoate dimethanesulfate.

*To whom reprint requests should be addressed. E-mail: gwss@bioeng.ucsd.edu.

The publication costs of this article were defrayed in part by page charge payment. This article must therefore be hereby marked "*advertisement*" in accordance with 18 U.S.C. §1734 solely to indicate this fact.

All groups except the sham group were subjected to 100 min of splanchnic arterial occlusion and reperfused for 30–120 min. Thirty minutes after reperfusion, arterial blood samples in heparin were collected from five rats in each group. To separate plasma, the blood samples were immediately centrifuged.

Activation of Naive Leukocytes by Experimental Plasma. Because activated leukocytes have high probability to become trapped in the microcirculation, there is a need to test the ability of plasma to activate with cells other than autologous circulating cells. Therefore, the activation produced by rat plasma aliquots was tested on naive granulocytes derived from nonischemic controls. Human leukocytes from laboratory volunteers without symptoms were used to minimize further use of donor animals. Pilot studies had shown that the levels of activation achieved by rat plasma applied to rat and to human granulocytes were not significantly different (10). Granulocyte-rich plasma was collected after about 40-min sedimentation of 40 ml of venous blood from a healthy human volunteer, layered onto 3.5 ml of Histopaque (Sigma), and centrifuged ($600 \times g$, 20 min). The leukocytes and erythrocytes (about 1 ml) were suspended in 1 ml of Krebs–Henseleit solution layered onto 2.5 ml of 55% and 74% isotonic Percoll solution (Sigma). After centrifugation at $600 \times g$ for 15 min, the intermediate purified granulocyte layer was removed and resuspended in 1 ml of 10 mM phosphate-buffered saline. These control cells are referred to as naive granulocytes.

The ability of plasma from the rats with splanchnic arterial occlusion to activate was determined by pseudopod formation on naive granulocytes. Suspended granulocytes (100 μl; about 10,000 per mm^3 in phosphate-buffered saline) were mixed with 100 μl of test plasma from shock rats. This mixture was incubated for 10 min at room temperature. Glutaraldehyde in phosphate-buffered saline (3%, 100 μl) was added to arrest the pseudopods. Granulocytes with cytoplasmic projections (pseudopodia) >1 μm were designated as activated cells. The fraction of activated granulocytes, of at least 200 cells, was counted.

Splanchnic Arterial Occlusion with Blockade of Pancreatic Enzymes in the Intestine. In this second set of experiments, the pancreatic blood supply was left intact, and instead the intestinal lumen was rinsed and portal venous blood samples were collected according to the following procedure. Besides femoral artery and vein catheters, a line (PE-50) was inserted into a cecal branch of the portal vein for blood collection. Because venous congestion was frequently encountered, the cecum was removed before cannulation of the portal vein.

The proximal duodenum and terminal ileum were cannulated (PE-280), and the initial intestinal contents were gently rinsed with 30 ml of saline and collected. Thereafter the intestinal lumen was gently rinsed with 3,000 ml of saline (37°C) at constant pressure (10–15 mmHg; 1 mmHg = 133 Pa) and then discarded. A closed-loop intestinal perfusion from the duodenum to the terminal ileum was set up with a peristaltic pump (MasterFlex, Cole–Parmer) with a priming volume of 50 ml.

Experimental Procedure. The intestinal contents were mixed and centrifuged ($500 \times g$, 10 min). The supernatant is referred to as intestinal fluid (IF). The small intestinal lumen was perfused at constant flow rate (4.0 ml/min) with the following three solutions: (*i*) 45 ml of diluted IF with 5 ml of 5% glucose (IF group, *n* = 5 rats); 45 ml of saline (*ii*) without, and (*iii*) with 10 mg of 6-amidino-2-naphthyl *p*-guanidinobenzoate dimethanesulfate (ANGD, 0.37 mM, Nafamostat Mesilate, Torii Pharmaceutical, Chiba, Japan) (14) in 5 ml of 5% glucose (SAL and ANGD groups, *n* = 5 rats, respectively). Five nontreated animals were used as nonischemic controls.

After 15 min of intestinal perfusion, the animals were subjected to 100 min of splanchnic ischemia, which was confirmed

by cyanotic organ discoloration and the loss of pressure pulsation in the mesentery. After 100 min of splanchnic ischemia, the celiac and superior mesenteric arteries were reperfused. One milliliter of circulating fluid in the intestine was collected from the reservoir before ischemia and 90 min after ischemia. Arterial and portal venous blood (0.3 ml each) was sampled before ischemia, after 90 min of ischemia, as well as 30, 60, and 120 min of reperfusion. Plasma Lyte A (0.6 ml; Baxter Scientific Products, McGaw Park, IL) was administered (i.v.) immediately after each blood withdrawal. Twenty microliters of arterial blood was used to measure the fraction of granulocytes with pseudopods in the circulation, and 20 μl was used for leukocyte counts. The remainder of the arterial and portal venous blood aliquot was centrifuged at 1,500 $\times g$ for 10 min, and supernatant plasma was stored at -70°C. After centrifugation, the blood cells were resuspended in 0.6 ml of Plasma Lyte A and reinjected (i.v.).

Leukocyte Counts and Activation. Leukocyte counts were made with a hemocytometer. Immediately after withdrawal, 20 μl of arterial blood was fixed in 3% glutaraldehyde (Fisher Scientific; in 10 mM phosphate-buffered saline), and stained with 20 μl of 0.02% crystal violet in phosphate-buffered saline. The fraction of activated granulocytes was counted (see above).

Organ Leukocyte Infiltration. Myeloperoxidase activity was used as marker for assessment of leukocyte infiltration into the small intestine (about 20 cm proximal to the ileocecal junction, 3 cm in length), liver, and right lung. Tissue myeloperoxidase levels were determined by a spectrophotometric method (15). Myeloperoxidase from human purulent leukocytes (Sigma) served as standard.

Protease Activity in Intestinal Fluid. Serine protease activity of any of the intestinal lumen perfusates was determined by a spectrofluorometric method (16). Trypsin (1440 BAEE units/mg, Sigma) was used as standard.

Intestinal Histology. A sample of small intestine was longitudinally dissected, fixed in 10% buffered formalin, and embedded in paraffin. Five-micrometer sections were made, stained with hematoxylin and eosin, and examined at ×200 magnification. Severity of intestinal injury was estimated by the length between the tip of the villi and the musculus mucosae, a measure of the mucosal layer thickness (17). In each specimen, the measurement was made at 10 randomly selected locations and averaged.

Bile Flow Rate. Bile flow rate was determined to indicate the severity of liver injury (18). The bile flow rate for 15 min was measured at each time point before splanchnic arterial occlusion and during reperfusion, and normalized for each animal with the flow rate before ischemia.

Lung Wet/Dry Weight Ratio. The left lung lobes were harvested, wet weight was determined, and samples were dried at 70°C for 72 hr. The residuum was weighed, and the ratio of wet to dried weight was computed.

***In Vitro* Production of Activator.** In three rats, the intestine was removed and homogenized in 0.25 mM sucrose solution at 1:5 weight ratio and centrifuged (1,000 $\times g$; 10 min). The sediment was discarded, and the supernatant was ultracentrifuged (15,900 $\times g$; 30 min; 4°C). The supernatant intestinal homogenate was mixed with 0, 1, 5, and 10 mg/ml trypsin (Sigma) and incubated for 0, 30, 60, 90, 120, and 150 min. At each time, the ability to activate naive leukocytes was tested by pseudopod formation (see above). Each test was completed in quadruplicate.

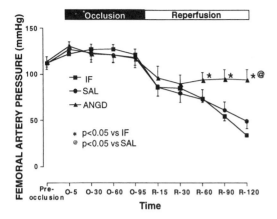

Fig. 1. Time course of femoral mean artery blood pressure. Numbers on the abscissa indicate the minutes after splanchnic arterial occlusion (O) and reperfusion (R). Squares refer to the intestinal fluid (IF) group, circles the saline rinse (SAL) group, and triangles to the ANGD group.

Statistics. Measurements are expressed as mean ± standard error of mean. Statistical significance in wet/dry ratio of the lung, tissue myeloperoxidase level, and histological measurements were compared by Fisher's protected least-squares difference. Values of mean arterial blood pressure, bile flow rate, activation of circulating leukocytes, serine protease activity in the intestinal fluid, and activation produced by plasma, and *in vitro* experimental data were tested by the analysis of variance with Bonferroni's correction. $P < 0.05$ was considered significant.

Results

Pancreatic Vascular Ligation in Splanchnic Artery Occlusion Shock. The femoral blood pressure during the occlusion period was not significantly different between group II without and group III with pancreatic vessel ligation, respectively. Thirty minutes after reperfusion, the arterial pressure dropped in both groups. Although the mean arterial pressure fell slightly less in the ligated group (63 ± 17 mmHg in group II vs. 47 ± 8 mmHg in group III with pancreatic ligation, $P > 0.05$), 60 min after reperfusion both groups reached 100% mortality. The ability of arterial plasma to activate naive leukocytes is significantly elevated 30 min after reperfusion in these groups, when compared with sham shock group I (22.6 ± 4% in splanchnic arterial occlusion group II and 5.3 ± 2% in sham shock group I) ($P < 0.05$). This sequence of events is unaffected by pancreatic vessel ligation (22.7 ± 5% in group III), indicating that the pancreas is less likely the predominant source of a leukocyte activator. Instead, the pancreas may serve as a source of enzymes that are discharged into the intestine as part of normal digestion. Therefore, our remaining study is focused on pancreatic enzymes in the intestine.

Significance of Proteolytic Reactions in the Ischemic Intestine. *Mean arterial pressure.* After splanchnic arterial occlusion, all animals exhibited a slight increase in mean arterial pressure and a sudden decrease after the reperfusion (Fig. 1). The groups with intestinal fluid in the small intestine (IF) and with a saline rinse of the intestinal lumen (SAL) did not recover from the hypotension after the reperfusion. In contrast, the mean artery pressure in the rats with saline rinse and ANGD in the intestinal lumen was significantly higher after reperfusion ($P < 0.01$ ANGD vs. IF and SAL groups). The mean arterial pressure in the ANGD group

remained on average at 84% of its initial value after 2 hr of reperfusion.

Serine protease activity in the ischemic intestine. Before splanchnic ischemia, the serine protease activity in the intestinal fluid in the IF group was 5.8 ± 1.8 × 10³ units/ml. Intestinal lavage with saline served to lower the activity to 2.0 ± 0.9 × 10³ units/ml ($P < 0.05$, IF vs. SAL group) and ANGD further decreased the activity to 1.2 ± 0.1 × 10³ units/ml ($P < 0.05$, SAL vs. ANGD). After 90 min of ischemia, the serine protease activity in the intestinal fluid of the IF, SAL, and ANGD groups was 5.6 ± 2.3 × 10³, 3.3 ± 2.3 × 10³, and 1.2 ± 0.2 × 10³ units/ml, respectively ($P < 0.05$, SAL vs. ANGD). Intestinal perfusion with ANGD kept the serine protease activity lower than in other groups ($P < 0.01$ ANGD vs. IF and SAL groups).

Leukocyte count and activation. In the preischemic period, the numbers of leukocytes in the arterial blood were the same (5.9 ± 0.4 per mm³ in the IF group, 5.4 ± 0.3 per mm³ in the SAL group, and 5.5 ± 0.2 per mm³ in the ANGD group). In the IF and SAL groups, the leukocyte count in arterial blood started to decrease after ischemia and reached its lowest value 120 min after reperfusion. Intestinal perfusion of protease inhibitor completely ameliorated the leukopenia (Fig. 2A).

Although no differences were detected among groups in the preischemic period (Fig. 2A), splanchnic ischemia and reperfusion led to an increased number of leukocytes with pseudopod projections. Intestinal perfusion of protease inhibitor, however, served to significantly lower the number of activated cells.

Activation potential of shock plasma. Arterial and portal venous plasma from nonischemic control animals only mildly activated naive leukocytes (8.0 ± 3.1% and 9.0 ± 2.0%, respectively). In the preocclusion period, the ability to activate naive leukocytes with arterial or portal venous plasma from either of the groups did not differ from the one in the plasma of nonischemic controls (Fig. 2B).

Ninety minutes of splanchnic ischemia served to increase the activation produced by portal venous plasma in the IF and SAL groups ($P < 0.01$ vs. nonischemic control, data not shown). At this time point, the portal venous plasma in both IF and SAL groups yielded higher activation values than did arterial plasma ($P < 0.01$). In the ANGD group, however, the activation produced by either the arterial or portal venous plasma was not significantly increased throughout the experiment, and remained much lower than in either the IF or SAL group ($P < 0.01$) (Fig. 2B).

Tissue myeloperoxidase levels. At 120 min after reperfusion, the intestinal, hepatic, and pulmonary myeloperoxidase levels were increased in all organs. This is an early indicator of leukocyte infiltration and organ failure in shock (19). Intestinal perfusion of protease inhibitor, however, significantly attenuated the myeloperoxidase activity (Fig. 3 A–C).

Intestinal injury after splanchnic ischemia reperfusion. Splanchnic arterial occlusion and reperfusion led to a reduction of the mucosal thickness, a phenomenon associated with morphological damage to the intestine. Intestinal perfusion of protease inhibitor served to maintain mucosal thickness and to reduce intestinal injury (Fig. 3D).

Bile flow rate. The baseline values for the bile flow rate were not different among groups (95 ± 12 μl/min in IF group, 75 ± 9 μl/min in SAL group, 85 ± 7 μl/min in ANGD group). Splanchnic arterial occlusion caused a reduction of bile production without significant differences among groups. A small degree of recovery was observed after reperfusion in the IF and SAL groups. Intestinal perfusion of protease inhibitor increased the recovery of bile flow rate (Fig. 3E).

Pulmonary wet/dry ratio. The level of edematous lung injury after splanchnic arterial occlusion and reperfusion is greatly increased in the IF and SAL groups. Perfusion of the intestinal lumen with protease inhibitor served to reduce the average lung wet/dry ratio to the level of the nonischemic control (Fig. 3F).

214

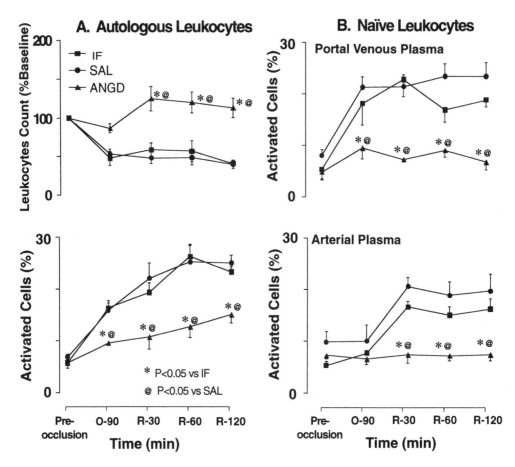

Fig. 2. (A) Activation of circulating leukocytes and leukocyte count. Numbers on the abscissa indicate the minutes after splanchnic arterial occlusion (O) and reperfusion (R). P < 0.01 at 30, 60, and 120 min of reperfusion. (B) Activation of naive leukocytes induced by rat plasma in splanchnic arterial occlusion shock. Numbers on the abscissa indicate the minutes after splanchnic arterial occlusion (O) and reperfusion (R). * and @, P < 0.05, vs. IF and SAL groups, respectively.

In vitro production of leukocyte activator. Without trypsin, a low level of activation was detected in intestinal homogenates incubated for 120 min at 37°C. Intestinal homogenates incubated with trypsin, however, increased the ability to activate naive leukocytes in a time- and trypsin concentration-dependent manner (Fig. 4).

Discussion

The current study indicates that pancreatic digestive enzymes in the intestinal lumen may form *in vivo* activators during ischemia and reperfusion of the intestine. These *in vivo* activators lead to up-regulation of cells in the circulation and compromise of microcirculatory functions. A key issue is the localization of pancreatic digestive enzymes. The pancreas serves as a source for digestive enzymes, and if homogenized under *in vitro* conditions, yields a large amount of potent activators (10). But as shown here, occlusion of the pancreatic blood supply during splanchnic arterial occlusion has no significant impact on cell activation or multiple organ failure. Instead, as part of normal digestion, the pancreas discharges into the intestine. In the presence of pancreatic enzymes, the ischemic intestine produces powerful *in vivo*

activators whose effect can be detected relatively early during ischemia in portal venous and central plasma. With release of the activators, the initial signs of multi-organ failure become apparent. These include arterial pressure reduction, leukopenia with infiltration of leukocytes into the microcirculation of the lung, liver, and other organs (19), pulmonary edema formation, morphological damage to the intestine, reduced liver function, and other cell and organ dysfunctions. The current results show that these processes are already attenuated by intestinal fluid lavage associated with reduced enzyme activity, and they are almost completely eliminated after protease blockade in the intestinal lumen.

Abdominal organs, such as intestine, have been the focus of shock research for several decades. Surgical excision of the small intestine increases survival of rats after hemorrhagic shock (20). But intestinal tissue by itself is not necessarily a source of a strong activator (Fig. 4). Our *in vitro* studies indicate that with the exception of the pancreas, homogenates of a variety of internal organs, including the intestine (after careful *in vitro* lavage), are less effective in activating leukocytes. Pancreatic enzymes or serine proteases *per se* (trypsin and chymotrypsin in saline after

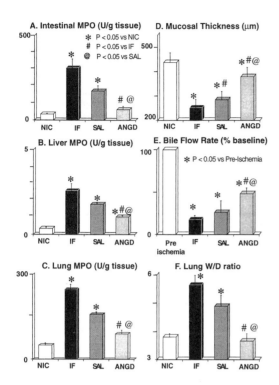

Fig. 3. (A–C) Tissue myeloperoxidase levels (MPO) in intestine (A), liver (B), and lung (C) after 90 min of reperfusion. (D) The level of intestinal injury as measured by mucosal thickness. (E) Bile flow rate as a measure of liver function. (F) Pulmonary dry/wet weight ratio. The abscissa indicates the time during splanchnic arterial occlusion (O) and reperfusion (R). NIC represents the nonischemic control. *, #, and @, $P < 0.05$, vs. NIC, IF, and SAL groups, respectively, A–D and F. *, $P < 0.05$ vs. preischemic control in E.

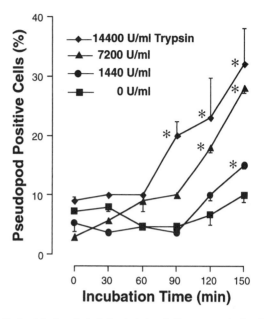

Fig. 4. Activation of naive leukocytes by intestinal homogenates incubated with increasing concentrations of trypsin. *, $P < 0.05$ vs. the values at 0 min.

incubation periods of 2 hr) are also ineffective in generation of leukocyte activation (10). Instead, incubation of tissue from different organs, including the intestine, with pancreatic proteases for 2 hr leads to the production of humoral activators. These activators have an impact on a variety of cell functions, including production of oxygen free radicals in endothelial cells and leukocytes, expression of leukocyte adhesion molecules, and initiation of the leukocyte adhesion cascade to the point of parenchymal cell apoptosis (10). The activators, derived from intestine incubated with pancreatic enzymes, may have different molecular masses and include materials <2 kDa (10). Such small products may serve as humoral activators by entry into the systemic circulation through the portal vein and the lymphatics (1, 12, 21), especially if the permeability in an ischemic and damaged intestine is increased (11, 22). Proteases (23) as well as oxygen free radicals (24) may be involved in production of morphological damage to the intestinal villi. Perfusion of the intestinal lumen with a combination of ANGD and allopurinol (10 mg/ml) to block xanthine oxidase during splanchnic artery occlusion and reperfusion does not lead to further improvement of blood pressure (Fig. 1), reduction of activators (Fig. 2), or organ function (Fig. 3) compared with ANGD alone, and ANGD does not inhibit the conversion of xanthine dehydrogenase into xanthine oxidase (results not shown).

Translocation of bacteria and endotoxins from the intestinal lumen into the lymphatics and vasculature has been proposed as a trigger mechanism for multiple-organ failure (25, 26). Endotoxin may induce shock, but there is no conclusive evidence that endotoxin translocation under experimental conditions serves as the main mechanism for inflammation (27, 28). The current evidence suggests that in addition to endotoxins, there are alternative mediators for cell activation. In light of the evidence that trypsin increases the ability of intestinal homogenates to activate, it is possible that activators are produced in the ischemic intestine by mechanisms that are influenced by protease activity in the lumen. A high protease activity in the ischemic intestine was detected even after intestinal lavage with saline (SAL group). In addition to an incomplete washout of pancreatic enzymes from the high number of minute crevices among intestinal villi, possible sources may be the conversion of trypsinogen, not removed by intestinal lavage (23), by enterokinase and/or lysosomal proteases from ischemic tissue (29, 30). Other sources may involve the complement or coagulation system, in which trypsin-like serine proteases are involved (31), and intestinal mast cells or activated leukocytes. Enzymes, which can be activated or released by proteolytic reactions, may be involved in the production of activators. For example, phospholipase A_2 may be an important enzyme in the production of stimulatory lysophospholipids, such as platelet-activating factor (32). Zymogen of pancreatic phospholipase A_2 is activated by trypsin. Inflammation caused by proteases or their products may also stimulate the secretion of phospholipase A_2 from the intestine.

The choice of enzyme inhibitor in the current study was governed by the fact that among five different inhibitors (Complete, phenylmethylsulfonyl fluoride, benzamidine, and aprotinin), ANGD was the most effective blocker for production of activators *in vitro* in the presence of pancreatic enzymes (10), providing ≈80% reduction of activation. ANGD is a broad-acting serine protease inhibitor (IC_{50} for trypsin = 1.3×10^{-8} M; IC_{50} for chymotrypsin = 5.9×10^{-4} M), which, at the concentrations used in the current experiments (0.37 mM) also blocks lipase ($IC_{50} = 8.4 \times 10^{-6}$ M) and phospholipase A_2 ($IC_{50} = 7.5 \times 10^{-5}$ M) (14) but has less effect on elastase or other enzymes. ANGD (at concentrations between 10^{-13} M and 10^{-3} M) does not significantly deactivate naive leukocytes after stimulation with plasma of rats after splanchnic arterial occlusion (collected at 30 min after reperfusion) or after stimulation with fMet-Leu-Phe (at 10^{-8} and 10^{-9} M; results not shown). Future use of enzyme blocker may involve even broader spectra to prevent activator production. Identification of the particular species constituting the plasma activator is the subject of current research (33).

The interaction between activated leukocytes and endothelial cells leads to accumulation of leukocytes in various organs, cytotoxicity, and cell death (34). Although such a process is mediated by humoral activators in the plasma of the systemic circulation, an inflammatory reaction is initiated in innocent-bystander organs that eventually may lead to multi-organ failure (35). When leukocytes are activated, the intracellular F-actin formation with pseudopod formation is up-regulated and several membrane adhesion molecules are expressed, a process that lowers cell deformability and leads to accumulation of leukocytes in the microcirculation (36). Not only may such leukocytes start inflammation but also the abnormal cellular entrapment in the microcirculation leads to immune suppression because of reduced numbers of circulating cells. Suppression of plasma activator production serves to maintain normal leukocyte counts in the circulation (Fig. 2). Furthermore, the plasma-derived activators may also have direct suppressive effects on the parenchymal cell function without the requirement for leukocyte involvement (10).

In summary, we provide evidence that the small intestine in splanchnic arterial occlusion shock plays a role in activation of circulating blood cells by a mechanism that is controlled by pancreatic enzyme activity. Activators can be produced in the ischemic intestine and then enter the systemic circulation, impair multiple cell functions, and trigger a cascade that results in multi-organ failure. The effect can be blocked by inhibitors of pancreatic enzymes in the lumen of the intestine, a procedure that may have clinical application.

National Institutes of Health Grant HL 43026 supported this research.

1. Deitch, E. A. (1992) *Ann. Surg.* **216,** 117–134.
2. Mazzoni, M. C. & Schmid-Schönbein, G. W. (1996) *Cardiovasc. Res.* **32,** 709–719.
3. Yao, Y. M., Redl, H., Bahrami, S. & Schlag, G. (1998) *Inflammation Res.* **47,** 201–210.
4. Xiao, F., Eppihimer, M. J., Young, J. A., Nguyen, K. & Carden, D. L. (1997) *Microcirculation* **4,** 359–367.
5. Barroso-Aranda, J., Zweifach, B. W., Mathison, J. C., Schmid-Schönbein, G. W. (1995) *J. Cardiovasc. Pharmacol.* **25,** S23–S29.
6. Rinaldo, J. E., Henson, J. E., Dauber, J. H. & Henson, P. M. (1985) *Tissue Cell* **17,** 461–472.
7. Palmer, R. M., Stepney, R. J., Higgs, G. A. & Eakins, K. E. (1980) *Prostaglandins* **20,** 411–418.
8. Camussi, G., Bussolino, F., Salvidio, G. & Baglioni, C. (1987) *J. Exp. Med.* **166,** 1390–1394.
9. Gerkin, T. M., Oldham, K. T., Guice, K. S., Hinshaw, D. B. & Ryan, U. S. (1993) *Ann. Surg.* **217,** 48–56.
10. Kistler, E. B. (1998) Ph.D. thesis (Univ. of California, San Diego).
11. Sun, Z., Wang, X., Deng, X., Lasson, A., Wallen, R., Hallberg, E. & Andersson, R. (1998) *Shock* **10,** 203–212.
12. Moore, E. E., Moore, F. A., Franciose, R. J., Kim, F. J., Biffl, W. L. & Banerjee, A. (1994) *J. Trauma* **37,** 881–887.
13. Upperman, J. S., Deitch, E. A., Guo, W., Lu, Q. & Xu, D. (1998) *Shock* **10,** 407–414.
14. Fujii, S. & Hitomi, Y. (1981) *Biochim. Biophys. Acta* **661,** 342–345.
15. Schierwagen, C., Bylund-Fellenius, A. C. & Lundberg, C. (1990) *J. Pharmacol. Methods* **23,** 179–186.
16. Aoyama, T., Ino, Y., Ozeki, M., Oda, M., Sato, T., Koshiyama, Y., Suzuki, S. & Fujita, M. (1984) *Jpn. J. Pharmacol.* **35,** 203–227.
17. Parks, D. A., Williams, T. K. & Beckman, J. S. (1988) *Am. J. Physiol.* **254,** G768–G774.
18. Karwinski, W., Husøy, A. M., Farstad, M. & Søreide, O. (1989) *J. Surg. Res.* **46,** 99–103.
19. Barroso-Aranda, J., Schmid-Schönbein, G. W., Zweifach, B. W. & Engler, R. L. (1988) *Circ. Res.* **63,** 437–447.
20. Chang, T. W. (1997) *J. Trauma* **42,** 223–230.
21. Upperman, J. S., Deitsch, E. A., Guo, W., Lu, Q. & Dazhong, X. (1998) *Shock* **10,** 407–414.
22. Granger, D. N., Sennett, M., McElearney, P. & Taylor, A. E. (1980) *Gastroenterology* **79,** 474–480.
23. Bounous, G., Menard, D. & de Medicis, E. (1977) *Gastroenterology* **73,** 102–108.
24. Parks, D. A., Bulkley, G. B., Granger, D. N., Hamilton, S. R. & McCord, J. M. (1982) *Gastroenterology* **82,** 9–15.
25. Wolochow, H., Hildebrand, G. J. & Lamanna, C. (1966) *J. Infect. Dis.* **116,** 523–528.
26. Bahrami, S., Yao, Y. M., Leichtfried, G., Redl, H., Schlag, G. & Di Padova, F. E. (1997) *Crit. Care Med.* **25,** 1030–1036.
27. Koike, K., Moore, E. E., Moore, F. A., Read, R. A., Carl, V. S. & Banerjee, A. (1994) *Crit. Care Med.* **22,** 1438–1444.
28. Schlichting, E., Grotmol, T., Kähler, H., Naess, O., Steinbakk, M. & Lyberg, T. (1995) *Shock* **3,** 116–124.
29. Leffer, A. M. & Barenholz, Y. (1972) *Am. J. Physiol.* **223,** 1103–1109.
30. Wildenthal, K. (1978) *J. Mol. Cell. Cardiol.* **10,** 595–603.
31. Nakayama, Y., Senokuchi, K., Nakai, H., Obata, T. & Kawamura, M. (1997) *Drugs Future* **22,** 285–293.
32. Shakir, K. M., Gabriel, L., Sundram, S. G. & Margolis, S. (1982) *Am. J. Physiol.* **242,** G168–G176.
33. Sofianos, A. (1999) M.S. thesis (Univ. of California, San Diego).
34. DeLano, F. A., Forrest, M. J. & Schmid-Schönbein, G. W. (1997) *Microcirculation* **4,** 349–357.
35. Barroso-Aranda, J., Chavez-Chavez, R. H., Mathison, J. C., Suematsu, M. & Schmid-Schönbein, G. W. (1994) *Am. J. Physiol.* **266,** H415–H421.
36. Ritter, L. S., Wilson, D. S., Williams, S. K., Copeland, J. G. & McDonagh, P. F. (1995) *Microcirculation* **2,** 315–327.

ENGINEERING

MEDICAL SCIENCES

CHAPTER 9

MOLECULAR BASIS OF CELL MEMBRANE MECHANICS

LANPING AMY SUNG

Department of Bioengineering, University of California, San Diego

1. Introduction

Molecular biology is an important part of bioengineering. This chapter will show how molecular biology is being integrated into research in biomechanics. One of the systems my laboratory is working on is the cell membrane. In this research, we address how individual components of the cell membrane — especially the protein molecules — affect the mechanical properties of cell membranes.

The plasma membrane is a 50-nm thick film that serves as a barrier between the contents of the cell and its surrounding medium. This film is composed mainly of a lipid bilayer, which is penetrated by channel and pump proteins that transport specific substances into and out of the cell. The lipid bilayer is also associated with other proteins, including those acting as sensors to detect the changes in its environment, and those function as infrastructures to provide mechanical stability for the membrane when the cell is being stressed or deformed.

Some of the mechanical properties of the membrane (e.g. membrane fluidity) are dependent on the composition and the structure of phospholipids, sterols and glycolipids that form the lipid bilayer of the membrane. Yet, other mechanical properties of the membrane (e.g. the resistance to elastic deformation or fragmentation) are primarily attributable to the membrane proteins. Cell membrane proteins can be classified into three categories on the basis of their relationships with the lipid bilayer: (A) transmembrane proteins that span the lipid bilayer once or more; (B) acetylated proteins (i.e. proteins having a fatty acid component) embedded, or partially embedded in the lipid bilayer; and (C) peripheral proteins that are noncovalently associated with the lipid bilayer or other proteins on either side of the membrane. Many of these membrane proteins interact with each other noncovalently. They form network structures that are integral with the lipid bilayer and provide the mechanical properties of the cell membrane.

We investigate the functional role of specific proteins in the organization of the protein network that is associated with the cell membrane, and how specific proteins affect mechanical properties and stability of the cell membrane. We especially want to know the stress-strain relationship of the cell membrane when a specific protein is absent, overexpressed, underexpressed, or mutated. In other words, we want to know the molecular basis of cell membrane mechanics.

Our research helps us to understand how the cell membrane deforms in response to stress, how much the cell membrane deforms before it breaks, how does a membrane skeletal protein affect the mechanics of the membrane, and what molecular and mechanical defects may be present in certain hereditary diseases. Our research also involves creating cell membrane models in which one specific membrane network protein is mutated or depleted by genetic engineering. Generating and analyzing these disease models helps us to understand and appreciate the importance of biomechanics in living tissues. Furthermore, restoring normal mechanical stability and properties to these model membranes, and rescuing other disease phenotypes in hereditary diseases by gene therapy will be among the challenges facing the new generation of bioengineers in the next century.

1.1. *The membrane skeletal network*

The human erythrocyte membrane is the best characterized cell membrane system in living cells. It is the simplest in that an erythrocyte has no complicated three-dimensional cytoskeletons connecting the plasma membrane to other parts of the cell. It has only a two-dimensional protein network underneath the lipid bilayer. Furthermore, an erythrocyte has no nucleus and other major organelles in the cytoplasm.

The mechanical properties of the erythrocyte membrane have been well characterized. The molecular structures of the major proteins and lipids and their organization in the membrane have also been well studied. Basically all major erythrocyte membrane skeletal proteins (Fig. 1A), and some minor ones, have been cloned and characterized in the past fifteen years. Electron microscopy and biochemical analyses of hypotonically expanded human erythrocyte membranes have shown that these major membrane proteins form a relatively regular, two-dimensional protein network (Fig. 1B) on the cytosolic side of the membrane (Fig. 1C). This hexagonal membrane skeletal network supports the mechanical stability of the lipid bilayer and contributes to the mechanical properties of the erythrocyte membrane. While the 200-nm long, spring-like spectrin tetramers between two junctional complexes are the thread of the hexagons (Fig. 1C), the short 37-nm stubs of actin protofilaments are the key components of the knobs. The knobs, or the junctional complexes, are where the tail ends of spectrin tetramers (usually six) meet. The head-to-head association of spectrin dimers occurs in the middle of the 200-nm long tetramers. The junctional complexes also contain several actin-associated proteins, such as tropomyosin and erythrocyte (E)-tropomodulin (3–10), that are involved in defining the uniform length (37 nm) of the actin protofilaments (Fig. 5).

This two-dimensional protein network is anchored to the endoface of the lipid bilayer in the erythrocyte membrane mainly through pairs of band 3/ankyrin/protein 4.2 complexes (Fig. 1C). The pairs of such complexes are located in mid regions of the spectrin tetramers. Each of the proteins in the complex plays an important role in attaching the membrane skeletal network to the lipid bilayer: band 3 is the major transmembrane protein that serves as the anion exchanger (a channel protein); Ankyrin is a peripheral membrane protein that binds strongly and noncovalently to the β subunit of spectrin; and protein 4.2, which binds to both band 3 and ankyrin, has a fatty acid component, making protein 4.2 and the entire complex tightly associated with the lipid bilayer.

Each triangle formed by three spectrin tetramers is hung perpendicularly to the lipid bilayer from underneath, by three pairs of the band 3/ankyrin/band 4.2 complexes, one pair on each tetramer, in the mid region, with each complex in a pair being 80 nm apart from each other (Fig. 1B and 1C). Sometimes, based on the electron microscopy, one or both sites

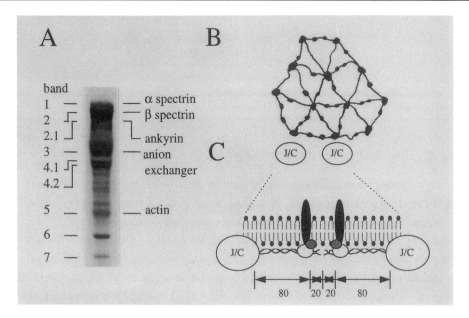

Fig. 1. The molecular organization of erythrocyte membrane skeletal network. (A) The protein profile of human erythrocyte membranes after being separated on the SDS-polyacrylamide gel. The specific names of the major proteins are on the right. The names of these proteins, based on their positions on the gel, are on the left. (B) The sketch showing the top view of the hypotonically expanded membrane skeletal network. J/C stands for junctional complex. The pair of smaller complexes between the two junctional complexes in mid regions of the spectrin tetramers, are protein 4.2/band 3/ankyrin complexes. (C) The sketch showing the side view of the erythrocyte membrane in between two junctional complexes. A spectrin tetramer is 200-nm long, consisting of two $\alpha\beta$ spectrin dimers, which are associated with a head-to-head fashion. The two tail ends of the spectrin tetramer joint the junctional complexes and the two head ends meet in the middle. A spectrin tetramer is associated with a pair of protein 4.2/band 3/ankyrin complexes, each is 20 nm away from the midpoint of the spectrin tetramer. The complex consisting of transmembrane proteins band 3 (in black), protein 4.2 (in gray), and ankyrin (in white), links the erythrocyte membrane skeletal network to the endoface of the lipid bilayer. Theses complexes are the major bridges that hang the spectrin-actin based membrane skeletal network to the lipid bilayer.

for the complexes on β spectrin are not occupied. In the native state, all the membrane skeletal proteins, while linked to the flexible skeletal network, may be pached closely with each other, forming a single layer of protein sheet under the lipid bilayer. What is the advantage of such network organization in providing the stability of the lipid bilayer? How does the network organization allow the elastic deformation of the erythrocytes? These are the important engineering questions because erythrocytes need to be strong in order to survive the flow dynamics of the cardiovascular system and yet highly deformable in order to negotiate through narrow capillaries, many of which have smaller diameters than that of the erythrocytes.

We have contributed to the better understanding of the molecular structures of the erythrocyte membrane components by cloning and characterizing two cDNA sequences of human reticulocytes (precursors of erythrocytes). One cDNA encodes for the major erythrocyte membrane skeletal protein 4.2 (1), and the other for a minor erythocyte membrane skeletal protein Eropomodulin (3). The paper describing the cloning of protein 4.2 is enclosed on p. 228.

Fig. 2. Deformation of a human erythrocyte in response to four different shear stresses in a flow channel. The wall shear stresses are 0, 0.5, 1.7, and 2.5 dynes/cm^2, from left to right.

1.2.   *The mechanical properties of erythrocyte membranes*

The mechanical properties of erythrocyte membranes have been characterized by several techniques using engineering principles and technologies. These include flow channel, micropipette aspiration, ektatocytometry, and microceiving techniques (11, 12, for review see Refs. 13 and 14). While the first three techniques analyze the stress-strain relationship of the erythrocyte membrane by observing the degree of erythrocyte deformation, the microceiving technique measures the resistance of the flow in relation to the deformability of erythrocytes. In addition, the osmotic fragility is also a measurement of the mechanical stability of the erythrocyte membrane when erythrocytes are subjected to low osmotic solutions, and the degrees of hemolysis of erythrocytes are quantified.

An example of analyzing the stress-strain relationship of an erythrocyte is demonstrated in Fig. 2. It shows the various degrees of cell deformation of a normal erythrocyte in response to increasing levels of shear stresses controlled by the fluid flow rate when the cell is attached to the floor of the channel with a point attachment (11).

To understand why the mechanical stability of the membrane or the deformability of diseased erythrocytes is altered in patients with hemolytic anemia or other disorders, and how to restore the mechanical stability and properties, we must elucidate the molecular basis of membrane mechanics of both normal and diseased erythrocytes.

2.   What Are the Studies Reported in the Enclosed Paper?

In order to understand how membrane skeletal proteins interact with each other in forming the protein network, and how the network organization effects the mechanical properties of the erythrocyte membrane, the amino acid sequences or the primary structures of these membrane skeletal proteins are needed. The amino acid sequences are also essential to identify the molecular defects in patients and to study the effects of mutations on the mechanical properties of erythrocyte membranes.

Recombinant DNA technologies are the technologies that have been used in recent years to reveal the primary structures of major membrane skeletal proteins. The recombinant DNA technologies include the use of restriction endonucleases, direct nucleotide sequencing, cDNA and genomic DNA cloning, site-specific mutagenesis, polymerase chain reaction (PCR), and generation of transgenic or knockout cells and animals.

Although our laboratory has used the above recombinant DNA technologies to clone and characterize both protein 4.2 and E-tropomodulin cDNAs and their corresponding genes, to generate a series of mutant recombinant proteins, and to knock out a specific gene in mouse models the paper reporting the cDNA cloning of human protein 4.2 (1) is enclosed in this chapter. It is chosen because it reports several important findings and basic analyses in one single scientific paper: (1) the technologies involved in the cDNA cloning, (2) the finding of two alternatively spliced cDNA isoforms of the same protein, (3) the analysis used for predicting a protein's secondary structure, (4) the hydropathy plot for predicting the possible association between a protein and the membrane lipid bilayer, (5) the homology search in protein and nucleotide databases, and the sequence alignment, and (6) the experience of a pleasant surprise in the world of scientific discovery.

2.1. The cloning of human erythrocyte protein 4.2

The cDNA cloning of human erythrocyte protein 4.2 was done by immunoscreening of a human reticulocyte cDNA library with a protein 4.2 specific antibody. The additional 5' sequences of the cDNA were obtained by PCR, a technique used to amplify a piece of DNA *in vitro* with pairs of sequence-specific primers. Nucleotide sequencing of these clones, and translating the open reading frame of the cDNAs, predicted a protein of 691 amino acids and a longer isoform with a 30-amino acid insertion. The authenticities of the cDNA sequences were verified by several lines of evidence. The translated amino acid sequence also allowed the prediction of protein properties, such as molecular mass, isoelectric pH, secondary structure, and hydropathy plot.

2.2. Sequence homologies between protein 4.2 and transglutaminases

The complete amino acid sequence of protein 4.2 allowed the search for its sequence homology among all published protein sequences collected in the databases. National Center for Biotechnology Information (NCBI, http://www.ncbi.nlm.gov), for example, provides such a service. By doing so we discovered, to our surprise, that human protein 4.2, a major skeletal protein in erythrocyte membranes, actually belongs to an enzyme family. Protein 4.2, in fact, has a significant homology with a group of transglutaminases, e.g. the coagulation factor XIIIa and liver transgulataminase. The sequence homology indicates that protein 4.2 and other transglutaminases evolved from a common ancestor gene from long ago!

Homologous proteins often have related (not necessarily similar) functions. Transglutaminases are a group of enzymes that generally function to catalyze covalent cross-linking of proteins and stiffen tissues and cells. Protein 4.2, however, is a pseudo-transglutaminase in that it cannot enzymatically cross-link proteins covalently. While all the functional transglutaminases have a Cys-Trp in their active sites to catalize the cross-linking, protein 4.2 has an Ala-Trp instead.

2.3. Why two transglutaminases (one pseudo- and one real) in erythrocytes?

Normal mature erythrocytes are highly deformable. The deformability is needed for erythrocytes to negotiate through some narrow capillaries whose diameters are smaller than that of erythrocytes. It is not clear why two transglutaminases (one pseudo-, one real) exist in erythrocytes. We hypothesized that protein 4.2 and erythrocyte transglutaminase, by coordinating their bindings to the major transmembrane protein band 3, regulate the degree of covalent cross-linking of the membrane protein network. The degree of cross-linking,

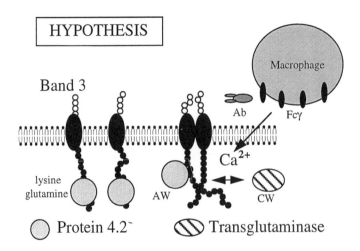

Fig. 3. A schematic drawing of hypothesis involving the function of protein 4.2 and erythrocyte transglutaminase. By coordinating their bindings to the major transmembrane protein band 3, these two proteins regulate the degree of covalent cross-linking of the membrane protein network. Ab: antibody; TcγR: Fcγ receptors on macrophages that undergo erythrophagocytosis; CW: Cys-Trp; AW: Ala-Trp

in turn, may regulate the elastic deformation, the surface topology, and the life span of erythrocytes (Fig. 3). We are designing experiments to validate (or to disapprove) the hypothesis, centering on the mechanical functions of these two interesting and homologous proteins in erythrocytes, especially near the end of their life spans.

3.   What Are the Related Researches Beyond the Enclosed Paper?

3.1.   *Protein binding site*

Recombinant DNA technologies have been used to clone more than a dozen of human erythrocyte membrane proteins and to characterize their interactions. Included is the cDNA for human E-tropomodulin, a tropomyosin- and actin-binding protein located in the junctional complexes of the erythrocyte membrane network (see Fig. 1B) (3).

E-tropomodulin is a capping protein that caps the actin filaments at the pointed end and regulates the length of the actin filaments (4). It is likely that the E-tropomodulin/tropomyosin complex plays an important role in there organization and the biomechanics of the membrane skeletal network during the terminal differentiation of erythrocytes. In nonmuscle cells, E-tropomodulin binds to the *N*-terminus of human tropomyosin isoform 5 (TM5), thus positions itself on the very pointed end of the actin filaments to function as a capping protein (Fig. 4) (5). We have also identified tropomyosin isoform 5 (5) and isoform 5b (6) to be the two major tropomyosin isoform incorporated in human erythrocyte membrane skeletons (5). Site-specific mutagenesis of human tropomoysin isoform 5 demonstrated that 3 hydrophobic residues isoleucine, valine, and isolencine located at the "*a*" and "*d*" positions of the *N*-terminal heptad repeats (between residue's 7–14) and a positively charged residue, arginine, at position 12, are critical for the E-tropomodulin-binding (Fig. 5 bottom panel) (7). Site-specific mutagenesis allows us to change one or more specific residues in a molecule in order to identify critical residues

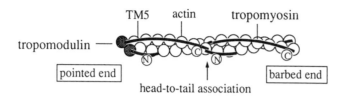

TM5   actin   tropomyosin

tropomodulin —

| pointed end | | barbed end |

head-to-tail association

Fig. 4. A molecular model of nonmuscle actin filament showing the binding of E-tropomodulin to the *N*-terminus of tropomyosin isoform 5 (TM5) at the pointed end and the head-to-tail association of tropomyosin molecules along the actin filament. In human erythrocytes, the actin protofilament filaments are only 6 or 7 actin monomers in length (∼ 33–37 nm), possibly associated with only two tropomyosin molecules, one in each groove of the actin filaments, and one E-tropomodulin at the pointed end. The actin protofilament, E-tropomodulin, tropomyosin, and other associated proteins form the junctional complex in the membrane skeletal network (J/C in Fig. 1, panels B and C).

involved in the protein interactions. The interactions among proteins are important for the network organization and the mechanical properties.

3.2. *Molecular ruler, protofilament, hexagon, and elastic deformation*

Although TM5 and TM5b are products of two different genes γ-TM and α-TM genes, respectively, they share several common features, including: (1) both are low molecular weight (LMW) isoforms, have 248 residues, and are ∼33–35 nm long, (2) both have a high actin-binding affinity, and (3) both have a high E-tropomodulin-binding affinity (6). The following discusses the actin affinity of TM and why it is important for the stability of protofilaments; the length of TM and how it defines the hexagonal geometry of the membrane skeleton; and the E-tropomodulin affinity of TM and how that modulates the function of TM isoforms and the length of the actin filaments (Fig. 5).

High actin affinity allows TM5 and TM5b to form more stable protofilaments. Erythrocytes are constantly subjected to the flow dynamics of the cardiovascular system and frequently deformed in negotiating their ways through narrow capillaries. Protofilaments located in the center of the junctional complexes must be strong enough to resist the pulling of spectrin in response to stresses, in order to maintain the integrity of the membrane skeletal network. The main function of TM is to coat and stiffen to actin filaments, making them more resistant to depolymerization and fragmentation. TM5 and TM5b, the two TM isoforms that have a high affinity toward actin, should function better than other LMW TM isoforms in stabilizing protofilaments.

The length of TM molecules defines the geometry of the membrane skeletal network. In general, high molecular weight (HMW) TM isoforms (284 residues, ∼40–43 nm long) may protect 7 G-actins in one strand and LMW TM isoforms (248 residues, ∼33–35 nm long) 6. As LMW isoforms, TM5 and TM5b stabilize six G-actins, and allow six αβ spectrin dimers to bind to one protofilament (presumably one αβ spectrin dimer binds to one pair of G-actin in the double helix). As a result, LMW isoforms, rather than HMW isoforms, favor the organization of hexagonal lattices in the membrane skeletal network. The mostly hexagonal arrangement of the lattices allows a seamless continuation of the spectrin-actin based skeletal network throughout the entire cell membrane. Such a 2-D membrane skeletal network is essential for the mechanical stability of a circulating erythroctye, as the

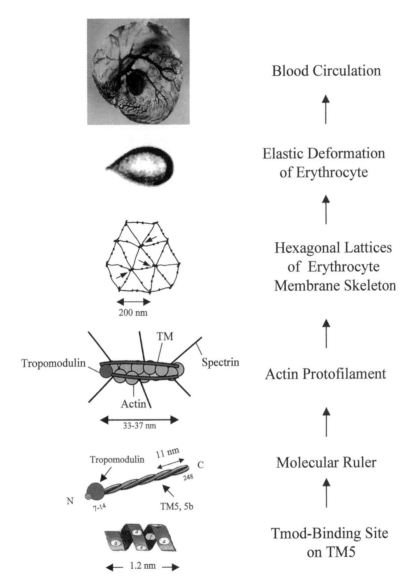

Fig. 5. A view on how E-tropomodulin and specific TM isoforms contribute to the erythrocyte membrane mechanics. E-tropomodulin binding site is described in Section 3.1. Molecular ruler, protofilament, and hexagon are described in Section 3.2. Elastic deformation of erythrocytes is described in Section 1.2. The top panel shows "blue" *E-Tmod* +/− erythrocytes circulating in a 9.5 day old mouse embryo. The yolk sac is outside and the embryo proper with a "blue" heart is inside. Blood cells and candiomyocytes turned blue after the X-gal staining because the "knocked in" *lacZ* reporter gene, which replaced the *E-Tmod* gene, was highly expressed as described in Section 3.6. From Ref. 15, reproduced by permission of the American Society of Hematology.

enucleated, biconcave mature erythrocyte has no longer a supporting 3-D actin cytoskeleton in the cytoplasm.

The high E-tropomodulin affinity of TM5 and TM5b makes the TM/E-tropomodulin complex an effective measuring device, or a molecular ruler, capable of metering off long actin filaments to short protofilaments. This is because TM complexed with E-tropomodulin is able to bind and block an actin filament at its pointed end, but not able to overlap with other TMs in a head-to-tail fashion along the actin filaments or at the barbed end (Fig. 4). As the result, only the first six G-actins (in one strand, or twelve in the double helix) located at the pointed end of the actin filaments are protected by TM. Given the stress and strain undergone in the erythrocyte membrane during circulation, segments of actin filaments not coated by TM are likely to be fragmented or depolymerized. Thus, only short segments of the actin filaments that are protected by 1 TM can survive. The complex, therefore, contributes to the formation and maintenance of protofilaments in erythrocytes. The fact that TM5 has a high initial binding affinity and a low cooperativity to actin filaments should also favor the formation of protofilaments with only 1 TM in length.

3.3. *Genomic organization*

Recombinant DNA technologies have also been used to clone and characterize genomic DNA of humans and other species. We have cloned the genomic DNA encoding protein 4.2 to understand how erythrocyte protein 4.2 isoforms are derived through the alternative splicing (2). We have also cloned human and mouse E-tropomodulin genes in order to understand their exon-intron organization (8). Studies on the genomic organization allow us to understand how the expression of a specific gene is regulated and how the splicing of the mRNA is achieved. For example, we have identified 9 exons encoding the 359 residues of E-tropomodulin in human and mouse erythrocytes (8). There is also a possibility that alternative promoters and/or alternative splicing may be involved in regulating the expression of *Tmod* or Tmod-like gene in different tissues. Genomic fragments are also needed to construct targeting vectors to knockout specific genes in creating disease models. We have also mapped the protein 4.2 and E-tropomodulin genes to human chromosome 15q15-20 and 9q22, respectively. Chromosomal mapping helps the localization and identification of disease genes in human patients. In addition to the *E-Tmod* gene, there are 3 more *Tmod* genes (*Sk-Tmod, N-Tmod,* and *Tmod 3*) recently found in human and mouse genomes. They are separate genes, encoding homologous but not identical tropomodulin proteins in overlapping or different tissues, such as in the skeletal muscle or the brain.

3.4. *Nonerythroid cells*

The gene encoding erythrocyte membrane skeletal proteins exist in all nucleated cells. They, however, are not always expressed. Meaning, the genomic DNA may not always be transcribed into mRNA, and used as a template to synthesize proteins. Some of the erythrocyte membrane skeletal proteins are tissue restricted (e.g. protein 4.2), some are more widely expressed among tissues. For example, the heart and skeletal muscle, which have the actin (thin) filament as an essential element in their sarcomeres, also express high levels of E-tropomodulin. It has been shown by other investigators that E-tropomodulin binds to the pointed end of the sarcomeric actin filaments and regulates the length of the actin filaments in muscular tissues (9). Furthermore, overexpression of E-tropomodulin is associated with the myofibril disarray and the dilated cardiomyopathy in mice. Therefore, studies of E-tropomodulin in erythrocytes have led to the findings of important mechanical functions of E-tropomodulin in nonerythroid cells and tissues. Figure 6 shows the distribution and

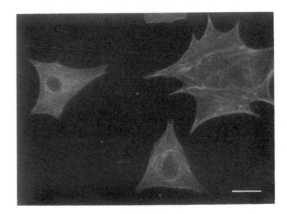

Fig. 6. The regular striated organization of E-tropomodulin in cultured neonatal rat cardiomyocytes. The cardiomyocytes were fixed, permealized, and incubated with a mouse monoclonal antibody made against human E-tropomodulin. After further incubation with a FITC-conjugated secondary antibody against mouse immunoglobulins, the cells were examined under a fluorescent microscope. The bar is 25 μm.

organization of E-tropomodulin in cultured neonatal rat cradiomyocytes. Note that E-tropomodulin exhibits a regular pattern of striation in cardiomyocytes as E-tropomodulin molecules are positioned at the pointed end (free end) of the well organized actin filament (thin filament) in sarcomeres.

3.5. *Mutants and diseases*

Mutations or deficiencies best reveal the function of a protein. Many hemolytic anemic patients have deficiencies or mutations in one of the major membrane skeletal network proteins in their erythrocytes, such as spectrins, ankyrin, band 3, protein 4.1, and protein 4.2 (see Fig. 1A). In many cases the protein deficiencies are associated with point mutations or frame shift mutations of these proteins. The erythrocytes of these patients exhibit sphere-ocytosis, elliptocytosis, or ovalocytosis, and the membranes demonstrate osmotic frigidity. The existence of these patients demonstrates the importance of membrane skeletal proteins in maintaining the mechanical stability and properties of the erythrocyte membrane.

3.6. *Knockout cells and mice: models to study the molecular basis of biomechanics*

When there is no disease model or the cause-effect relationship remains to be confirmed between a mutation and a disease, creating a knockout cell line or animal model deficient of a specific protein is a clean approach to demonstrate the role of this protein. This is accomplished by first constructing a targeting vector, which is a modified genomic fragment, and then introducing, by electroporation, the targeting vector into a cell line. The cell lines will be embryonic stem (ES) cells if the intention is to generate knockout mouse models. Through the rare probability of homologous recombination between the targeting vector and the endogenous gene, the endogenous normal gene is replaced by the targeting vector that carries a disrupted gene (Fig. 7). As a result, the expression of this gene, or the synthesis of this protein is specifically blocked. Investigating such cells or cells derived

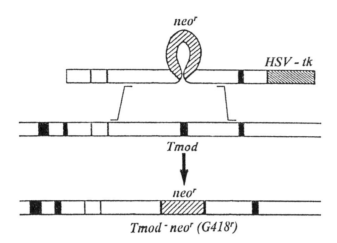

Fig. 7. A simplified schematic drawing showing the procedure used to enrich the mouse embryonic stem cells containing a targeted disruption of the E-tropomodulin gene. Top: targeting vector; Middle: endogenous gene; Bottom: disrupted gene after homolgous recombination between te targeting vector and the endogenous gene. Drawing modified from Mansour *et al.*, Ref. 16.

from such genetically engineered cell lines allows us to establish the functional role for this protein.

The embryonic stem cells, in which a targeted disruption of a specific gene has occurred, allow us to create knockout animal models. This is achieved by microinjecting the genetically altered mouse embryonic stem cells into early embryos at the stage of tropoblasts and then transplanting these early embryos into pseudopregnant mice. The resulting chimeric mice may give rise to heterozygous knockout mice if some of their germ cells are derived from the genetically altered embryonic stem cells. Breeding between the heterozygous knockout mice may give rise to homozygous knockout mice if the homozygous null mutation is not a lethal mutation. The homozygous knockout mice are completely deficient of the protein encoded by that specifically disrupted gene. Several knockout mouse models or cell lines deficient of erythrocyte membrane skeletal proteins have thus been created. We have contributed to these studies by creating embryonic stem cells and mice that are deficient of E-tropomodulin (10).

We first characterized the exon-intron organization of the mouse E-tropomodulin gene *E-Tmod* (8). We then constructed a targeting vector in which the exon containing the ATG initiation codon was replaced by a neomycine resistant (*Neo^r*) gene and a bacterial *lacZ* gene. Such targeting vector has been used to create embryonic stem cell lines heterozygous and homozygous for E-tropomodulin null mutation (10). The expression of the *lacZ* reporter gene, which encodes a β-galactosidase, can be detected by its substrates, e.g., X-gal or FDG (a fluorescent substrate), allowing us to identify cells or tissues that normally express E-tropomodulin. These embryonic cells can be driven into erythroid and other lineage in culture to demonstrate the effect of E-tropomodulin deficiency on the mechanical properties of erythroid and nonerythroid cells.

The embryonic stem cells, whose one copy of the E-tropomodulin gene is disrupted, can also be used to generate chimeric, heterozygous, and homozygous knockout mice. While the heterozygous knockout mice (E-$Tmod$ +/−) survive to the adulthood, homozygous knockout mice (E-$Tmod$ −/−) die during the early embryonic development. The results indicate that E-tropomodulin is essential for the embryogenesis, and that there are no other genes in the mouse genome to compensate for the function of E-tropomodulin when both copies of the E-tropomodulin gene are disrupted. The disease phenotypes in E-$Tmod$ −/− embryos include arrests of heart development, vasculogenesis, and hematopoiesis (10). Our initial analysis using micropipette aspiration technique (12), in collaboration with K.-L. Paul Sung, shows that erythrocytes and erythroid cells isolated from E-$Tmod$ −/− yolk sacs have less mechanical strength in their membranes in response to a negative pressure, exhibit partial hemolysis, and are easier to be fragmented as compared to the normal (E-$Tmod$ +/+) cells. We will soon begin to analyze the organization of the membrane skeletal network and to complete the characterization of the mechanical properties of E-$Tmod$ −/− erythroid cells, so that the structural and mechanical functions of E-tropomodulin in the erythrocyte membrane can be established.

Since E-tropomodulin is normally highly expressed in cardiac, skeletal, and smooth muscles as reported by the *lacZ* reporter gene, the mechanical functions of E-tropomodulin in these tissues will also be studied *in vivo*. These will be done by analyzing the structures, mechanical properties, and functions of the heart, the blood vessel, and the skeletal muscle of the wild type, the heterozygous and homozygous knockout mice or embryos. Furthermore, the phenotypes observed in knockout mice will be used as a guideline to screen human patients with similar disease phenotypes blood, cardiovascular system, and muscular tissues. The studies of knockout mice will, therefore, help us in identifying human diseases whose molecular defects are associated with mutations or abnormal expression of E-tropomodulin.

3.7.   Cell membranes: A target and a barrier for gene transfer

After the molecular basis of cell membrane mechanics is established with the use of mutant or knockout models, we will want to see whether the disease phenotypes such as the arrests of heart development, vasculogenesis, and hematopoiesis in E-$Tmod$ −/− mice can be rescued by the strategies of gene therapy. Delivery of a normal cDNA into the knockout cells e.g., ES cells, in culture by electroporation can be performed. Introducing a normal gene to knockout mice by viral vectors, lipofusion, or other methods can be carried out *in vivo*. Although these techniques have been utilized to transfer genes into living cells and tissues, their efficiencies are far from satisfactory and some safety concerns have not been ruled out. We hope, in the future, after introducing a normal copy of the E-tropomodulin cDNA or gene to the mutated cells or individual, the cell membrane will regain the normal mechanical strength, exhibit no premature hemolysis, and resist to fragmentation in response to a normal physiological shear stress.

While the cell membrane is the target of the treatment in several mouse models and some hemolytic anemic patients, the cell membrane itself is an important physical barrier for all techniques involved in gene transfer. Bioengineers working on the molecular basis of the biomechanics of cell membranes may help in improving strategies of gene therapy. One of the greatest challenges for biomedical researchers, including bioengineers, at the turn of this century is developing technologies to deliver transgenes with high efficiencies.

A transgene needs to be delivered with a high efficiency to the right tissues, at the right time, not only to cross the physical barriers of the plasma membrane, but also that of the nuclear membrane, not only to replace the disease genes but also to express (transcribe into mRNA) under the normal control, and on a long-term basis.

It is interesting that the cell membrane is a target and, at the same time, a barrier for gene transfer. For successful gene transfer, understanding both the biochemistry and biomechanics of the cell membrane is important. Thus, knowledge and skills in both molecular biology and engineering are in order to bring about innovative solutions to meet this important challenge.

4. Conclusion

This chapter demonstrates that the two seemingly unrelated fields, biomechanics and molecular biology, can be integrated. Together they can address questions neither fields alone can do. The example given here is our research work designed to address the molecular basis of cell membrane mechanics. While we have made much progress, more need to be done.

By integrating biomedical sciences and engineering, bioengineers have improved our basic understanding of biological systems. Having analyzed biomedical problems in this light, they have designed practical treatments for disease. Molecular biology and engineering are a powerful combination. The new generation of bioengineers capable of executing recombinant DNA technologies, will clearly be at the frontier, making more original and exciting contributions at the interface of medicine and engineering.

Acknowledgments

The work was supported by grants from the National Institute of Health. I thank students, postdoctoral fellows, research associates, and collaborators who have made significant contributions in these research projects. Some of the recent works were conducted in the Molecular Biology Common Facility and was established in the Department of Bioengineering with the generous support from the Whitaker Foundation.

References

1. Sung, L. A., Chien, S., Chang, L.-S., Lambert, K., Bliss, S. A., Bouhassira, E. E., Nagel, R. L., Schwartz, R. S. and Rybicki, A. C., Molecular cloning of human protein 4.2: A major component of the red cell membrane, *Proc. Natl. Acad. Sci. USA* **87**, 955–959 (1990).
2. Sung, L. A., Chien, S., Fan, Y.-S., Lin, C. C., Lambert, K., Zhu, L., Lam, J. S. and Chang, L.-S., Human erythrocyte protein 4.2: Isoform expression, differential splicing and chromosomal assignment, *Blood* **79**, 2763–2770 (1992).
3. Sung, L. A., Fowler, V. M., Lambert, K., Sussman, M., Karr, D. and Chien, S., Molecular cloning and characterization of human fetal liver tropomodulin: A tropomyosin-binding protein, *J. Biol. Chem.* **267**, 2616–2621 (1992).
4. Weber, A., Pennise, C. R., Babcock, G. G. and Fowler, V. M., Tropomodulin caps the pointed ends of actin filaments. *J. Cell Biol.* **127**, 1627–1635 (1994).

5. Sung, L. A. and Lin, J. J.-C., Erythrocyte tropomodulin binds to the *N*-terminus of hTM5, a tropomyosin isoform encoded by the γ-tropomyosin gene, *Biochem. Biophys. Res. Commun.* **201**, 627–634 (1994).

6. Sung, L. A., Gao, K.-M., Yee, L. Y., Temm-Grove, C. J., Helfman, D. M., Lin, J. J.-C. and Mehrpouryan, M., Tropomyosin isoform 5b is expressed in human erythrocytes: Implication of tropomodulin-TM5 or tropomodulin-TM5b complexes in the protofilament and hexagonal organization of membrane skeletons. *Blood* **95**, 1473–1480 (2000).

7. Vera, C., Sood, A., Gao, K.-M., Yee, L. J., Lin, J. J.-C. and Sung L. A., Tropomodulin-binding site mapped to residues 7–14 at the *N*-terminal heptad repeats of human tropomyosin insoform 5. *Arch. Biochem. Biophys.* **378**, 16–24 (2000).

8. Chu, X., Thompson, D., Yee, L., and Sung, L. A., Genomic organizations of mouse and human erythrocyte tropomodulin genes. Gene (in press).

9. Gregorio, C. C., Weber, A., Bondad, M., Pennise, C. R. and Fowler, V. M., Requirement of pointed-end capping by tropomodulin to maintain actin filament length in embryonic chicken cardiac myocytes. Nature **377**, 83–86 (1995).

10. Chu, X., Chen, J., Chien, K. R., Vera, C. and Sung, L. A., Tropomodulin null mutation arrests cardiac development, vasculogenesis and hematopoiesis during embryogenesis. *Mol. Cell. Biol.* **10**, 153a (1999).

11. Chien, S., Sung, L. A., Lee, M. M. L. and Skalak, R., Red cell membrane elasticity as determined by flow channel technique, *Biorheology* **29**, 467–478 (1992).

12. Sung, K.-L. P., Dong, C., Schmid–Schönbein, G. W., Chien, S. and Skalak, R. Leukocyte relaxation properties. *Biophy. J.* **54**, 331–336 (1988).

13. Chien, S., Principles and techniques for assessing erythrocyte deformability, *Blood Cells* **3**, 71–99 (1977).

14. Fung, Y. C., Mechanics of erythrocytes, leukocytes, and other cells, *Biomechanics: Mechanical Properties of Living Tissues*, 2nd edition, Springer-Verlag, New York (1995).

15. Sung, L. A., A view on the molecular basis of erythrocyte membrane mechanics, Blood **96**, 781–782 (2000).

16. Mansour, S. L., Thomas, K. R., Capecci, M. R., Disruption of the protooncogen int-2 in mouse embryo-derived stem cells. A general strategy for targeting mutations to non-selective genes. Nature **336**, 348–352 (1988).

Proc. Natl. Acad. Sci. USA
Vol. 87, pp. 955–959, February 1990
Biochemistry

Molecular cloning of human protein 4.2: A major component of the erythrocyte membrane

(band 4.2/membrane skeleton/factor XIII/transglutaminase/cDNA)

Lanping Amy Sung*†, Shu Chien*, Long-Sheng Chang‡, Karel Lambert*, Susan A. Bliss‡, Eric E. Bouhassira§, Ronald L. Nagel§, Robert S. Schwartz§, and Anne C. Rybicki§

*Departments of Applied Mechanics and Engineering Sciences-Bioengineering and Medicine, and Center for Molecular Genetics, University of California at San Diego, La Jolla, CA 92093; ‡Department of Biology, Princeton University, Princeton, NJ 08544; and §Division of Hematology, Albert Einstein College of Medicine/Montefiore Medical Center, Bronx, NY 10467

Communicated by Russell F. Doolittle, November 16, 1989

ABSTRACT   Protein 4.2 (P4.2) comprises ≈5% of the protein mass of human erythrocyte (RBC) membranes. Anemia occurs in patients with RBCs deficient in P4.2, suggesting a role for this protein in maintaining RBC stability and integrity. We now report the molecular cloning and characterization of human RBC P4.2 cDNAs. By immunoscreening a human reticulocyte cDNA library and by using the polymerase chain reaction, two cDNA sequences of 2.4 and 2.5 kilobases (kb) were obtained. These cDNAs differ only by a 90-base-pair insert in the longer isoform located three codons downstream from the putative initiation site. The 2.4- and 2.5-kb cDNAs predict proteins of ≈77 and ≈80 kDa, respectively, and the authenticity was confirmed by sequence identity with 46 amino acids of three cyanogen bromide-cleaved peptides of P4.2. Northern blot analysis detected a major 2.4-kb RNA species in reticulocytes. Isolation of two P4.2 cDNAs implies existence of specific regulation of P4.2 expression in human RBCs. Human RBC P4.2 has significant homology with human factor XIII subunit a and guinea pig liver transglutaminase. Sequence alignment of P4.2 with these two transglutaminases, however, revealed that P4.2 lacks the critical cysteine residue required for the enzymatic crosslinking of substrates.

Membrane skeletal proteins play an important role in regulating the viability and mechanical properties of erythrocytes (RBCs) (1, 2). All major RBC membrane proteins have been identified, most have been characterized, and several have been cloned (3–6). Isoforms of many RBC membrane proteins have been identified in various types of cells and tissues (1, 2). Protein 4.2 (P4.2), which represents ≈5% of the protein mass of human RBC membranes, is one of the last major membrane proteins to be characterized. P4.2 has an apparent molecular mass of ≈72 kDa and associates with the cytoplasmic domain of the anion exchanger, band 3 (7). Recent evidence suggests that P4.2 interacts with ankyrin and may function to stabilize ankyrin in the membrane (8). Individuals whose RBCs are severely deficient in P4.2 experience various levels of anemia, further indicating an important functional role for this protein (8).

We now report the molecular cloning and characterization of the full-length cDNA for human RBC P4.2.¶ Two cDNA sequences have been identified that differ only by a 90-base-pair (bp) insert located near the 5′ end of the coding region. The presence of two P4.2 cDNAs resembles the transcript heterogeneity found in membrane skeletal protein 4.1 (9–11) and nonerythroid α-spectrin (6) and suggests that regulation of P4.2 expression exists in human RBCs, possibly by alternative splicing.

MATERIALS AND METHODS

Screening of λgt11 cDNA Library. Affinity-purified rabbit anti-human P4.2 IgG prepared by Rybicki *et al.* (8) was used to screen a cDNA expression library in λgt11 constructed from human reticulocyte mRNA, kindly provided by J. G. Conboy and Y. W. Kan (5). Immunoscreening of the λgt11 expression library was performed according to Huynh *et al.* (12), except that positive clones were identified with goat anti-rabbit IgG conjugated with horseradish peroxidase (Bio-Rad). For each 150-mm Petri dish, 5×10^4 plaque-forming units were used.

Subcloning and Sequence Analysis. cDNA inserts from positive phage clones were subcloned into pBS(+) plasmids (Stratagene). Unidirectional deletion clones were generated by using BAL-31 exonuclease (13), and cDNA fragments were sequenced with T3 and T7 primers by the dideoxynucleotide chain-termination method (14). Sequence analysis and GenBank data base searches were performed by IBI Pustell sequence analysis software (International Biotechnologies).

5′-End Extension of cDNA. Three oligonucleotides were prepared in a technique based on the polymerase chain reaction (PCR) to synthesize the missing 5′ sequence of the partial cDNA clone: p1 was composed of nucleotides (nt) 7–23 of clone 7 (c.7); p2 was complementary to nt 36–52 of c.7 plus an *Eco*RI restriction site at its 5′ end; p3 was composed of the *Eco*RI polylinker and the poly(dC) originally used for the first-strand cDNA synthesis when the library was constructed. The sequences of p1, p2, and p3 were, respectively, 5′-dTGAGGATGCTGTGTTCC-3′, 5′-dTCGAATTCGTACTC-CATGCGCTGAG-3′, and 5′-dGCGGAATTCCCCCCCCCCC-CCCC-3′, with the *Eco*RI sites underlined. p2 and p3 were used as PCR primers, and the reticulocyte cDNA library (5 μl with 10^6 phages per μl) was used as a template. The reaction product was electrophoresed and stained with ethidium bromide. The major band was excised and subcloned into pGEM 3zf plasmids (Promega). From >500 transformants, 24 colonies were randomly chosen to make minipreparations of plasmid DNA, of which 80% were positive when hybridized with ³²P-labeled p1.

Western Blot Analysis of Fusion Proteins. Recombinant lysogens of three positive λgt11 clones were prepared and induced for expression of β-galactosidase fusion proteins (12). Clear lysate containing the fusion proteins was separated by SDS/PAGE (7.5% polyacrylamide gel) (15), transferred to

Abbreviations: LTG, guinea pig liver transglutaminase; nt, nucleotide(s); ORF, open reading frame; PCR, polymerase chain reaction; P4.2, protein 4.2 of human erythrocytes; RBC, erythrocyte; XIII$_a$, subunit a of human factor XIII.
†To whom reprint requests should be addressed.
¶The sequences reported in this paper have been deposited in the GenBank data base (accession nos. M30646 and M30647).

nitrocellulose filters (16), and immunostained with the affinity-purified rabbit anti-human P4.2 IgG and a goat anti-rabbit IgG conjugated with alkaline phosphatase (Promega).

Generation of Cyanogen Bromide-Cleaved Fragments. Purified P4.2 (50 μg) prepared according to Korsgren and Cohen (7) was incubated with cyanogen bromide (17). The cleaved P4.2 mixture was concentrated in a Speed Vac Concentrator (Savant), and the peptide fragments were separated by HPLC (Aquapore C03-GU, 30 × 4.6 mm; Brownlee Lab) with a 0.1% trifluoroacetic acid/acetonitrile linear gradient (0–90% acetonitrile in 13 min) at a flow rate of 1 ml/min. The peptide peaks detected at 210 nm were collected, concentrated, and sequenced using an ABI 1470 gas phase protein sequenator (Applied Biosystems).

RNA Isolation and RNA Blot Analysis. Human reticulocyte RNA was prepared (**5**) from peripheral blood of an anemic individual with paroxysmal nocturnal hemoglobinuria. The RNA was electrophoresed on a 1% agarose/formaldehyde gel (18), transblotted onto nitrocellulose paper, and hybridized with ^{32}P-labeled probes generated by random-primer extension (Pharmacia LKB).

RESULTS

Isolation of P4.2 cDNA Clones. Immunoscreening of $5 × 10^5$ recombinant phages with anti-human P4.2 antibody yielded 12 potentially positive clones, 6 of which contained large-sized inserts ranging from 1.2 to 1.8 kilobases (kb). The 1.8-kb insert of c.7 was found to cross-hybridize with those of c.4 (1.4 kb), c.8 (1.7 kb), and c.9 (1.2 kb), suggesting that these four clones contained overlapping nucleotide sequences.

Sequence analysis indicated that c.7 had a poly(A) tail and one long open reading frame (ORF) at the 5' end. Restriction analyses and partial nucleotide sequences showed that these cross-hybridizing clones contained the 3' portion of the cDNA with varying lengths from the 5' end (Fig. 1). Since P4.2 has an apparent molecular mass of 72 kDa, its cDNA is expected to have a coding region of ≈2.2 kb. Hence, the longest c.7 of 1.8 kb was not long enough for the entire coding region.

5'-End Extension of cDNA. To obtain the missing 5' end of the cDNA, we performed PCR 5'-end extension by using primers synthesized according to the sequences of c.7 insert and the λgt11 clones in the cDNA library. Sequencing of the four largest PCR-extended cDNA clones (including c.12 and c.16 in Fig. 1) showed that they had identical sequences, except that the largest one (c.12) had a 90-bp insert near the 5' end of the coding region (Fig. 1). The combined length of c.7 and the 5'-extended cDNA is 2.4 kb (or 2.5 kb with the 90-bp insert). Sequence analyses showed that they contained ORFs of 2.1 and 2.2 kb, respectively, and thus were capable of encoding a protein of ≈72 kDa.

Expression of Fusion Proteins and Amino Acid Sequence Analysis of Cyanogen Bromide Fragments. In addition to the coding capacities of the cDNAs, two other lines of evidence support the identity of these cDNAs as P4.2. An ≈175-kDa β-galactosidase fusion protein encoded by recombinant phage c.7 was detected by anti-P4.2 antibody (data not shown). Since β-galactosidase contributes 114 kDa to the fusion protein, c.7 insert encodes a peptide of ≈61 kDa. The combined protein size of c.7 and c.16 (15 kDa) or c.12 (18 kDa) is close to the apparent molecular mass of P4.2. The most convincing evidence for the authenticity of the cDNAs was the complete match of 46 amino acids from three independent cyanogen bromide peptides of P4.2 with the amino acid sequence deduced from the cDNA (Fig. 2, boxes).

Sequence Analysis of the P4.2 cDNAs. The complete nucleotide sequence of P4.2 cDNA and the deduced amino acid sequence of P4.2 are shown in Fig. 2. The cDNA has a 227-nt untranslated region upstream from the putative ATG start codon. The nucleotide sequence CAACC ATG G around this initiation site is similar to the consensus sequence for initiation found in higher eukaryotes (19), except that the second nt in the P4.2 cDNAs is A rather than C. This ATG initiation site is followed by an ORF through the c.7 cDNA. There is another ATG at nt −179 to −177, but it is followed by an in-frame termination codon 19 nt downstream. The presence or absence of the 90-nt insert (underlined in Fig. 2) gives rise to two P4.2 cDNA sequences. The 2.4-kb cDNA contains 2382 nt with an ORF of 691 amino acids, predicting an ≈77-kDa protein; the 2.5-kb cDNA contains 2472 nt with an ORF of 721 amino acids, predicting an ≈80-kDa protein. The cDNA ends in a poly(A) tail, and the 3' untranslated region is relatively short, containing only 82 nt. There is no polyadenylylation signal sequence AATAAA, but a sequence AATCTAAA is located at nt 2204–2211.

RNA Blot Analysis of Human Reticulocytes. Northern blot analysis using c.7 insert as a probe detected a 2.4-kb RNA species in human reticulocytes (Fig. 3). This result indicates that the cDNAs obtained in this study (2382 and 2472 bp) are apparently the full-length cDNAs for P4.2. Overexposure of the blot showed two additional minor bands of 3.9 and 1.7 kb.

Structural Analysis of P4.2. The amino acid sequence derived from the 2.5-kb cDNA contains ≈43% nonpolar, ≈35% polar, ≈10% acidic, and ≈12% basic amino acid residues. The most abundant amino acids are leucine (82 residues) and alanine (60 residues). There are 49 serine and 43 threonine residues (potential sites for O-glycosylation), representing 13% of the total residues. There are 16 cysteine residues, 6 potential N-glycosylation sites (Asn-Xaa-Ser/Thr) at Asn-103, -420, -447, -529, -604, and -705, 1 potential cAMP-dependent phosphorylation site (basic-basic-Xaa-Ser) at Ser-278 (20), and 9 potential protein kinase C phosphorylation sites (Ser/Thr-Xaa-Arg/Lys) at Ser-7, -57, -58, -154, -222, -449, -455, and -666, and Thr-287 (21). There is one Arg-Gly-Asp sequence at 518–520. Secondary structure analysis using the Chou and Fasman method (22) predicted that P4.2 contains ≈33% β-sheet, ≈24% α-helix, and ≈45% reverse turns.

Hydropathy analysis of the deduced amino acid sequence using the algorithm and hydropathy values of Kyte and Doolittle (23) revealed a major hydrophobic domain (residues 298–322; Fig. 4, b). This hydrophobic region was predicted to be mainly a β-sheet structure with a possible turn. There is a strongly hydrophilic region (residues 438–495; Fig. 4, c). Toward the C terminus of this region, there is a highly charged segment predicted to be an α-helix (residues 470–492; underlined in Fig. 5) and containing a large number of both positively and negatively charged residues, especially glutamic acid.

Homology searches of GenBank 59 (released March 1989) and NBRF-PIR Protein Sequence Database (release 19,

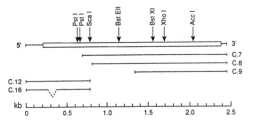

Fig. 1. Schematic diagram of human RBC P4.2 cDNAs. Horizontal open bar represents the coding region. Lines flanking it represent the 5' and 3' untranslated regions. Three clones isolated by immunoscreening (c.7, c.8, and c.9) and two clones obtained by PCR extension (c.12 and c.16) are shown below. The dashed V-line on c.16 indicates the absence of the 90-bp insert found in c.12.

233

Biochemistry: Sung *et al.* *Proc. Natl. Acad. Sci. USA* 87 (1990) 957

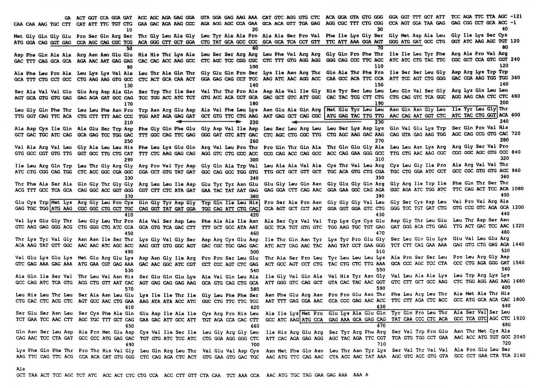

FIG. 2. Composite nucleotide sequence of P4.2 cDNA and deduced amino acid sequence. The first nucleotide of the putative ATG initiation codon is designated as nucleotide position 1. The underlined (without arrow) sequence represents the 90-bp insert in the larger cDNA. Boxed residues are those matched with the amino acid sequences obtained from three independent cyanogen bromide-cleaved peptides of purified P4.2. Positions where the primer sequences were derived for PCR extension are underlined with double arrowheads (p1) and a single arrowhead (p2).

updated December 1988) revealed that the amino acid sequence of human RBC P4.2 has significant homology with subunit a of human placental factor XIII (XIII$_a$, plasma transglutaminase; refs. 24, 26, 27) and also with the guinea pig liver transglutaminase (LTG; ref. 25) (Fig. 5). While the two transglutaminases are themselves 35.3% identical, P4.2 has 24.6% and 33.5% identity with XIII$_a$ and LTG, respectively. Fig. 5 shows the alignment of these three proteins by the method of Feng and Doolittle (28). In this alignment, a total of 122 residues are identical in all three proteins (Fig. 5, asterisks). The strongest homology is found in residues

293–325 of P4.2 (Fig. 5, large box), in which 20 of the 33 residues, or 60.6%, are identical for all three proteins. This highly conserved region contains the catalytic thiol sites of XIII$_a$ and LTG where Cys-Trp is required for transglutaminase activity (24, 26). P4.2, however, despite the high homology around the active site, has an alanine (residue 298 of P4.2; Fig. 5, arrow) instead of cysteine at the corresponding position.

The hydropathy plots for XIII$_a$, LTG, and P4.2 show similarities (Fig. 4), especially with regard to the hydrophobic region labeled b (residues 298–322 in P4.2), which overlaps with the 60.6% identity area (residues 293–325 in P4.2) in the larger box of Fig. 5. This region begins with the critical cysteine residue in XIII$_a$ and LTG and the corresponding alanine in P4.2. In the region marked c in Fig. 4, P4.2 has a strongly hydrophilic region; it is less hydrophilic in LTG and least in XIII$_a$. Near the center of this hydrophilic region in P4.2 lies a potential Ca^{2+} binding site (Fig. 5, small box), as identified by its alignment and homology with XIII$_a$ and other Ca^{2+} binding proteins (24).

DISCUSSION

In this communication, we report the cloning and sequencing of cDNA encoding the human RBC membrane P4.2. Two P4.2 cDNA sequences were obtained: one contains 2382 nt with an ORF of 691 residues; the other, with a 90-nt insert, contains 2472 nt with an ORF of 721 residues. These predictions are in good agreement with the apparent molecular mass of P4.2 (≈72 kDa). The finding of the two P4.2 cDNAs is consistent with the immunostaining pattern of P4.2 on SDS/

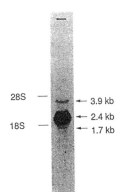

28S —

18S —

← 3.9 kb
← 2.4 kb
← 1.7 kb

FIG. 3. Northern blot analysis of human reticulocyte RNA. A major 2.4-kb hybridizing RNA was detected with c.7 probe (see Fig. 1). The other two arrows indicate the minor bands (3.9 and 1.7 kb) that appeared after overexposure. The standards used for size estimation were 18S rRNA (≈1.8 kb) and 28S rRNA (≈4.8 kb). The line at the top of the gel represents the origin.

Proc. Natl. Acad. Sci. USA 87 (1990)

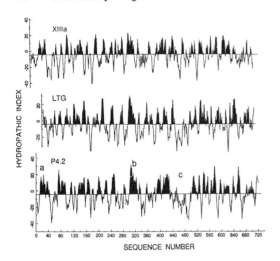

FIG. 4. Hydropathy plot of the deduced amino acid sequence of human P4.2 (with regions marked a, b, and c) and its comparison with subunit a of human factor XIII and guinea pig liver transglutaminase LTG. The hydropathic index was obtained from windows of seven amino acids. The three plots are aligned according to the highly conserved hydrophobic region (designated b: residues 298–322 in P4.2), which contains the transglutaminase active site in XIII$_a$ (24) and LTG (25). The catalytic cysteine of the active sites of the transglutaminases and the corresponding alanine in P4.2 (see arrow in large box in Fig. 5) are all located at the point of transition from hydrophilic to hydrophobic regions—i.e., the beginning of area b. In the P4.2 panel, area a (residues 4–33) shows the hydrophobic characteristics of a 30-amino acid insert in the 2.5-kb cDNA, and area c (residues 438–495) shows the strongly hydrophilic region of P4.2.

PAGE, which showed a diffuse band slightly higher than the major 72-kDa band (data not shown).

PCR-extended sequences contain 44 nt (including the 17-nt primer) overlapping with the 5′ end of c.7 obtained directly from the cDNA library. Furthermore, both the PCR-extended sequences (after removal of the nucleotides overlapping with c.7) and c.7 insert hybridized strongly with a P4.2 partial genomic DNA clone (unpublished observation). All of the above show that the PCR products were part of the P4.2 gene. The c.12, however, does not have the first 7 nt (ACAAACT) at the 5′ end of c.7. These 7 nt may represent the end of another insertion/deletion sequence that was not amplified and subcloned during PCR extension. Interestingly, an ACAAACT sequence is found further upstream at nt 279–285.

The mechanism by which the two P4.2 isoforms arise is unknown. Alternative splicing is an attractive possibility, especially in light of recent findings that isoforms of protein 4.1 mRNAs are generated by such a mechanism (9–11). The junction sequence around the 90-bp insert are G/G and T/G, which have been reported as junction nucleotides between exons (29, 30). This 30-amino acid insert has the characteristics of signal peptides containing a stretch of hydrophobic residues and shows homology with the internal sequences of a group of tyrosine kinase-related transforming proteins— e.g., c-src (31). This 30-amino acid insert may represent an imported exon. The expression and possible function of this insert warrants further study. RNA blot analysis indicates that the cDNAs obtained in this study represent the full-length message for P4.2 in reticulocytes, although the 90-nt difference of these two isoforms cannot be resolved. Whether the two minor hybridizing RNA species (3.9 and 1.7 kb) represent additional isoforms or messages of related proteins in reticulocytes needs to be investigated.

P4.2 binds to the cytoplasmic pole of band 3 in membranes (32, 33) and to ankyrin and protein 4.1 in solution (33). The availability of the P4.2 cDNAs has allowed us to conduct preliminary studies on the functional domains of P4.2, including those involved in its binding to other proteins. P4.2 contains 43% hydrophobic amino acid residues and shows at

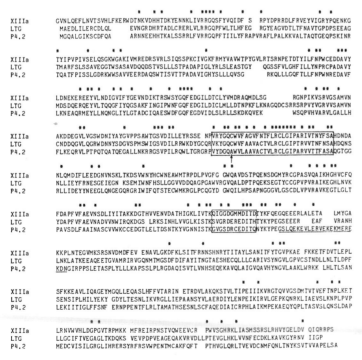

FIG. 5. The alignment of human P4.2 with XIII$_a$ (694 amino acids) and LTG (689 amino acids). Asterisks indicate residues that are identical in all three proteins. The alignment necessitates removal of the 30-amino acid insertion from the long isoform of P4.2—i.e., the shorter isoform (691 amino acids) is used in this plot. Gaps are allowed for maximizing identity. The region with the greatest identity is enclosed in the larger box (residues 293–325 of P4.2), which contains the highly conserved active site with Cys-Trp (CW) in XIII$_a$ and LTG (24). Arrow indicates the presence of alanine in P4.2 instead of cysteine at this site. Smaller boxed area (residues 452–463 of P4.2) is the potential Ca^{2+} binding site of XIII$_a$. Underlined sequence (residues 470–492 of P4.2) is the highly charged region in P4.2 with a predicted α-helical structure flanked by nearby proline residues. Amino acids are designated by the single-letter code.

least one domain of high hydrophobicity (Fig. 4, b). The hydrophobic region(s) might interact with membrane lipids or allow P4.2 to fold within itself. Since the hydrophobic domain labeled b is highly conserved in P4.2 and the two transglutaminases, it may be important in forming the active site itself and/or positioning the sites in cells.

It is interesting that P4.2 contains one potential cAMP-dependent phosphorylation site, since Suzuki *et al.* (34) have previously reported that P4.2 was phosphorylated by a cAMP-dependent protein kinase and that phosphorylation was stimulated by heavy metal ions.

P4.2 has significant homology with XIII$_a$ (24, 26, 27) and LTG (25), especially around their active sites. XIII$_a$, the final component in the coagulation pathway, plays an important role in the stabilization of fibrin clots by covalently crosslinking fibrin monomers through γ-glutamyl-ε-lysine bridges and by preventing proteolysis (35). Liver transglutaminase has some of the activity associated with the plasma membrane and may be responsible for forming covalently crosslinked matrices of proteins at sites of cell-to-cell contact (36). P4.2, however, has alanine instead of the cysteine indispensible for transglutaminase activity (37) in the active site area. It is possible that P4.2 may use this site to bind other RBC membrane proteins without forming covalent crosslinks. P4.2, along with protein 4.1, has been proposed to be one of the last membrane proteins synthesized during RBC maturation (38). P4.2 may contribute to the stabilization of the membrane skeleton through its binding with membrane proteins and thus protects them from being degraded (e.g., by proteases) or crosslinked (e.g., by cytoplasmic transglutaminase). Many proteins that are labile in the cytosol become resistant to degradation once assembled into the skeletal network (39).

The phylogenetic history of the three proteins can be inferred from their relative similarities. The human P4.2 is more similar to LTG than XIII$_a$, but it appears to have undergone a faster rate of change, suggesting that P4.2 is an offshoot of a tissue enzyme.

Our finding of P4.2 cDNAs in the human reticulocyte cDNA library indicates that circulating reticulocytes retain intact P4.2 mRNAs. Immunoreactive analogs of P4.2 are also present in nonerythroid cells and tissues, including platelets, brain, and kidney (40, 41). These results raise the possibility that P4.2, like protein 4.1, may be a ubiquitous component of cell membranes, although its function in other cells may differ from that in RBCs. The availability of the cDNA for P4.2 should aid considerably in the study of structure–function relationships of this protein, including investigations on the mechanisms responsible for P4.2 deficiency in human patients and the expression of P4.2 in different tissues and during differentiation.

Note. Korsgren *et al.* (42) have also obtained a cDNA sequence of human RBC membrane P4.2. Their nucleotide sequence, except for the following, is identical to our shorter isoform: (*i*) Absence of the first 39 nucleotides at the 5' end; (*ii*) C instead of A at −188 and absence of T at −37; (*iii*) four synonymous differences: G instead of A at 420, T vs. G at 531, and C vs. G at both 1137 and 1215; (*iv*) two nonsynonymous differences: G vs. C at both 1138 and 1216; (*v*) absence of AG at 1094–1095 and presence of extra CC after 1109, causing a frameshift in the intervening nucleotides. *iv* and *v* result in seven amino acid differences.

We thank Drs. John G. Conboy, Y. W. Kan, and Narla Mohandas for their generous gift of the cDNA library, Drs. Al Smith and John Gardner for amino acid sequence analysis of P4.2 peptides, Dr. Thomas Shenk for his laboratory facilities, Dr. Michael G. Rosenfeld for his valuable advice and help throughout this study, Dr. John Trombold for blood samples for RNA preparations, and Dr. Russell F. Doolittle for his expert advice and help in alignment and hydropathy analysis of proteins. We appreciate the technical expertise of Sylvia Musto, June Wang, Eugene Leung, and Gerard Norwich. This work was supported by Research Grants HL19454, HL21016, HL33084, HL38655, and HL44147 from the National Heart, Lung and Blood Institute.

1. Bennett, V. (1985) *Annu. Rev. Biochem.* **54,** 273–304.
2. Becker, P. S. & Benz, E. J. (1990) in *Molecular Biology of the Cardiovascular System,* ed. Chien, S. (Lea & Febiger, Philadelphia), in press.
3. Curtis, P. J., Palumbo, A., Ming, J., Fraser, P., Cioe, L., Meo, P., Shane, S. & Rovera, G. (1985) *Gene* **36,** 357–362.
4. Kopito, R. R. & Lodish, H. F. (1985) *Nature (London)* **316,** 234–238.
5. Conboy, J., Kan, Y. W., Shohet, S. B. & Mohandas, N. (1986) *Proc. Natl. Acad. Sci. USA* **83,** 9512–9516.
6. McMahon, A. P., Giebelhaus, D. H., Champion, J. E., Bailes, J. A., Lacey, S., Carritt, B., Henchman, S. K. & Moon, R. T. (1987) *Differentiation* **34,** 68–78.
7. Korsgren, C. & Cohen, C. M. (1986) *J. Biol. Chem.* **261,** 5536–5543.
8. Rybicki, A. C., Heath, R., Wolf, J. L., Lubin, B. & Schwartz, R. S. (1988) *J. Clin. Invest.* **81,** 893–901.
9. Ngai, J., Stack, J. H., Moon, R. T. & Lazarides, E. (1987) *Proc. Natl. Acad. Sci. USA* **84,** 4432–4436.
10. Tang, T., Leto, T. L., Correas, I., Alonso, M. A., Marchesi, V. T. & Benz, E. J. (1988) *Proc. Natl. Acad. Sci. USA* **85,** 3713–3717.
11. Conboy, J. G., Chan, J., Mohandas, N. & Kan, Y. W. (1988) *Proc. Natl. Acad. Sci. USA* **85,** 9062–9065.
12. Huynh, T. V., Young, R. A. & Davis, R. W. (1985) in *DNA Cloning: A Practical Approach,* ed. Glover, D. M. (IRL, Oxford, U.K.), Vol. 1, pp. 49–78.
13. Maniatis, T., Fritsch, E. F. & Sambrook, J. (1982) *Molecular Cloning: A Laboratory Manual* (Cold Spring Harbor Lab., Cold Spring Harbor, NY), pp. 135–139.
14. Sanger, F., Nicklen, S. & Coulson, A. R. (1977) *Proc. Natl. Acad. Sci. USA* **74,** 5463–5467.
15. Laemmli, U. K. (1970) *Nature (London)* **227,** 680–685.
16. Towbin, H., Staehelin, T. & Gordon, J. (1979) *Proc. Natl. Acad. Sci. USA* **76,** 4350–4354.
17. Drickamer, L. K. (1977) *J. Biol. Chem.* **252,** 6909–6917.
18. Ausubel, F. M., Brent, R., Kingston, R. G., Moore, D. D., Smith, J. A., Seidman, J. G. & Struhl, K. (1987) *Current Protocols in Molecular Biology* (Wiley, New York), p. 4.1.2.
19. Kozak, M. (1987) *J. Mol. Biol.* **196,** 947–950.
20. Krebs, E. G. & Beavo, J. A. (1979) *Annu. Rev. Biochem.* **48,** 923–959.
21. Woodget, J. R., Gould, K. L. & Hunter, T. (1986) *Eur. J. Biochem.* **161,** 177–184.
22. Chou, P. Y. & Fasman, G. D. (1974) *Biochemistry* **13,** 222–245.
23. Kyte, J. & Doolittle, R. F. (1982) *J. Mol. Biol.* **157,** 105–132.
24. Takahashi, N., Takahashi, Y. & Putnam, F. W. (1986) *Proc. Natl. Acad. Sci. USA* **83,** 8019–8023.
25. Ikura, K., Nasu, C., Yokota, H., Tsuchiya, Y., Sasaki, R. & Chiba, H. (1988) *Biochemistry* **27,** 2898–2905.
26. Ichinose, A., Hendrickson, L. E., Fujikawa, K. & Davie, E. W. (1986) *Biochemistry* **25,** 6900–6906.
27. Grundmann, U., Amann, E., Zettlmeissl, G. & Kupper, H. A. (1986) *Proc. Natl. Acad. Sci. USA* **83,** 8024–8028.
28. Feng, D.-F. & Doolittle, R. F. (1987) *J. Mol. Evol.* **25,** 251–260.
29. Yamada, Y., Avvedimento, V. E., Mudryi, M., Ohkubo, H., Vogeli, G., Meher, I., Pastan, I. & deCrombrugghe, B. (1980) *Cell* **22,** 887–892.
30. Mount, S. M. (1982) *Nucleic Acids Res.* **10,** 459–472.
31. Hanks, S. K., Quinn, A. M. & Hunter, T. (1988) *Science* **241,** 42–52.
32. Steck, T. L. (1974) *J. Cell Biol.* **62,** 1–19.
33. Korsgren, C. & Cohen, C. M. (1988) *J. Biol. Chem.* **263,** 10212–10216.
34. Suzuki, K., Ikebuchi, H. & Terao, T. (1985) *J. Biol. Chem.* **260,** 4526–4530.
35. Lorand, L., Downey, J., Gotoh, T., Jacobson, A. & Tokura, S. (1968) *Biochem. Biophys. Res. Commun.* **31,** 222–230.
36. Slife, C. W., Dorsett, M. D. & Tilotson, M. L. (1986) *J. Biol. Chem.* **261,** 3451–3456.
37. Folk, J. E. (1980) *Annu. Rev. Biochem.* **49,** 517–531.
38. Chang, H., Langer, P. J. & Lodish, H. F. (1976) *Proc. Natl. Acad. Sci. USA* **73,** 3206–3210.
39. Moon, R. T. & Lazarides, E. (1984) *J. Cell Biol.* **98,** 1899–1904.
40. Schwartz, R. S., Rybicki, A. C., Heath, R., Shew, R. & Lubin, B. (1987) *Blood* **70,** Suppl. 1, 42a (abstr.).
41. Friedrichs, B., Koob, R., Kraemer, D. & Drenckhahn, D. (1989) *Eur. J. Cell Biol.* **48,** 121–127.
42. Korsgren, C., Lawler, J., Lambert, S., Speicher, D. & Cohen, C. M. (1990) *Proc. Natl. Acad. Sci. USA* **87,** 613–617.

CHAPTER 10

BIOMECHANICS OF INJURY AND HEALING

PIN TONG

Department of Mechanical Engineering,
Hong Kong University of Science and Technology,
Visiting Professor, University of California, San Diego

Y. C. FUNG

Department of Bioengineering University of California, San Diego

1. Introduction

Injuries may occur in many different ways. While many injuries result from sporting events or falls, the more severe impacts occur at higher speed impact associated with transportation related accidents. There are over 3 million people injured per year in traffic crashes in the United States alone, in which more than 40,000 people were killed. The economic cost alone of motor vehicle crashes in 1994 was estimated to be more than 150 billion (1). Among the injuries, brain injury is the leading killer and cause of disability in children and young adults. A conservative estimate puts the total number of traumatic brain injuries at over 500,000 per year in the United States severe enough to require hospitalization, and about 75,000 to 100,000 will die as a result of a traumatic brain injury. Of those who survive their initial injury, approximately 70,000 to 90,000 will endure lifelong debilitating loss of function. A survivor of a severe brain injury typically faces 5 to 10 years of intensive services and estimate cost in excess of \$4 million (2). Motor vehicle crashes cause about half of all brain injuries. There are approximately 1,000 bike-related deaths per year, of which about 75% are the results of head injuries. Brain injury is the leading cause of death in bicycle incidents.

Injuries are abundant in life. We are interested in healing and reduction of injury through engineering. Each of us knows somebody injured in some unfortunate accidents. Each of us could think of ways to improve our chance of survival. Thinking concretely of ways to implement our wishes is our engineering opportunity.

2. Historical Perspective

Society makes laws to assure safety of people. Engineers design their vehicles and structures with safety as a primary consideration. A typical approach taken by engineers with regard to head injury in automobile transportation can be quoted. In this approach, an injury criterion is defined as a measure to indicate injury potential of a particular body region in response to mechanical inputs such as applied forces, pressure and/or kinematics. Injury criterion is intended for a broad range of applications. With the great variability in

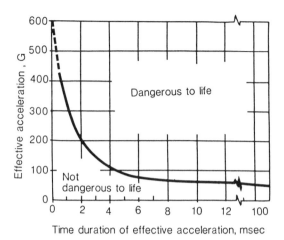

Fig. 1. The "Wayne State University" tolerance curve for the human brain in forehead impact against plane, unyielding surface.

biological systems, a criterion must be statistical in nature. The current US Federal Motor Vehicle Safety Standard 208 specifies the '*Head Injury Criterion*' as a safety criterion that new vehicles introduced to US highways must comply with. The criterion is used as an indicator for head injury potential in motor vehicle crash testing. This requirement has led to better vehicle and restraint system designs, which provide improved occupant protection for selected impact situations. The criterion evolved from the *Wayne State Tolerance Curve*, which is shown in Fig. 1. This curve was based on the clinical observation of frequent concomitant concussions in skull fracture (80% of all concussion cases involved skull fractures, (3). With the observed data, Gurdjian *et al.* (4) suggested that the tolerance of the skull to fracture is an effective indicator for tolerance to brain injury. In this curve shown in Fig. 1, the ordinate is the "effective" acceleration (which is an average front-to-back acceleration of the skull measured at the occipital bone over a "duration" T) for impacts of the forehead against a plane, unyielding surface; the abscissa is the duration of the effective part of the pulse. Lissner *et al.* (5) later developed a correlation between the magnitude of the translational anterior-posterior acceleration and the load duration that can potentially lead to skull fracture and head injury. Lissner's result was based on a variety of empirical data from cadaveric, animal, and human tests. Gadd (6) later used the tolerance curve to develop the *Gadd Severity Index*. These eventually lead to the concept of specifying a specific maximum number of an integral called the *Head Injury Criterion* (HIC). The HIC that relates the severity of head injury potential to the amplitude and duration characteristics of the resultant translational acceleration at the center of gravity of a dummy head is defined as follows:

$$HIC = \max \left\{ \frac{1}{(t_2 - t_1)} \int_{t_1}^{t_2} [a(t)]dt \right\}^{2.5} (t_2 - t_1)$$

where $a(t)$ = resultant *translational acceleration* at the center of gravity (CG) of the head in multiples of g, the gravitational acceleration on earth at sea level; $(t_2 - t_1)$ = time interval maximizing the *Head Injury Criterion* in milliseconds.

The U.S. Federal Standard 208 (effective 9/1/91) specifies that with $t_2 - t_1$ equal approximately to 36 ms (milisecond), HIC should be less than 1000. While this *Head Injury Criterion* has been used as an indicator for both skull fracture and brain injury potential, its efficacy in detecting and evaluating rotationally induced injuries appears to be limited. However, its long and popular use by the automobile industry shows nevertheless the importance of acceleration and the period of impact in causing head injury.

3. Basic Biomechanics of Injury and Healing

In the following, we shall discuss the basic features of the dynamics of our bodies in resisting an impact load and the healing afterwards. The material in Sections 3–9 are reproduced from articles written by the present authors, (7,8,9), by permission.

First, we must realize that when a load hits a person, the stress in the person depends on how fast the body material moves. This is because our body reacts to the load by setting up internal elastic waves. An understanding of this point is very important.

The impact load that may cause injury to a person frequently comes as a moving mass (e.g. a bullet, a flying object, a car), or as an obstruction to a moving person (e.g., falling to the ground, running into a tree). The impact causes the material of the human body in contact with the load to move relative to the rest of the body. The initial velocity induced in the body material particles that come into contact with the load has a decisive influence on the stress distribution in the body following the impact. The velocity can be supersonic transonic, subsonic, or so slow as to be almost static. The body reacts differently to these speeds. This is of central importance to the understanding of trauma, and is explained below.

If a load comes like a bullet from a gun, it sets up a shock wave. The shock wave will move in a person's body with a speed faster than the speed of sound in the body. At supersonic speed, the shock wave carries energy that is concentrated at the shock-wave front. Thus in a thin layer in the body, a great concentration of strain energy exists, which has a high potential for injury. This is analogous to the sonic boom coming from a supersonic airplane. People are familiar with the window-shaking and roof-shattering thunder to sonic booms. In these booms the shock energy of the airplane is transmitted down to the house. Similarly, the shock wave created by a load that hits at a supersonic load can cause damage to a human body. A fast-moving blunt load that does not penetrate can nevertheless cause shock-wave damage.

If the body material moves at a transonic or subsonic velocity, stress waves will move in the body at sonic speed. These stress waves can focus themselves into a small area and cause concentrated damage in that area. They can also be reflected at the border of organs and cause greater damage in the reflection process.

The complex phenomena of shock- and elastic-wave reflection, refraction, interference, and focusing are made more complex in the human body by the fact that different organs have different damping characteristics and different sound speed. There are little data available on the damping characteristics of organs. The sound speeds in different organs are listed in Table 1 (from Fung, (8), p. 473, in which original references are listed) and Fig. 2. It is seen that the cortical bone has a sound speed of 3,500 m/sec. This may be

Table 1. Velocity of sound in various tissues, air, and water.

Tissue	Density (g/cm^3)	TPP* (kPa)	Sound speed mean ± S.D. (m/sec)	Reference
Muscle	1		1580	Ludwig (1950), Frucht (1953), von Gierke (1964)
Fat	1		1450	Ludwig (1950), Frucht (1953)
Bone	2.0		3500	Clemedson and Jönsson (1962)
Collapsed lung	0.4		650 (ultrasound)	Dunn and Fry (1961)
Collapsed lung pneumonitis	0.8		320 (ultrasound)	Dunn and Fry (1961)
Lung, horse	0.6		25	Rice (1983)
Lung, horse	0.125		70	Rice (1983)
Lung, calf			24–30	Clemedson and Jönsson (1962)
Lung, goat		0	31.4 ± 0.4	Yen *et al.* (1986)
		0.5	33.9 ± 2.3	
		1.0	36.1 ± 1.9	
		1.5	46.8 ± 1.8	
		2.0	64.7 ± 3.9	
Lung, rabbit		0	16.5 ± 2.4	Yen *et al.* (1986)
		0.4	28.9 ± 3.3	
		0.8	31.3 ± 0.9	
		1.2	35.3 ± 0.8	
		1.6	36.9 ± 1.7	
Air			340	Dunn and Fry (1961)
Water, distilled, 0°C			1407	Kaye and Labby (1960)
Air bubbles (45% by vol.) in glycerol and H$_2$O			20	Campbell and Pitcher (1958)

*TPP =Transpulmonary pressure = airway pressure − pleural pressure.
1 kPa = 10^3 N/m^2 ∼ 10.2 cm H$_2$O.

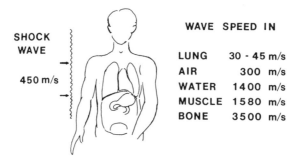

Fig. 2. The speed of propagation of sound waves (elastic waves of small amplitude) in human organs.

compared with the sound speed of 4,800 m/sec in steel, aluminum, copper, etc. The speed of elastic waves in the lung is of the order of 30–45 m/sec. This is much lower than the sound speed in the sir, almost ten times slower. The lung has such a low sound speed because it has a gas-filled, foamy structure. Roughly speaking, the lung tissue has the elasticity of the gas and the mass of the tissue. Sound speed being proportional to the square root of elastic modulus divided by the mass density, the lowering of sound speed by the tissue mass is understandable. The lung structure is so complex that several types of stress waves can exist. Table 1 lists several sound speeds in the lung. The speeds given by Yen *et al.* (10) were measured from the lungs of man, cat, and rabbit under impact pressures and wall velocities comparable with those induced by shock waves of an air blast (e.g. due to a bomb explosion, or a gun fired not too far away, or a gasoline tank explosion). The wave speeds found by Yen *et al.* depend on the transpulmonary pressure of the lung (i.e. on how large the lung is inflated) and the animal species. The sound speed given by Rice (11) was measured by a microphone picking up a sound made by an electric spark. The speed given by Dunn and Fry (12) was measured by ultrasound waves, which appear to be quite different from the waves measured by Yen *et al.* (10) and Rice (11).

At an impact speed like that of a fast moving automobile, the impact on the lung can be supersonic. The impact coming to the lung from a high speed deployment of an air bag is usually supersonic. Thus the risk of air bag injury to child passenger can be anticipated. One can easily understand also a military fact: In a pressure wave impact of an explosive, the lung is the first organ at risk for injury.

In general, for the same force, the stress induced in the body is the smallest if it is applied very slowly. The stress will generally be larger when the rate at which the load is applied is increased. As the rate increases, first the induced vibration may cause additional stress. Then the elastic waves may cause stress concentration. The spatial distribution of stress is different in static, vibratory, and elastic wave regimes. Damping can have a significant influence on the dynamic stress.

In the following section we shall discuss the strength of the materials, i.e. the maximum stress a material can bear without failure. It will be seen that the strength depends on the rate of change of strain. Therefore, the effects of the speed of loading are twofold: the strain rate influences the maximum stress induced by the impact, and it influences the strength of the material. Thus the limit of safety, defined by having the maximum stress staying below the critical limit of strength, depends on the rate of loading.

4. The Strength of a Tissue or Organ Is Expressed by a Tolerable Stress, Which May Vary with the Condition with Which the Load Is Applied

We spoke in the preceding section about the stresses induced in the tissues and organs due to an external load. Now let us consider the strength of the tissues and organs in greater detail. We have to define strength and tolerance very carefully, because they depend on what we mean by failure (8).

To discuss strength, failure and tolerance, let us consider first the simple experiments shown in Fig. 3 (from Ref. 8, p. 456):

1. A piece of twine is to be cut by a pair of dull scissors. The cutting is more difficulte when the twine is relaxed. But if one pulls the string tight and then cut it, it breaks easily, Why?

2. A stalk of fresh celery breaks very easily in bending. An old, dehydrated one does not. Practice on carrots, also!

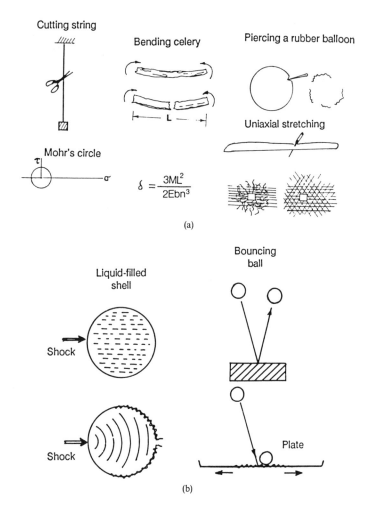

(a)

(b)

Fig. 3. Several experiments demonstrating that the meaning of the term "strength" depends on the type of stress, whether it is uniaxial, biaxial, or triaxial. Strength also depends on the fluid content of a tissue, whether a body contains fluid or not, whether elastic and shock waves cause concentration of stress, and whether the propagating kinetic energy is carried away by elastic waves upon impact. The rate of change of strain affects the critical stress at which rupture occurs. From Fung (8). Reproduced by permission.

3. A balloon is inflated. Another is not inflated but is stretched to a great length. Prick them with a needle. One explodes. The other does not. Why?

4. A thin-walled metal tube is filled with a liquid, as an Atlas missile. Strike it on one side. Sometimes the shell fails on the other side. This is known as contre coup. How can this happen?

5. Take a small nylon ball, or a pearl, or a ball bearing, and throw it onto a hard surface. It bounces. Throw it onto a thin metal plate such as that used in the kitchen for baking, and it won't bounce. Why?

We can explain the first example by computing the maximum principal stress in the string due to the action of shear and the pulling. Pulling the string taut increases the maximum principal stresses. When a limit is exceeded the string breaks.

In the second example, the specimen fails by bending. The bending stress in the fibers of the celery is higher in the fresh and plump celery, lower in the dehydrated specimen.

To explain the phenomenon shown in the third example, we have to think of the long-chain molecules of the rubber membrane. These molecules or fibers are bent and twisted randomly in every direction. When the balloon is inflated, fibers in every direction are stretched taut. A prick of the pin breaks fibers in every direction passing through the hole, and the membrane breaks with an explosion. On the other hand, when the rubber membrane is stretched uniaxially, only the fibers in the direction of the stretching are pulled taut. Those in other directions are still relaxed. The pin prick breaks only the fibers in one direction; those in other directions remain intact. Hence no explosion.

The fourth example shows what focusing of stress waves can do. The compression wave in the fluid initiated by the impact moves to the right. The flexural wave of the metal shell also moves to the right along the curved surface of the tube wall. If the flexural wave and the compression wave arrive at the other side simultaneously, a concentration of stress may occur that may exceed the ultimate stress of the materials and cause fracture on the far side. This situation may occur also in head injury.

The fifth example shows the effect of energy transfer between two bodies in impact. When a ball impacts a hard surface, it bounces. When it impacts a thin plate, it does not bounce. In the first case the kinetic energy of the ball is not transfered to the hard surface. In the second case the kinetic energy of the ball is transfered to the plate.

There are many biological analogs of these examples. Altogether they tell us that in answering questions of strength and tolerance, we must consider the magnitude of the maximum principal stress and the rate at which the stress varies with time in a biological material. We also have to consider the molecular configuration of biological material, which depends on the nature of the stress, whether it is uniaxial, biaxial, or triaxial, and the stress waves around the point of concern in the biological material. The fourth example shows that the stress concentration due to the elastic waves may result in a weakness. In all cases, however, for human safety, the focus of attention is on dynamic stress.

In summary, we have explained that the stress in the body in response to external force depends on the speed at which the force is applied onto the human body. In making a stress analysis, we first obtain the static stress distribution in the body under the external load (including the inertia force due to deceleration of a car). Then determine the dynamic amplification due to vibrations, then assess the stress concentration due to elastic waves and shock waves.

We see that the strength of an organ or a tissue in our body depends not only on the magnitude of the stress, static or dynamic, but also on the type of the stress: whether it is uniaxial, biaxial, or triaxial.

In the following sections, we consider in greater detail the meaning of the strength of material relative to injury, repair, growth, and resorption.

5. Injury of Organs and Tissues

In human society, the concept of injury is largely subjective. To bring some order into this subjective world, objective clinical observations and tests are desirable, but not necessarily easy. For example, consider head injury. The brain can be injured by fracture, impingement, excessively high localized pressure or tensile stress, high localized shear stress and strain, and cavitation in high-tension regions. The regions where the maximum normal stress occurs are usually different from where the maximum shear stress occurs; and they are affected significantly by the flow through foramen magnum (opening at the base of the skull) during impact. The brain tissue can be contused, leading to axonal damage; and the blood vessels may be ruptured, causing brain hematomas.

A well-known trauma is brain concussion, which is defined as a clinical syndrome characterized by immediate transient impairment of neural function, such as loss of consciousness, and disturbances of vision and equilibrium due to mechanical forces. Normally, concussion does not cause permanent damage. It is the first functional impairment damage of the brain to occur as the severity of head impact increases. It is reproducible in experimental animals. It has been studied with respect to rotational acceleration and flexion-extension of the upper cervical cord during motion of the head-neck junction. A detailed discussion of the influence of angular kinematics and its effects on induced strains in the brain is contained in Ommaya (13), Tong *et al.* (14) and Bandak and Eppinger (15).

Logically, one could correlate lesions, immunochemical changes and neurological observations with the stress in the brain. But today, a thorough correlation still does not exist. There are a number of difficulties. First of all, the evaluation of stress distribution in the brain is difficult. Not only is the computation of stress distribution in a specific boundary-value problem difficult, but the identification of the boundary conditions in known automobile collisions or aircraft accidents is nearly impossible today. And the neurological data must be correlated with the stress distribution.

Injury to other organs also needs attention. Each organ has its own characteristics, and its injury potential varies with the rate of application of the load (static, subsonic, or supersonic), and the type of critical stress (uniaxial, biaxial, or triaxial).

A publication edited by Woo and Buckwalter (16) presents a detailed discussion of injury and repair of the musculoskeletal soft tissues, including tendon, ligament, bone-tendon and myotendinous junctions, skeletal muscle, peripheral nerve, peripheral blood vessel, articular cartilage, and meniscus. A detailed mechanical analysis of vibration and amplification of elastic waves is presented by Fung (8, Chapter 12). The trauma of the lung due to impact load and the cause of subsequent edema are also examined in detail in Fung (8, pp 475–482). Existing data on the tolerance of organs to impact loads are quite extensive. See References listed in (8, pp. 493–498).

Obviously, the world needs basic research to obtain data on organ and tissue morphology, histology, biochemistry, material density, mechanical properties, stress distribution, and signal pathways. We need accurate mathematical models and efficient computing programs to evaluate injury and repair potentials, and devices to help healing and rehabilitation.

6.   Biomechanical Analysis

In assessing human tolerance to impact loads and in designing vehicles for crashworthiness, it is necessary to calculate the stress and strain at specific points in various organs, and this is best done by mathematical modeling. A complete dynamic simulation of a crash event requires detailed mechanical descriptions of both the vehicle and occupant structures, as well as descriptions of forces and kinematics resulting from vehicle/occupant interaction. Numerical methods are generally necessary for analyzing such complex nonlinear problems. The finite element method (17) is most suitable. This may be done by employing finite element representations for both vehicle and occupants (18), and using lumped parameter models to represent occupants and contact surfaces with a deceleration pulse applied to the surrounding structure. In the case of head-impact simulation, one can replicate the measured-momentum field that describes the inertial loads applied to a finite element representation of the soft tissue (19).

The finite element model is capable of predicting brain responses under dynamic loading. The distribution of strain fields and their time history are useful for visualizing the evolution of injury potential during impact, identifying the regions of the brain that may be more vulnerable than others under particular loading conditions, and assessing effectiveness of countermeasures for injury mitigation. It also provides a means of comparing analytical predictions with studies that map the distribution of damaged tissue.

Through mathematical modeling, one can connect pieces of information on anatomy, physiology, and clinical observations with people, vehicle, and accident. A validated model can then become a foundation of engineering. Biomechanical modeling is being developed vigorously. A selected bibliography is given in Fung (8).

7.   Healing and Rehabilitation are Helped by Proper Stress

Living organisms are endowed with certain ability to heal when damaged. Orthopedic surgeons were the first to pay attention to the role played by biomechanics in the healing of bone fracture. In 1866, G. H. Meyer (20) presented a paper on the structure of cancellous bone and demonstrated that "the spongiosa showed a well-motivated architecture which is closely connected with the statics of bone." A mathematician, C. Culmann, was in the audience. In 1867, Culmann (21) presented Meyer a drawing of the principal stress trajectories on a curved beam similar to a human femur. The similarity between the principal stress trajectories and the trabecular lines of the cancellous bone is remarkable. In 1869, Wolff (22) claimed that there is a perfect mathematical correspondence between the structure of cancellous bone in the proximal end of the femur and the trajectories in Culmann's crane. In 1880, W. Roux (23) introduced the idea of "functional adaptation." A strong line of research followed Roux. Pauwels, beginning in his paper in 1935 and culminating in his book of 1965 (24), turned these thoughts to precise and practical arts of surgery. Vigorous development continues. The biology of bone cells is advancing rapidly.

Thus both the motivation and the knowledge and art of providing suitable stress to the bone and cartilage to promote healing and rehabilitation are clear and advancing rapidly.

8.   Remodeling of Soft Tissues in Response to Stress Changes

The best-known example of soft-tissue remodeling due to change of stress is the hypertrophy of the heart caused by a rise in blood pressure. Another famous example was given by Cowan and Crystal (25), who showed that when one lung of a rabbit was excised, the remaining lung expanded to fill the thoracic cavity, and it grew until it weighted approximately the initial weight of both lungs. On the other hand, animals exposed to the weightless condition of space flight have demonstrated skeletal muscle atrophy. Leg volumes of astronauts are diminished in flight. In-flight vigorous daily exercise is necessary to keep astronauts in good physical fitness over a longer period of time.

Immobilization of muscle causes atrophy. But there is a marked difference between stretched immobilized muscle vs. muscle immobilized in the resting or shortened position. Fundamentally, growth is a cell-biological phenomenon at molecular level. Stress and strain keep the cells in a certain specific configuration. Since growth depends on cell configuration, it depends on stress and strain.

How fast do soft tissues remodel when stress is changed? That question is discussed in Chapter 2 of this book. It is shown that in the case of the lung the active remodeling of blood-vessel wall proceeds quite fast. Histological changes can be identified within hours. The maximum rate of change occurs within a day or two. See discussions on p. 28 et seq.

Morphometric changes are not the only changes occurring in blood vessel wall when the blood pressure changes as a step function. The zero-stress state of the vessel and the mechanical properties of the vessel wall also change with time as the remodeling proceeds. This is expected because the mechanical properties follow material composition and structure. So the properties will change when the composition and structure changes. Data on zero-stress state changes during tissue remodeling in pulmonary and system arteries and in veins are given in papers by Fung and Liu (26) and Liu and Fung (27). Data on changes in mechanical properties due to tissue remodeling in arteries are presented in Liu and Fung (28, 29).

A corresponding program of research on the hypertrophy of the heart leads also to many new findings. Increased stress in the heart also leads to hypertrophy, and morphometric structural and mechanical properties change.

What these experiments reveal is that other things being equal, tissue growth is related to stress. The growth-stress relationship plays a role in the healing of the tissues and in rehabilitation to normal life.

9.   Tissue Engineering

Looming large in the future is tissue engineering, which can be defined as engineering the improvement of natural tissues of man, or creating living artificial tissue substitutes with human cells. One example is the skin substitute made with a patient's own cells cultured in a scaffold of biodegradable or nondegradable material. Another example is the blood-vessel substitute seeded with the patient's own endothelial cells. Work on cartilage, bone, and other organs is in progress. These new techniques became possible because major advances were made in the art of tissue culture in recent years as a consequence of the discovery of growth factors and various culture media. These discoveries make tissue engineering thinkable. With the feasibility established, we can now think of practical applications of the stress-growth laws to engineering living tissues. Success in this area will have great impact on healing and rehabilitation of injuries.

10.   Conclusion

Biomechanics is a key factor in understanding accidental injury and healing. This chapter presents an overview and an introduction to some terminology and basic concepts. A large factor in accident prevention is not biomechanics, not medicine, not surgery, not rehabilitation, not law. It is culture. It is good for the professional people to think of the big picture. However, the professional side must not be minimized. The injured persons need care. The society needs injury prevention. At the level of emergency care, medical and surgical treatment, physical rehabilitation, designing of a safer vehicle, providing better protection of man, and establishing as better public policy, we need the best of scientific knowledge. A thorough understanding of the fundamental topics discussed in this article will help.

This discussion points out the directions to look for opportunities to make a contribution to reduce injury and promote healing. Specific opportunities are not always obvious. Significant findings are reserved only for the alert and careful observer. The editorial article by Savio Woo on pp. 249–251 makes a good point for biomechanical research in the new millenium.

References

1. 'A Compilation of Motor Vehicle Crash Data from the Fatality Analysis Reporting System and the General Estimates System,' National Highway Traffic Safety Administration, report No. DOT HS 808 806.
2. Interagency Head Injury Task Force Reports, National Institute of Neurological Disorders and Stroke, cited in http://www.shepherd.org/abi/facts.htm
3. Melvin, J. W., Lighthall, J. W., and Ueno, K., Brain injury biomechanics. In *Accidental Injury: Biomechanics and Prevention*, A. M. Nahum and J. W. Melvin, (eds.) New York: Springer-Verlag (1993), pp. 268–291.
4. Gurdjian, E. S., Webster, J. E., and Lissner, H. R., Observations on the mechanism of brain concussion, contusion, and laceration. *Surg. Gynec. Obstet.* **101**, 680–690 (1955).
5. Lissner, H. R., Lebow, M., and Evans, F. G., Experimental studies on the relation between acceleration and intracranial pressure changes in man. *Surgical Gynecol. Obstet.* **111**, 329–338 (1960).
6. Gadd, C. W., Criteria for injury potential. In Impact Acceleration Stress Symposium. Washington, D. C.: National Academy of Sciences (1961).
7. Fung, Y. C., The application of biomechanics to the understanding of injury and healing. In *Accidental Injury: Biomechanics and Prevention*, edited by Alan M., Nahum, and John W. Melvin. New York: Springer (1993), pp. 1–11.
8. Fung, Y. C., *Biomechanics: Motion, Flow, Stress and Growth*, New York: Springer (1990).
9. Fung, Y. C., In *the Biomechanics of Trauma*, edited by Nahum, A. M. and Melvin, J., Norwalk, CT: Appleton-Century-Crofts (1985).
10. Yen, R. T., Fung, Y. C., Ho, H. H., and Butterman, G., Speed of stress wave propagation in the lung. *J. Appl. Physiol.* **61**(2), 701–705 (1986).
11. Rice, D. A., Sound speed in pulmonary parenchyma. *J. Appl. Physiol.* **54**(1), 304–308 (1983).
12. Dunn, F. and Fry, W. J., Ultrasonic absorption and reflection of lung tissue. *Phys. Med. Biol.* **5**, 401–410 (1961).

13. Ommaya, A. K., Faas, F. and Yarnell, P., Whiplash injury and brain damage: An experimental study. *JAMA*, **204**(4), 285–9 (1986).

14. Tong, P., Galbraith, C. and DiMasi, F. P., Three dimensional modeling of head injury subjected to inertial loading. *Proc. of the 12th International Techn. Conf. on Experimental Safety Vehicles*, Gothenburg, Sweden, June (1989).

15. Bandak, F. A. and Eppinger, R., A three dimensional finite element analysis of the human brain under combined rotational and translational acceleration. *38th Stapp Car. Crash Biomechanics Conference Proceedings*, paper 279, 145–163 (1994).

16. Woo, S. L. Y. and Buckwalter, J. A., (eds) *Injury and repair of the musculoskeletal soft tissues*. American Academy of Orthopedic Surgeons. Park Ridge, IL (1988).

17. Fung, Y. C. and Tong, P., *Classical and Computational Solid Mechanics*, World Scientific Publisher, Inc. Singapore (2000).

18. Tong, P., Crashworthiness. *Proceedings of the 10th Annual Meeting of Society of Engineering Science* (1975).

19. DiMasi, F. P., Eppinger, R. and Bandak, F. A., Computational analysis of head impact response under car crash loadings. *39th Stapp Car Crash Biomechanics Conference Proceedings*, Paper 299, 425–438 (1995).

20. Meyer, G. H., Die Architektur der spongiosa. *Archiv für Anatomie, Physiologie, und wissenschaftliche Medizin. (Reichert und wissenschaftliche Medizin, Reichert und Du Bois-Reymonds Archiv)* **34**, 615–625 (1867).

21. Culmann, C., *Die graphische Statik*, Zurich: Meyer und Zeller (1866).

22. Wolff, J., Über die bedeutung der Architektur der spondiosen Substanz, *Zentralblatt für die medizinische Wissenschaft.* **6**, 223–234 (1869).

23. Roux, W., *Gesammehe Abhandlungen über die entwicklungs mechanik der Organismen*. W. Engelmann, Leipzig (1895).

24. Pauwels, F., *Biomechanics of the locomotor apparatus*, German ed. (1965). English translation by P. Maqnet and R. Furlong. Springer-Verlag, Berlin, New York (1980).

25. Cowan, M. J. and Crystal, R. G., Lung growth after unilateral pneumonectomy: Quantitation of collagen synthesis and content. *Am Rev Respir Disease* **111**, 267–276 (1975).

26. Fung, Y. C. and Liu, S. Q., Change of residual strains in arteries due to hypertrophy caused by aortic constriction. *Circ. Res.* **65**, 1340–1349 (1989).

27. Liu, S. Q. and Fung, Y. C., Relationship between hypertension, hypertrophy, and opening angle of zero-stress state of arteries following aortic constriction. *J. Biomech. Eng.* **111**, 325–335 (1989).

28. Liu, S. Q., and Fung, Y. C., Influence of STZ-induced diabetes on zero-stress states of rat pulmonary and systemic arteries. *Diabetes* **41**, 136–146 (1992).

29. Fung, Y. C., and Liu, S. Q., Strain distribution in small blood vessels with zero-stress state taken into consideration. *Am J Physiol: Heart & Circ.* **262**, H544–H552 (1992).

J Orthop Sci (2000) 5:89–91

Journal of

Orthopaedic Science

© The Japanese Orthopaedic Association

Editorial

The importance of biomechanics for the new millennium

Savio Lau-Yuen Woo

Director, Musculoskeletal Research Center, Department of Orthopaedic Surgery, University of Pittsburgh Medical Center, Pittsburgh, PA, USA

Key words: biomechanics, ligaments, knee, injury

"Biomechanics is the middle name of biological structure and function"

When Professor Y.C. Fung made this statement in 1998, he indicated that biomechanics research constitutes the investigation of structures at all levels, ranging from molecules, to cell membranes, to tissues, to organs, to the body. At the beginning of the new millennium, it is indeed an exciting time to reflect on the potential of biomechanics. In this editorial, I have chosen to write about my particular area of interest — ligaments. Nevertheless, I do believe the concepts presented are applicable to other areas of orthopedic research.

The medial collateral ligament (MCL) of the knee is known to heal spontaneously after injury. However, one problem is that the morphology, biochemistry, and mechanical properties of the healed MCL have failed to return to their uninjured condition, even after long periods of time.[4] The cruciate ligaments of the knee have limited healing capability, and surgical reconstruction with tissue grafts is needed. Although outcomes have been reasonably successful for the anterior cruciate ligament (ACL), the same cannot be said for the posterior cruciate ligament (PCL). Significant research efforts are being made to improve the outcome of such knee ligament injuries, and future directions point towards the integration of biological and biomechanical approaches.

"Functional tissue engineering" is a concept which offers the potential to improve the quality of ligaments during the healing process. This concept is based on the manipulation of cellular and biochemical mediators to affect protein synthesis. Novel techniques — such as the application of growth factors, gene transfer technology for growth factor delivery, mesen-chymal stem cell (MSC) therapy, biomatrix scaffolding coupled with seeding of cells, and the mechanical loading of cells — have shown some promise. Among these applications are growth factors such as platelet-derived growth factor (PDGF)-BB, epithelial growth factor (EGF), and transforming growth factor (TGF) β1, TGF β2, insulin-like growth factor (IGF), and bone morphogenic proteins (BMPs). Investigators have successfully introduced marker and therapeutic genes into ligaments, using retroviral and adenoviral techniques. Adeno-associated virus is also being explored, as it infects both dividing and nondividing cells, with almost no immune response. Other investigators have taken advantage of non-viral and non-toxic gene transfer techniques, such as hemagglutinating virus of Japan (HVJ)-conjugated liposomes, with which there are few constraints on the size of gene to be delivered. Frank et al.[2] hypothesized that decorin, a small leucine-rich protein, may play a role in fibrillogenesis. Antisense decorin oligodeoxynucleotides were introduced, using an in-vivo liposome method, to decrease decorin RNA expression and protein synthesis. As a result, large collagen fibrils were formed, and the mechanical properties of the healing ligament were enhanced. Another current area of research uses silicon microgroove surfaces to align fibroblasts in a parallel fashion, aiming to generate a new matrix that resembles the uninjured ligament.

It should be noted that the above avenues are all in the early stages of development. By combining appropriate engineering mechanics with other basic science, we believe that the process of ligament healing can be functionally engineered to produce better outcomes. It is hoped that the knowledge gained can be extended to enhance the healing of: (1) the ACL, PCL, and all ligaments, (2) soft tissues in the bone tunnel, and (3) ligament attachment at the insertion sites.

From a bioengineering perspective, the robotics/universal force moment sensor testing system has been a unique development; with it, measurement of the in-situ forces in ligaments can be done with greater accuracy.[3] In the coming years, the focus will be to obtain kinematics data in vivo. These data can be used experimentally to determine the in-situ forces of knee ligaments during the same in-vivo activities. Also, the data are needed to validate mathematical models (Fig. 1). These models, in turn, can be used to calculate stresses and strains in ligaments and to examine the mechanical effects on cellular responses. This combined experimental and mathematical modeling approach offers the opportunity to establish a database on intact ligaments and ligament replacement grafts, based on factors such as sex, age, and size. Hence, database optimization of the variations that occur during ligament reconstruction (ie, graft tension, placement, and position) can be performed on an individual basis. Further, surgical preplanning and postoperative rehabilitation protocols can be customized. Ultimately, the outcome for patients should be improved.

Bioengineers and biologists have a reasonably good history of collaboration with orthopedic surgeons. However, in the future, even more integration of disciplines, including biology, biochemistry, biomechanics, and surgery, is necessary. Investigators need to work together in the same laboratory.[1] To do this, one must learn to truly respect disciplines other than one's own, and major efforts are needed to learn the languages of the other fields. I have long believed that it is uncommon for a good biomechanician to also be a good biologist, and vice versa. Finding willing collaborators and developing methods of communication are efficient ways of accomplishing our common goals (see Fig. 2).

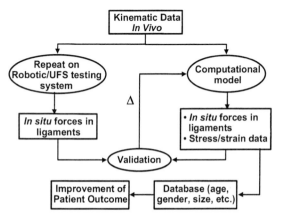

Fig. 1. Flow chart detailing the general approach to gain data on the in situ forces, as well as stresses and strains, in the ligaments of the knee and shoulder under loading conditions that involve in-vivo activities. *UFS*, Universal force moment sensor

References

1. Andriacchi TP, Alexander EJ, Toney MK, et al. A point cluster method for in vivo motion analysis: applied to a study of knee kinematics. J Biomech Eng 1998;120:743–49.

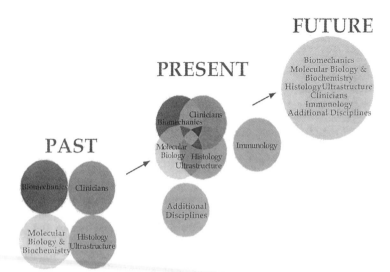

Fig. 2. Collaboration among scientists is an efficient and ideal way to accomplish common goals

2. Frank C, Shrive N, Hiraoka H, et al. Optimization of the biology of soft tissue repair. J Sci Med Sport 1999;2:190–210.

3. Rudy TW, Livesay GA, Woo SL-Y, Fu FH. A combined robotic/universal force sensor approach to determine in situ forces of knee ligaments. J Biomech 1996;29:1357–60.

4. Woo SL-Y, Smith DW, Hildebrand KA, et al. Engineering the healing of the rabbit medial collateral ligament. Med Biol Eng Comput 1998;36:359–64.

Speed of stress wave propagation in lung

R. T. YEN, Y. C. FUNG, H. H. HO, AND G. BUTTERMAN

Department of AMES-Bioengineering, University of California, San Diego, La Jolla, California 92093

YEN, R. T., Y. C. FUNG, H. H. HO, AND G. BUTTERMAN. *Speed of stress wave propagation in lung.* J. Appl. Physiol. 61 (2): 701–705, 1986.—The speed of stress waves in the lung parenchyma was investigated to understand why, among all internal organs, the lung is the most easily injured when an animal is subjected to an impact loading. The speed of the sound is much less in the lung than that in other organs. To analyze the dynamic response of the lung to impact loading, it is necessary to know the speed of internal wave propagation. Excised lungs of the rabbit and the goat were impacted with water jet at dynamic pressure in the range of 7–35 kPa (1–5 psi) and surface velocity of 1–15 m/s. The stress wave was measured by pressure transducer. The distance between the point of impact and the sensor at another point on the far side of the lung and the transit time of the stress wave were measured. The wave speed in the goat lung was found to vary from 31.4 to 64.7 m/s when the transpulmonary pressure Pa − Ppl was varied from 0 to 20 cmH₂O where Pa represents airway pressure and Ppl represents pleural pressure. In rabbit lung the wave speed varied from 16.5 to 36.9 m/s when Pa − Ppl was varied from 0 to 16 cmH₂O. Using measured values of the bulk modulus, shear modulus, and density of the parenchyma, reasonable agreement between theoretical and experimental wave speeds were obtained.

lung parenchyma; compression wave; elastic modulus; density of lung parenchyma

IN ADDITION to scientific curiosity, there are other reasons why people want to study wave propagation in the lung. Some people want to use waves to probe the interior of the lung for changes and diseases (6, 11) much as the geophysicists set off blasts to induce elastic waves for oil prospecting (8). Other persons want to know if there are ill effects from exposure to ultrasound waves (7). Still others want to use lung sound for diagnosis (5). For the present paper, our motivation is trying to understand why, among all internal organs, the lung is the most easily damaged when a human or animal is subjected to a blast load. The stress wave speed is much smaller than that in other organs. The features of wave reflection, focusing, and interaction, therefore, often dominate the scene of lung dynamics in response to impact loading. There exists a large literature in which the edema and/or hemorrhage in the lung were identified as the cause of death when humans or animals were exposed to explosive overpressure (1–4, 14, 15). To explain this phenomenon, we thought the stress-wave aspect must be looked into, because by wave reflection, refraction, and focusing, high stress regions may exist in the lung where trauma may be initiated. As a first step, the speeds of stress

waves must be determined.

The lung is a complex structure. Approximately 10% of the lung volume is occupied by the larger conducting airways and blood vessels, and 90% is occupied by the parenchyma, which consists of the alveoli, alveolar ducts and sacs, and capillary blood vessels, arterioles, and venules. Stress waves can travel through the lung in many pathways, either through the larger airways and blood vessels, or through the parenchyma. In impact problems the critical trauma is edema in the parenchyma; hence, we shall be concerned mainly with wave propagation in the lung parenchyma. The lung parenchyma is a two-phase composite structure of soft tissues and gas. In the parenchyma the characteristic dimension is alveolar diameter and the Reynolds number of gas motion is less than one. The no-slip condition for the relative motion of gas and tissue implies that the parenchyma can be treated as a continuum. Many types of progressive waves can propagate in such a continuum. The tissue motion may principally be shearing (transverse to the direction of wave propagation). The wave front may be planar or curved. The amplitude of motion may be spatially uniform, or varied as in surface waves. The wave length may be very long compared with the alveolar diameter, (as in waves in the kHz range), or of the same order of magnitude, (as in waves in the MHz range), or even much shorter, (higher frequencies). The type of motion an alveolar wall or an alveolar mouth in the duct can execute depends on the ratio of the wave length to the alveolar diameter. Thus, in principle, there can be many types of waves, each of which has a characteristic speed. In different types of waves the wave speeds are different because the partition of energy between gas and tissue is different. Which one is excited depends on the mode of excitation and the amplitude of the loading. It follows that one should design an experiment according to the intended application, so that the type of the impact load and its amplitude level can be properly chosen. In the present study, we are interested in the dynamics of the lung in response to impact or shock wave acting on the lung surface; hence the impact load level should be appropriate to the problem of possible lung injury. In the following, we present the results of our measurement of wave speed in the lungs of the rabbit and goat, by using a water jet to impose an impact load on the lung. The impact pressure level on the surface was in the range of 7 to 35 kPa (1–5 psi), and the impact velocity at the lung surface was on the order of 1–15 m/s. According to Clemedson and Jönsson (4), severe hemorrhages in the lung are highly probable for a maximum impact velocity at the lung surface exceeding about 10 m/s. A pressure transducer was used to record the pressure wave produced by the jet hitting the lung. The transducer was also placed at a point on the far side of the lung to sense the arrival of the stress waves. A photosensor was placed at various places between the nozzle of the water jet and the lung to measure the arrival of the water. From these signals the velocity of water jet and the speed of the stress wave in the lung were determined.

APPARATUS

Figure 1 is a schematic diagram of the testing apparatus. The apparatus consists of three subgroups, the first of which is the water jet for imposing impact loading on the surface of the lung. To obtain a well defined impulse loading covering a circle of 2-mm diameter on the surface of the lung, a water jet was constructed. A fluid reservoir, driven by a pressure regulated air source, charged a small hydraulic accumulator through a normally open electronic flexure poppet valve. This valve, manufactured by Clippard Minimatic, has an activating time of < 3 ms due to the very low mass of the flexure poppet. The valve was designed to operate at 5 V DC at 100 mA. When a normal trigger was applied to the valve sequencer, valve A would close, isolating the accumulator from the fluid reservoir. After 30 ms valve B, an identical flexure poppet valve, would open for 5 ms, discharging the accumulator through the nozzle. The nozzle itself consisted of 5 holes of 0.3-mm diameter each, grouped in a 2 mm circle.

The second subgroup consisted of the photosensor system, which was used to sense the arrival of the water jet. The photosensor could be moved with precision along the axis of the jet to calculate the velocity of the fluid in

the air. The conditioning electronics incorporated a Hi-pass filter with a time constant of 100 μs to eliminate ambient light effects.

The third subgroup consisted of the pressure transducer used to record the response due to water impact of the lung. The arrival time and the pressure wave were recorded. The pressure transducer was connected to a Neff amplifier and the output was shown on a Tektronix 5440 oscilloscope. The pressure transducer was an Endevco model 8510-2 m21 with a calibrated sensitivity of 151.6 mV/psi.

ANIMAL PREPARATION

The animal (goat or rabbit) was first anesthetized with pentobarbital sodium at 7 mg/kg. Heparin (145 U/kg) was then injected in the bloodstream to prevent blood coagulation. The trachea was cannulated. Through a midline incision of the chest, the lung was exposed and excised and an isolation of the lung lobes was made.

EXPERIMENTAL METHOD

The excised lung was placed firmly on a pressure transducer in a supporting basin. A water jet was placed directly above the lung, with a distance of 1.8 cm from the bottom of the nozzle to the surface of the lung. The nozzle has five holes to allow five small jets converging onto one area on the lung surface. A photosensing tube was placed between the lung surface and the nozzle. In each experiment, the reservoir air pressure was kept constant (50–60 psi). With a touch of the control unit key, a water jet was fired. The screen of the oscilloscope showed the output. For each constant reservoir air pressure, at least three photosensor positions were used in the experiment. Then another reservoir air pressure of

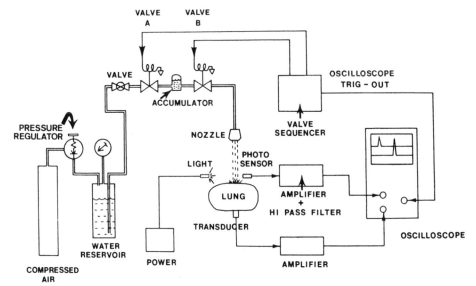

FIG. 1. A schematic diagram of wave propagation testing apparatus. Apparatus consists of 3 subgroups: water jet, photosensor system, and pressure response measuring system.

the lung was set at specific levels. The pleural pressure was zero (atmospheric).

We then plotted on an X-Y coordinate paper (Fig. 2) the distance between the sensing elements of the pressure transducer and the photosensor (d) vs. the difference in time (t) between the leading edges of the pressure wave and the photosensor wave. For each water ejection setting (pressure in the accumulator) a straight line was obtained. By changing the setting, different straight lines were obtained. These lines could intersect at one point, which marked the time used and distance traveled by the wave in the lung. In this way we obtained the lung thickness d and the corresponding time, t, of travel of the stress wave through the lung tissue. The wave speed in the lung tissue is $c = d/t$. A cathetometer was used to measure d. The cathetometer uses a telescope with a cross-hair reticle to measure the thickness between two points without actually touching the tissue. When obtained in this way, d agreed within 5% with the other measurement.

DATA ANALYSIS

A typical set of experimental curves is shown in Fig. 3, A and B. Figure $3A$ shows the pressure wave due to impact of the water jet on the transducer located at the place where the jet hits the lung (with the lung removed). The particular wave in Fig. $3A$ shows a peak pressure of 1.7 psi. Figure $3B$ shows the oscilloscopic traces of the photosensor output (*upper curve*) and the pressure sensor output (*lower curve*). The photosensor indicates the arrival of the water jet. The pressure transducer indicates the stress wave at the place where the transducer was located. The first high peaks in these records were used to correlate the impact wave and the transmitted wave. The time difference between the first arrival of the water jet in the line of sight of the photosensor and the first arrival of the stress wave at the pressure transducer was measured and designated as t_1. The corresponding distance between these sensors is designated as d_1. These data are plotted as shown in Fig. 2. From these plots the wave speed is calculated by a method presented earlier.

Rice (12) showed that the speed of a plane compression wave in an elastic continuum is

$$c = \left(\frac{B}{\rho}\right)^{1/2} \qquad (1)$$

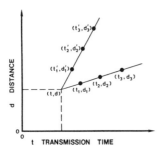

FIG. 2. X-Y plot of distance d vs. transmission time t.

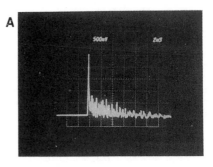

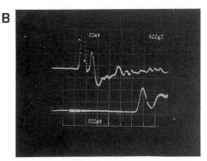

FIG. 3. A: pressure wave due to impact of water jet on transducer located at place where jets hit lung (with lung removed). This particular wave shows a peak pressure of 1.7 psi. B: usual format of outputs recorded on screen of oscilloscope. *Upper curve* is response of photosensor indicating arrival of water jet. *Lower curve* is response of pressure transducer, which indicates stress wave in lung at location of transducer. First sharp wave is used to correlate impact wave imposed on lung. Time difference between arrival of water jet at photosensor and first arrival of wave by impact was measured. Results are used to obtain water jet speed in air as well as distance and traveling time in lung tissue.

where c is the sound speed in m/s, B is an elastic constant[1] in N/m² (Pa) and ρ is the density in kg/m³.

In our experiment, we used the first arrival time of the stress wave to compute the wave speed. We may identify the wave as a compression wave, for which we can use Eq. 1 with $B = \lambda + 2G$, where λ and G are the Lame constants. Since λ is related to the bulk modulus, K, and the shear modulus, G, by the formula

$$\lambda = K - \frac{2}{3} G \qquad (2)$$

we have

$$B = K + \frac{4}{3} G \qquad (3)$$

We have also measured the elastic moduli K and G for the lungs of the goat, rabbit, and human (R. T. Yen, Y.

---

[1] Rice (12) identified B as the "volumetric stiffness." Actually, according to the theory of elasticity, B should be the shear modulus G for a shear or transverse wave, or $\lambda + 2G$ for a dilatational or longitudinal wave, λ and G being the Lame constants. The volumetric stiffness or bulk modulus, K, is related to λ and G by the relation $K = \lambda + (2/3)G$. See, for example, Timoshenko and Goodier (13) or Fung (9).

C. Fung, and C. Artaud, unpublished observations). We measured K first with the airway open at the mouth, and then with the airway closed at the mouth. The former is much smaller than the latter. On the other hand, the shear modulus of elasticity, G, is related to deformation without volumetric change; hence it is independent of whether the airway is open or closed at the mouth. So the values of G, measured with airway open at the mouth, can be used. Using these experimental values of K and G, and using Eq. 1, we obtained the theoretical values of the wave speed.

RESULTS

Table 1 gives the experimental results of the goat lung for several transpulmonary pressures (airway minus pleural pressures, pa − ppl). The values of c lie in the range of 30–70 m/s, and have a general pattern of increasing with increasing transpulmonary pressure. Here n is the number of lungs tested, and SD is the standard deviation.

Table 2 summarizes the experimental wave speed in the rabbit lung. Table 3 is a comparison of the calculated wave speed in the rabbit lung with the experimental value. The table lists the measured values of lung parenchyma density, the bulk modulus, the shear modulus,

and the theoretical wave speed computed from Eq. 1. The bulk modulus K in Table 3 is obtained with the airway closed. These results will be discussed below.

DISCUSSION

The lung has a complex structure and the lung parenchyma is a composite material with a highly nonlinear stress-strain relationship. Many types of waves can exist in it. Which one is excited depends on the mode of excitation and the amplitude of the loading. Data in the literature support this concept. Dunn and Fry (7), using the reflection coefficient of ultrasound energy (1 MHz) and the density of the lung tissue ($\sim$0.4 g/ml), calculated a wave speed of 650 m/s in excised dog lung tissue in which $\sim$⅔ air remained. The wave length at 1 MHz is 0.66 mm. Rice (12), using a microphone as a sensor and an electric spark discharge as a sound source, measured the transit time of sound through the lung parenchyma of a horse and estimated the sound speed in air-filled excised horse lung to be between 25 and 70 m/s at 20–25°C. Rice varied the ambient pressure of the air, and also substituted air with He and SF_6, and showed that the trend of his results agree with the theoretical formula for the speed of a plane compression wave in an elastic continuum

$$c = (B/\rho)^{1/2}$$

where c is the sound speed in m/s, B is an elastic constant in N/m^2 (Pa), and ρ is the density in kg/m^3. Recently, Kraman (10) measured the transit time of sound waves filtered between 125 and 500 Hz from a point on the neck beneath the larynx and another at each of eight locations on the chest wall, and obtained the result that at functional residual capacity the average wave speed is on the order of 30 m/s. Although this cannot be interpreted as sound speed in the parenchyma, it is certainly much lower than that in the trachea [350 m/s according to Rice (12)] and chest wall ($\sim$10^3 m/s, our estimation).

Relevant to the air shock problem, Clemedson and Jönsson (3) measured the wave speed in the lung by inserting two pressure transducers (of diameter ³/₃₂ in. or 2.38 mm) into the airways of the right and left lungs of an anesthetized rabbit or calf and then subjected the animal to a shock by chemical explosive. By measuring the difference of the arrival times of the stress waves at the probes and the distance between the probes by au-

TABLE 1. *Stress wave speed in goat lung*

Transpulmonary Pressure (Pa − Ppl), cmH$_2$O	n	Measured Wave Speed, m/s
0	3	31.4±0.4
5	4	33.9±2.3
10	6	36.1±1.9
15	6	46.8±1.8
20	4	64.7±3.9

Values are means ± SD. Pa, airway pressure; Ppl, pleural pressure.

TABLE 2. *Stress wave speed in rabbit lung*

Transpulmonary Pressure (Pa − Ppl), cmH$_2$O	n	Measured Wave Speed, m/s
0	6	16.5±2.4
4	6	28.9±3.3
8	5	31.3±0.9
12	4	35.3±0.8
16	7	36.9±1.7

Values are means ± SD. Pa, airway pressure; Ppl, pleural pressure.

TABLE 3. *Comparison of calculated wave speed in rabbit lung with experimental value*

Transpulmonary Pressure (Pa − Ppl), cmH$_2$O	Density (ρ), g/ml	Bulk Modulus (K), cmH$_2$O	Shear Modulus (G), cmH$_2$O	Calculated Wave Speed $\left[c = \left(\dfrac{K + 4G/3}{\rho} \right)^{1/2} \right]$, m/s	Experimental Wave Speed (c), m/s
0	0.596±0.056	1,425±27	2.2±0.16	15.3	16.5±2.4
4	0.145±0.010	1,327±27	4.7±0.16	30.0	28.9±3.3
8	0.118±0.007	1,291±25	7.3±0.25	32.9	31.3±0.9
12	0.109±0.007	1,280±24	9.9±0.35	34.1	35.3±0.8
16	0.103±0.006	1,279±25	12.5±0.35	35.1	36.9±1.7

Values are means ± SD. The values of ρ are obtained from measurements of 3 rabbit lungs. Weight of rabbits ranges from 2 to 2.5 kg. K are closed glottis values.

topsy, they obtained the velocity of the main part of the pressure pulse in the calf lungs inflated to approximately intravital size to be between 24 and 30 m/s. For rabbit lungs they found speeds of 15, 32, and 69 m/s in several animals but considered the data to be uncertain. In collapsed rabbit lung, they found a wave speed of 15 m/s.

The 10- to 20-fold difference between the data quoted above reflects the fact that different kinds of motion were excited in different experiments. The type of impact load and the amplitude of the loading should be chosen according to the intended application. Our choice of the water jet is based on the following consideration. Clemedson and Jönsson (4) used the denotation of a high explosive in a "wave guide" (constituted of two concentric cylinders) to impose a planar shock wave on the animal. By their method the determination of the wave arrival time is difficult, as can be seen from the cathode ray traces in their paper. In earlier stages of our own experiment, we designed and used several shock tubes (using compressed air and paper membrane) to produce a loading in an overpressure range of a few pounds per square inch. We found that the weak supersonic shock was followed by bulk flow. The lung responded to both the weak supersonic shock wave and the large bulk flow, and the determination of wave propagation characteristics was often difficult. This difficulty was minimized by using the water jet because the leading edge of the water jet produced a well-defined impact wave.

The impact velocity and the overpressure of the water jet used in our experiment were in the range that can cause lung injury. The target size used in our experiment, however, was very small. The wave front, therefore, will likely be spherical and the wave amplitude will decrease with the inverse square of the distance from the point of impact. This is why the amplitude of the stress wave shown in Fig. 3B was so small. Thus, a few centimeters away from the impact area, the waves were sound waves of small amplitude. Since they are longitudinal waves and there is no intereference by reflection waves, it is not surprising that the speed we found is of the same order of magnitude as that found by Rice (12).

Wave speeds obtained from experiments and theoretical calculations are listed in Table 3. It is seen that the theoretical wave speeds are fairly close to the experimental results when the value of K corresponding to the case of airway closed at the mouth is used. This is interpreted as indicating that the wave proceeds so fast that the gas in the alveoli and ducts at the wave front have no time

to escape to seek equilibrium. The shock process is so fast that the bulk movement of gas through the glottis is negligible in the first few moments. Hence the gas is compressed locally when the compression wave arrives. This is entirely reasonable in view of the very low Reynolds number of the gas flow in the alveoli and alveolar ducts.

Support of this research, given by the Walter Reed Army Institute of Research, Walter Reed Medical Center, under contract with Jaycor, is gratefully acknowledged. The views, opinions, and/or findings contained in this report are, however, solely those of the authors.

The design and fabrication of the impact machine was mainly due to Eugene Mead, development engineer at the University of California, San Diego.

Received 20 May 1985; accepted in final form 3 March 1986.

REFERENCES

1. BOWEN, I. G., A. HOLLADAY, E. R. FLETCHER, D. R. RICHMOND, AND C. S. WHITE. A Fluid-Mechanical Model of the Thoracoabdominal System with Applications to Blast Biology. Albuquerque, NM: Lovelace Found. Med. Ed. Res. 1965. (Rep. DASA-1675).
2. CLEMEDSON, C.-J., AND A. JÖNSSON. Transmission and reflection of high explosive shock waves in bone. Acta Physiol. Scand. 51: 47–61, 1961.
3. CLEMEDSON, C.-J., AND A. JÖNSSON. Distribution of extra- and intrathoracic pressure variations in rabbits exposed to air shock waves. Acta Physiol. Scand. 54: 18–29, 1962.
4. CLEMEDSON, C.-J., AND A. JÖNSSON. Dynamic response of chest wall and lung injuries in rabbits exposed to air shock waves of short duration. Acta Physiol Scand. Suppl. 233: 31, 1964.
5. DEGOWIN, E. L., AND R. L. DEGOWIN. Bedside Diagnostic Examination. New York: Macmillan, 1976, p. 339–342.
6. DONNERBERG, R. L., C. K. DRUZGALSKI, R. L. HAMLIN, G. L. DAVIS, R. M. CAMPBELL, AND D. A. RICE. Sound transfer function of the congested canine lung. Br. J. Dis. Chest. 74: 23–31, 1980.
7. DUNN, F., AND W. J. FRY. Ultrasonic absorption and reflection by lung tissue. Phys. Med. Biol. 5: 401–410, 1961.
8. EWING, W. M., W. S. JARDETZKY, AND F. PRESS. Elastic Waves in Layered Media. New York: McGraw-Hill, 1957.
9. FUNG, Y. C. A First Course in Continuum Mechanics. Englewood Cliffs, NJ: Prentice-Hall, 1977.
10. KRAMAN, S. S. Speed of low-frequency sound through lungs of normal humans. J. Appl. Physiol. 55: 1862–1867, 1983.
11. PLOY-SONG-SANG, Y., R. R. MARTIN, W. R. D. ROSS, R. G. LONDON, AND P. MACKLEM. Breath sounds and regional ventilation. Am. Rev. Respir. Dis. 116: 187–193, 1977.
12. RICE, D. A. Sound speed in pulmonary parenchyma. J. Appl. Physiol. 54: 304–308, 1983.
13. TIMOSHENKO, S., AND J. N. GOODIER. Theory of Elasticity, New York: McGraw-Hill, 1951.
14. WHITE, C. S., R. K. JONES, E. G. DAMON, E. R. FLETCHER, AND D. R. RICHMOND. The Biodynamics of Airblast. Albuquerque, NM: Lovelace Found. Med. Ed. Res., 1971. (Rep. DNA 2738T).
15. WHITE, C. S., AND D. R. RICHMOND. Blast Biology Albuquerque, NM: Lovelace Found. Med. Ed. Res., 1959. (Tech. Prog. Rep. TID 5764).

Reprinted from February 1988, Vol. 110, Journal of Biomechanical Engineering

A Hypothesis on the Mechanism of Trauma of Lung Tissue Subjected to Impact Load

Y. C. Fung

R. T. Yen[2]

Z. L. Tao[1]

S. Q. Liu

Department of AMES/Bioengineering,
University of California, San Diego,
La Jolla, CA 92093

When a compressive impact load is applied on the chest, as in automobile crash or bomb explosion, the lung may be injured and show evidences of edema and hemorrhage. Since soft tissues have good strength in compression, why does a compression wave cause edema? Our hypothesis is that tensile and shear stresses are induced in the alveolar wall on rebound from compression, and that the maximum principal stress (tensile) may exceed critical values for increased permeability of the epithelium to small solutes, or even fracture. Furthermore, small airways may collapse and trap gas in alveoli at a critical strain, causing traumatic atelectasis. The collapsed airways reopen at a higher strain after the wave passes, during which the expansion of the trapped gas will induce additional tension in the alveolar wall. To test this hypothesis, we made three new experiments: (1), measuring the effect of transient overstretch of the alveolar membrane on the rate of lung weight increase; (2) determining the critical pressure for reopening collapsed airways of rabbit lung subjected to cyclic compression and expansion; (3) cyclic compression of lung with trachea closed. We found that in isolated rabbit lung overstretching increases the rate of edema fluid formation, that the critical strain for airway reopening is higher than that for closing, and that these critical strains are strain-rate dependent, but independent of the state of the trachea, whether it is open or closed. Furthermore, a theoretical analysis is presented to show that the maximum principal (tensile) stress is of the same order of magnitude as the maximum initial compressive stress at certain localities of the lung. All these support the hypothesis. But the experiments were done at too low a strain rate, and further work is needed.

1 Introduction

This article is aimed at explaining the mechanism of impact injury of the lung. The lung is a soft organ capable of large deformation, yet under rapidly applied compressive load it is strangely susceptible to edema following the impact, and injury by hemorrhage. Before formulating an explanation, we shall review the literature in order to learn about the phenomena involved. Then we shall present a hypothesis and the evidences that support it. Three new experiments and a new theoretical analysis are presented. Yet the strain rate used in our experiment is far slower than that occur in real impact. Hence our hypothesis remains hypothetical at this time.

2 A Review of the Literature on Trauma and a Summary of Facts

Since the lung has the elasticity of the air but the mass of the tissue, in the lung the speed of sound is low (about 1/10 of the

speed of sound in air, see Clemedson and Jönsson, 1962; Rice, 1983, Kraman, 1983, Yen et al. 1986), hence under a rapidly applied loading the response of the lung is dominated by stress waves (Fung, 1985). Through reflection, refraction, and focusing, these waves may cause large stresses and strains in some small regions of the lung and induce severe injury. It is known that when an animal is subjected to bomb explosion the most vulnerable organ is the lung (Clemedson, 1956, 1964a; White et al., 1971; Jönsson, 1979). Lung injury occurs also in automobile crash (Viano, 1978, Lau and Viano, 1981), and pulmonary edema in highway traffic victims is a frequent clinical finding. Although the rate of loading in automobile crash is much lower than that in bomb explosion, in both cases the first shock is a compression wave. It is well known that the reflection of a compression wave at a rigid boundary intensifies the compressive stress, whereas one reflected from a free boundary decreases it. If compression is the cause of injury then a rigid boundary will be conducive to injury. We tested the effect of boundary conditions of an excised isolated rabbit lung on the rate of edema fluid formation following an impact load. Our results show that edema was more severe when the lung was supported on a rigid plate, than when it was resting on a net of nylon stockings (Yen et al., 1987a). Furthermore, it has been observed that in animals subjected to blast load, rib markings occur on the lungs (Frisoli and Cassen, 1950;

[1] Presently, Biomechanics Laboratory, Institute of Mechanics, Chinese Academy of Science, Beijing, People's Republic of China.
[2] Presently, Department of Mechanical Engineering, Memphis State University, Memphis, Tenn., 38152.

Contributed by the Bioengineering Division for publication in the JOURNAL OF BIOMECHANICAL ENGINEERING. Manuscript received by the Bioengineering Division May 6, 1987; revised manuscript received November 23, 1987.

Clemedson 1964b; Yen et al., 1985), and edema and hemorrhage usually follow, (Clemedson 1964a; Jönsson 1979; Yen et al. 1987a). In our experiments on the weight change of isolated lungs of the rabbit subjected to impact load and perfused with saline (Macrodex), (Yen et al., 1987a), we found that the maximum surface velocity of the lung under the impact load correlates well with the rate of increase of the lung weight following impact. For isolated lungs, the critical surface velocity was 11 m/s (24.6 mph). If the surface velocity due to impact was less than 11 m/s, no edema followed the impact. If the velocity exceeded 11 m/s, edema followed, and increased in severity with increasing surface velocity.

We tested also anesthetized intact rabbit impacted on the chest. The statistical mean of the wet lung weights of these animals remained equal to that of intact animals 1.5 hr after impact if the velocity of the chest wall under the impact load was less than 2 m/s. Above 2 m/s the wet/dry lung weight ratio after 1.5 hr significantly increased with increasing surface velocity. This can be explained by the fact that in the intact animal, the stress concentration due to the complex elastic waves is greater than that in the isolated lung. An additional feature was uncovered by these experiments: hemorrhage followed by coagulation is helpful in preventing massive edema. Hemorrhage occurred when the maximum chest wall velocity exceeded 12 m/s.

3   Literature on Alveolar Edema

It seems evident that hemorrhage is a result of fracture of the blood vessel. But long before the occurrence of fracture the blood vessels may be so stretched that the pores in the cell membranes or the cell-to-cell junctions in the endothelium are so changed as to affect the movement of the solutes. Similar changes may occur in the epithelium that separates the interstitium of the interalveolar septa from the alveolar gas. Movement of fluid is assumed to obey Starling's law. The rate of movement of fluid per unit area of a membrane is proportional to the difference of static pressures on the two sides of the membrane minus the osmotic pressure difference. Edema is conceived as caused by a change of the distributions of the concentrations of the solutes, or the static pressures, in such a way that fluid will move from the interstitium to the alveolar space. For example, if the epithelium becomes permeable to a certain small solute, then that solute in the interstitium will cross the epithelium into the alveolar side, increase the osmotic pressure there, and pull fluid from the interstitium into the alveolus. The movement of fluid and solutes can be studied by isogravimetric method, indicator dilution method, electron microscopy, lymph measurement etc. The books edited by Crone and Lassen (1970), Fishman and Hecht (1969), Giuntini (1971), and Staub (1978) provide a convenient introduction to the literature. The papers by Bo et al. (1977), Brigham (1978), Effros et al. (1982), Egan et al. (1975, 76a, 76b, 80), Nicolaysen and Hauge (1982) and Staub (1978) are particularly relevant to the problem of pulmonary edema. Effros, Egan, Nicolaysen et al. have shown that the epithelium is less permeable to small solutes than the endothelium, and that the change of permeability of the epithelium to small solutes is likely the reason for alveolar edema.

4   Hypothesis

The facts listed in Section 2 show that a compressive impact load can cause trauma. Since biological tissues are quite strong in compression, why does a compression wave cause edema or hemorrhage in the lung?

Our hypothesis is that the damage is caused by the tensile principal strains induced in the pulmonary alveolar walls. The tensile principal strain may cause the epithelial membrane of the alveolar wall to change its permeability with respect to

solutes and lead to edema, or to injure or stretch the pores of the endothelium and epithelium to such a degree as to leak larger solutes or even blood.

As dilatational and shear waves coexist in an elastic solid, the existence of shear stress is expected in lung tissue under impact. Is permeability of the epithelial membrane to small solutes a function of "distorsion"? or "dilatation"? or "stretching"? Mathematically, to which of the following is permeability correlated: maximum shear, biaxial or triaxial octahedral shear, biaxial or triaxial mean strain, areal strain, or the maximum principal strain? At the present time this is unknown. It is an important question awaiting an answer. In the hypothesis above, we made a tentative choice for the maximum principal strain. This choice may have to be modified in the future. The effect of shear stresses is of course included in this choice.

There are two ways to induce tensile strains in alveolar walls. One is by macroscopic dynamic response of the lung-chest system to the impact load. The other is the micromechanical response of the alveoli to compression waves. The dynamics of the chest and lung subjected to compression waves has been analyzed by several authors. Bowen et al. (1965) and White et al. (1971) treated the chest and lung as a single-degree-of-freedom elastic shell enclosing a gas which has a uniform pressure. Chuong (1985) freed the uniform pressure hypothesis by analyzing the stress waves with the finite-elements method in a two dimensional model of the lung-chest system. He showed that the pressure distribution is very nonuniform when the lung is subjected to a traveling shock wave, and that in some locations tensile strains comparable with the maximum initial compressible strains are obtained.

The micromechanical response is explained as follows. When a lung is subjected to a compression of sufficient magnitude and rate, some airways may be collapsed while the alveoli peripheral to them remain open. Gas is then trapped in these alveoli. Later, when the compression wave has passed or expansion wave arrives, the pressure outside the trapped unit would be reduced, and the gas in the trapped unit would expand. This expansion puts the alveolar walls in tension. Thus tensile stress can be induced in the alveolar walls even when the strain wave was compressive.

5   Experiments and Analysis to Test the Hypothesis

To test our hypothesis, we performed the following experiments. First, we used isolated rabbit lung perfused with saline at the *isogravimetric condition* (without change of lung weight with time), and measured the rate of lung weight increase due to a transient increase of stretch of the alveolar membrane. The stretch lasted for 1 min, then the lung was returned to the normal condition to measure the rate of weight change. A positive rate will lead to edema. This test reveals whether edema is related to the transient stretching of the alveolar walls.

Next, we determined the stress-strain relationship and hysteresis of the lung when it is subjected to compression and re-expansion. This is done by measuring the pressure-volume relationship of the lung. The result tells us also how much gas can be expected to be trapped in the alveoli due to closure of small airways when the lung is subjected to compression, and how much tension can be sustained when expansion wave arrives.

Then we measured the pressure-volume relationship of lung with trachea closed. The purpose is to show that the conventional pV relationship really should be transformed into and interpreted as a tissue stress-strain relationship. Then the relationship can be used in the dynamic analysis of the lung subjected to impact load.

Finally, we present a brief theoretical analysis of the lung

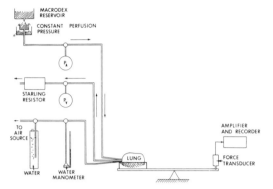

Fig. 1 A schematic diagram of the experiment to measure the rate of the wet lung weight increase following a stretch of the alveolar wall in a perfused rabbit lung. The lung was perfused with Macrodex (saline with dextran) and was isogravimetric (constant weight condition) initially under a transpulmonary pressure of 10 cm H_2O. Then the transpulmonary pressure acting on the lung was increased to stretch the lung for 1 min. In the resting period following the stretch the pressure was returned to 10 cm H_2O, and the rate of weight increase of the lung was measured. After a rest the lung was stretched again to a higher level, and the measurement repeated, and so on.

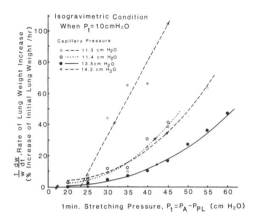

Fig. 2 The rate of increase of the wet lung weight due to edema following successive steps of increased stretching of rabbit lung. Each stretch lasted 1 min. Each resting period was 5 min.; during which the transpulmonary pressure was 10 cm H_2O and the rate of weight increase was measured. Four rabbits; each symbol represents a rabbit. The initial isogravimetric condition is stated in the inset at top left. The rate of increase of lung weight is expressed as percent of the initial lung weight per hour. The regression lines and the initial weight are, from top down, in the range of $20 < x < 60$,

(1) $y = 4.58x - 104.30$, $(r = 0.9151)$, $w = 13.30$g.
(2) $y = 0.06x^2 - 2.40x + 25.75$, $(r = 0.9882)$, $w = 26.55$g.
(3) $y = 0.05x^2 - 2.25x + 29.14$, $(r = 0.9884)$, $w = 23.40$g.
(4) $y = 0.03x^2 - 1.33x + 16.04$, $(r = 0.9961)$, $w = 41.50$g.

tissue subjected to a transient compression wave and rebound in tension. Details of these experiments and their interpretation are presented in the succeeding sections.

6 Experiments on the Rate of Lung Weight Increase Due to Transient Stretch of the Lung

The distribution of solutes and fluid in the lung has been studied extensively in physiology (see Section 3). Most authors try to keep the lung as close to normal physiological condition as possible. Eagan et al. (1975, 76a, 76b, 80) and Nicolaysen and Hauge (1982) have also studied the effect of steady elevated alveolar pressure. But no data exists, as far as we are aware, on the rate of increase of the wet lung weight (caused by edema or hemorrhage) due to transient overstretching of the alveolar wall membranes. To obtain the desired data, we measured this rate in excised lungs of the rabbit by inflating the lungs to various stretching pressures for one minute, then returning to the normal condition to measure the rate of edema fluid formation. The key to the measurement is that the lung must be perfused and initially edema free. This was obtained by establishing an *isogravimetric condition* initially. Rabbits of either sex weighing between 2 to 2.5 kg were anesthetized with intravenous pentobarbital (40 mg/kg) and heparinized. After a midline chest incision, cannulas were tied into the pulmonary artery, trachea and left atrium. The lung was then excised and placed on a screen made of nylon stocking material stretched over a double-ring frame used in embroidery, which provided an excellent support that minimizes stress concentrations. See Fig. 1. The specimen and the support system were then put on a weighing platform, and the weight of the lung was recorded continuously during the experiment to assess the edema development. All connections to the lung were made of soft rubber tubing passing through the axis of rotation of the balance to minimize the effect of tubing weight.

The lung was perfused at a constant pulmonary arterial pressure from an open reservoir. The left atrial pressure was controlled by a Starling resistor whereas the airway pressure was controlled by a water immersion manostat and monitored by a water manometer. Pulmonary arterial and venous pressures were measured simultaneously during the experiment using variable resistance pressure transducers with a

range of ± 60 cm H_2O. At the start of the experiment, the lung was perfused with Macrodex (6 percent dextran in normal saline) with the pulmonary venous pressure fixed at 11 cm H_2O, the airway pressure held at 10 cm H_2O, and the pleural pressure atmospheric (0). The pulmonary artery pressure was adjusted to attain the first isogravimetric condition (the rate of change of total weight equals zero). The arterial pressure was then lowered in steps by lowering the reservoir and the venous pressure adjusted after each step until an isogravimetric condition was reached. At the final isogravimetric step, the arterial pressure and venous pressure was the same and therefore no flow occurred, the system was in static equilibrium. In this condition, we cyclically inflated and deflated the lung by changing the airway pressure with a one minute "inflation" period and 5 minute resting period (during which the airway pressure was returned to 10 cm H_2O). The lung weight was monitored continuously in the resting period to assess the rate of edema formation in that period. The one-minute "inflation" pressure was fixed initially at 10 cm H_2O, and then increased by 5 cm H_2O in successive cycles to an ultimate maximal value of 60 cm H_2O. The airway pressure in the resting period was always 10 cm H_2O. In one case, after reaching the 60 cm H_2O maximum pressure, we decreased the inflation pressure at 5 cm H_2O steps in repeated cycles. In all cases, we also examined the effect of transient decrease of airway pressure, and found that the rate of lung weight change was zero to less than 10 cm H_2O.

The results are given in Figs. 2 and 3. The rate of increase of lung weight in the resting period, dw/dt, is expressed as percent of the initial lung weight per hour, and is plotted against the 1 minute stretching pressure preceding that period. Figure 2 shows the measurements of 4 rabbits with each symbol representing a rabbit. It is seen that as the lung was increasingly stretched, the rate of increase of lung weight rises. Considerable difference exists between individuals. The curve can be fitted by a parabola with coefficients listed in the legends. Figure 3 shows the measurements in a rabbit lung subjected first to a stepwise increase in stretching (the lower dots), then

260

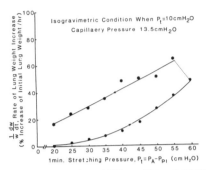

Fig. 3 The rate of edema in a rabbit lung which was subjected to a stepwise increase in stretching up to a transpulmonary pressure of 60 cm H_2O, (lower curve with dots), and then to a stepwise decrease in stretching (upper curve with stars). The cumulative damage is seen. The regression line of rising stretching is $y = 0.03x^2 - 1.33x + 16.04$, ($r = 0.9961$). That of continued stretching with decreasing amplitude is $y = -0.001x^2 + 1.39x - 10.80$, ($r = 0.9437$), for $20 < x < 60$.

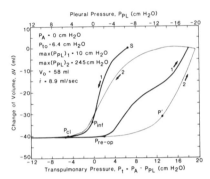

Fig. 4 The pressure-volume relationship of rabbit lung with the regime of small and negative transpulmonary pressure emphasized. The airway pressure was atmospheric (zero). At the starting point, S, the transpulmonary pressure was 6.4 cm H_2O, and volume was 58 ml. The rate of volume reduction was 8.9 ml/s. In application, these curves may be read as a stress-strain relationship of the lung because the tissue stress is equal to the transpulmonary pressure (1 cm H_2O = 98 N/m^2), and the change of volume divided by initial volume is the volumetric strain.

to a stepwise decrease in stretching (the upper stars). The effect of cumulative stretching is seen. The damage is accumulative, as shown by the increased dw/dt even while the stretching pressure was subsequently and successively lowered. These results support our hypothesis.

7 Cyclic Compression and Expansion of the Lung

The airway closure and gas trapping problem has been studied extensively in the literature (see reviews in Anthonisen, 1977, Bates et al., 1977, Hoppin and Hildebrandt, 1977). The existence of a critical lung volume at which some small airways become closed to trap gas in the alveoli is well known. It is related to the maximal expiratory flow phenomena, and in pulmonary funtion tests it furnishes a convenient index for certain lung disease such as silicosis and black lung disease of coal miners. Kooyman (1981) has shown the remarkable fact that diving animals such as Weddell Seals have plenty of cartilage in their bronchioles, all the way to the alveoli, presumably to prevent the closure of the bronchioles before the closure of the alveoli. This may explain why seals do not get bends or decompression sickness whereas human divers do in deep sea diving.

In extending the concept to the trauma problem, however, we need to know how the collapsed airways are reopened, how are the critical strains at closure and reopening affected by the strain rate, and whether very high compression, which occurs in traumatic impact but not in normal physiology, influences reopening. No reference to these aspects exist in the literature. Hence we performed a new experiment to obtain the desired data. We compressed isolated rabbit lungs with trachea open to the atmosphere and recorded the relationship between the lung volume and pleural pressure. While details are presented in Tao and Fung (1987), the results are summarized below. We found that for the rabbbit lung a range of pleural pressure exists in which gas will be trapped in the alveoli. At a critical pleural pressure a limiting lung volume will be reached beyond which the lung behaves like a closed balloon obeying Boyle's law. In the rabbit this limiting volume was roughly one-quarter to one-half of the initial lung volume which was about 60 percent of the total lung capacity.

On decreasing the pleural pressure again, there exists another critical pressure at which the lung volume begins to increase again. This critical pressure for reopening is higher than that for closing, and varies with the initial lung volume, the rate of strain, and the maximum compression imposed on the lung.

A typical example is shown in Fig. 4. An isolated rabbit lung

hanging in a lucite box was compressed, beginning at a state indicated by the point S in Fig. 4, which corresponds to an initial volume of 58 ml and a transpulmonary pressure of 6.4 cm H_2O. Following the curve number 1 a point of inflection is reached when the transpulmonary pressure is p_{inf}. Further down, at a "closing pressure" P_{cl} and volume V_{cl}, the slope of the PV curve became that of the gas law, PV = constant. The volume did not change appreciably with further increase in the pleural pressure in the range examined here. Gas was trapped in the alveoli.

Curve 1 stops at the left end at a transpulmonary pressure $p_t = -10$ cm H_2O. Then the transpulmonary pressure p_t was increased again. At $p_t = p_{re-op}$, the "reopening" pressure, the PV curve begins to rise again, following the curve marked No. 1 for the first cycle. In the second cycle, curve No. 2, dotted in Fig. 4, the compression was carried to a maximum of $p_t = -245$ cm H_2O. Returning stroke followed the dotted curve No. 2. At the point P' the slope of the PV curve in expansion is equal to that in compression at the same volume; the PV curve bends somewhat sharply at this point.

If the strain rate changes, the pressure at the inflection point, P_{inf}, and the closing pressure, P_{cl}, remains relatively unchanged, but the reopening pressure, P_{re-op}, and the characteristic pressure p', where the PV curve bends upward sharply on reinflation will change to the extent of several cm H_2O. If the lung were compressed to a much higher pressure, the reopening pressure is increased.

If the transpulmonary pressure remains smaller than the reopening pressure, then the collapsed airways will not be reopened and traumatic atelectasis results.

These results show that gas trapped in compressed lung is a reality, in support of our hypothesis. However, the strain rate used in our experiment is orders of magnitude slower than that occurring in a lung subjected to a shock wave. Since the effect of high strain rate is still unknown, our hypothesis cannot be considered fully supported.

8 Experiments on Lungs With Trachea Closed

The third experiment is designed to further clarify the meaning of the pressure-volume relationship explored in the preceding section. We would like to show that it is the compressive stress and strain of the lung tissue that determines the critial trapping of the gas, not the pleural and airway pressures themselves. The latter are important only insofar as they in-

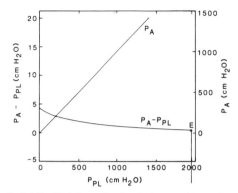

Fig. 5 Relationship between the pleural, airway, and transpulmonary pressures (p_{PL}, p_A, $p_A - p_{PL}$, respectively) in a rabbit lung with closed trachea. Note that the airway pressure follows the pleural pressure, and the difference $p_A - p_{PL}$ remains positive but small even though p_{PL} reaches 2000 cm H_2O. At the point E the lung volume control was released, the airway pressure decreased as the gas in the lung expanded into a syringe. The lung collapsed, trapping about one-third of the gas in it.

fluence the stress and strain in the lung tissue. With this interpretation made clear, then the results and implications of the preceding section can be used *locally* in three-dimensional dynamic analysis of the lung subjected to impact load. Furthermore, when lungs are impacted by supersonic shock waves the process is so fast that the amount of gas escaping from the mouth and nose is negligible compared with the total amount of gas in the lung at the time when the maximum stress is reached. In that case the trachea may be regarded as closed.

We used the same apparatus and the same procedure mentioned in the preceding section except that after the lung was hung in the air chamber with the airway pressure p_A equal to 0 (atmospheric) and the pleural pressure equal to $-p_{to}$ the valve at the top was closed, making the airway noncommunicative with the atmosphere. Then the pleural pressure p_{PL} was increased progressively, while the airway pressure p_A and the transpulmonary pressure $p_t = p_A - p_{PL}$ were recorded. Figure 5 shows a record of p_A and $p_A - p_{PL}$ versus p_{PL}. It is seen that the change of the transmural pressure $p_A - p_{PL}$ was quite small despite the large changes in p_A and p_{PL}. Furthermore, $p_A - p_{PL}$ remained positive. As expected, the lung did not collapse in this condition.

To demonstrate the collapse of the lung we connected a large syringe to the trachea beyond the valve at the top. At the condition signified by the point E, the valve was opened so that the trachea became communicative with the syringe. If the syringe was allowed to move slowly, the airway pressure decreased and the lung collapsed to a volume of 33.3 ml. If the syringe was allowed to move rapidly without significant restraint, the lung collapsed to a volume of 35.5 ml. When the syringe stopped the airway pressure is nearly atmospheric (0). The initial volume at $p_A = 4.2$ cm H_2O, $p_{PL} = 0$, was 103 ml. Hence 32-34 percent of gas was trapped in the alveoli.

This experiment shows that a lung with closed airway is much more rigid than one with open airway. The reason is that when the airway is closed the PV relation follows Boyle's law, the bulk modulus is two orders of magnitude larger than that of the lung tissue alone (Yen et al. 1987). It shows further that p_A or p_{PL} alone does not cause the alveoli or small airways to collapse. It is the transpulmonary pressure $p_A - p_{PL}$ that is relevant. But $p_A - p_{PL}$ is approximately equal to the lung tissue stress; and the change of lung volume per unit volume is the volumetric strain of the lung tissue. Hence it is the stress and strain of the lung tissue that characterizes the stability of the airway and the alveoli.

9 Theoretical Analysis of the Response of a Group of Alveoli With Trapped Gas to a Tension or Compression Wave

Recognizing the highly nonuniform stress distribution as the major feature of the stress waves in the lung in response to shock loading, we considered the dynamics of those alveoli in which gas is trapped by the collapse of small airways when the lung is subjected to an impact load.

In the Appendix, an analysis is presented under the assumptions that the alveoli with trapped gas behave like an elastic shell embedded in a continuum. The inertial effect of the tissue surrounding these alveoli is accounted for by an "apparent mass." The gas enclosed in the shell obeys gas law. The initial pressure of the trapped gas can be computed by a macroscopic theory such as Chuong's (1985). The gas trapping mechanism is as described in Section 7 supra. For simplicity it is assumed that the shell is a cylinder of finite length with spherical ends. It is shown that when a traveling stress wave arrives and passes, the shell responds and moves past the equilibrium condition by the inertial force; and an oscillation follows. In the oscillation the shell contracts and expands. In expansion tensile stress is generated in the wall of the shell.

10 Conclusion

The inflation experiments show that overstretching of the alveolar wall upsets isogravimetric condition and causes edema or hemorrhage. The compression-expansion experiments show that under a compression of sufficient magnitude, about 1/4 to 1/2 of the gas in the lung will be trapped in the alveoli. Decompression will cause the alveoli to reopen at a higher critical stress.

We learned that the pressure-volume curve of lung with open airway can be interpreted as the stress-strain relationship of the lung parenchyma in uniform compression and expansion. Whereas it is well known that the lung tissue has hysteresis due to surface tension, our experiment shows that for lungs which are compressed to collapse the hysteresis is much larger. The critical stress for reopening the collapsed airway is larger than that for closing by 5 to 15 cm H_2O in the range of strain rate tested.

We learned that the elasticity of the lung tissue is much smaller than that of the air. Hence in the analysis of the dynamic response of the lung to impact load the lung elasticity can be considered essentially the same as that of the air. A chest-lung model should recognize the mass and elasticity of the chest wall and diaphragm, the compressibility of the lung gas, and the mass of the lung tissue. In evaluating the effect of the compressibility of the lung gas the Mach number must be based on the very low velocity of sound in the lung. Lung strains computed from such a model are the strains of the tissue. The tissue stress can then be computed from the tissue stress-strain relationship.

Dynamic analysis by Chuong (1985) shows that in response to shock waves the stress and strain are transient and very nonuniform. In some localities the maximum tensile strain is comparable to the maximum compressive strain. Superposed on this macroscopic tensile strain is the microscopic mechanism of gas trapping in alveoli and re-expansion, which augments the tensile stress.

These facts taken together support our hypothesis that lung trauma is caused by overstretching the alveolar membrane. However, we cannot yet claim confirmation because the strain rate effect has not been fully examined, and the locations of the collapsed airways have not been identified. Further, a word in our hypothesis, "the maximum principal strain," remains tentative. Basic research in the future must identify the particular strain measure that best correlates with the change of permeability of the epithelium of the aveolar wall to small solutes, whether it is the maximum principal strain, or the maximum shear strain, or areal strain, or the octahedral strain. Finally, we learned in our recent impact tests on pigs that there are other patho-physiological factors involved in impact trauma, such as change of blood sedimentation rate, immune reaction, and activation of complement. The problem is more complex than it appears at first sight. As a hypothesis, however, our proposal is specific, consistent, and conducive to new directions of research and new understanding. It is not only relevant to the understanding of trauma, but also important for trauma management.

Acknowledgment

This work was done with support from grants No. NIH HL 26647 and NSF EET 85-18559.

References

Anthonisen, N. R., 1977, "Closing Volume," In: *Regional Differences in the Lung*, edited by J. B. West, Academic Press, New York, pp. 451-482.

Bates, D. V., Macklem, P. T., and Christie, R. V., 1971, *Respiratory Function in Disease*, W. B. Saunders, Philadelphia.

Bø, G., Hauge, A., and Nicolaysen, G., 1977, "Alveolar Pressure and Lung Volume as Determinants of Net Transvascular Fluid Filtration," *J. Appl. Physiol.*, Vol. 42, pp. 476-482.

Bowen, I. G., Holladay, A., Fletcher, E. R., Richmond, D. R., and White, C. S., 1965, "A Fluid-Mechanical Model of the Thoraco-Abdominal System With Applications to Blast Biology," Report No. DASA-1675, Lovelace Foundation for Medical Education and Research, Albuquerque, NM.

Brigham, K. L., 1978, "Lung Edema Due to Increased Vascular Permeability," In *Lung Water and Solute Exchange*, ed. by N. Staub, Marcel-Dekker, New York, pp. 235-276.

Chuong, C. J., 1985, "Biomechanical Model of Thorax Response to Blast Loading," Report on Contract No. DAMD 17-82-C-2062. Jaycor, 11011 Torreyana Rd., San Diego, CA, 92138.

Clemedson, C. J., 1956, "Blast Injury," *Physiol. Rev.*, Vol. 36, pp. 336-354.

Clemedson, C. J., and Jönsson, A., 1962, "Distribution of Extra- And Intrathoracic Pressure Variations in Rabbits Exposed to Air Shock Waves," *Acta Physiol. Scand.*, Vol. 54, pp. 18-29.

Clemedson, C. J., and Jönsson, A., 1964a, "Dynamic Response of Chest Wall and Lung Injuries in Rabbits Exposed to Air Shock Waves of Short Duration," *Acta Physiol. Scand.*, Vol. 62, Supp. 233, 31 pp.

Clemedson, C. J., and Jönsson, A., 1964b, "Differences in Displacement of Ribs and Costal Interspaces in Rabbits Exposed to Air-Shock Waves," *Amer. J. Physiol.*, Vol. 207, pp. 931-934.

Crone, C., and Lassen, N. A. (editors), 1970, *Capillary Permeability*, Proc. of a Sym., Academic Press, New York.

Effros, R. M., Mason, G., Uszler, J. M., and Chang, R. S. Y., 1982, "Exchange of Small Molecules in the Pulmonary Microcirculation," In *Mechanisms of Lung Microvascular Injury, Annals of the New York Acad. of Sci.*, Vol. 384, pp. 235-245.

Egan, E. A., Oliver, R. E., and Strang, L. B., "Changes in Non-Electrolyte Permeability of Alveoli and the Absorption of Lung Liquid at the Start of Breathing in the Fetal Lamb," *J. Physiol.* (*London*), Vol. 244, pp. 161-179.

Egan, E. A., Nelson, R. M., and Oliver, R. E., 1976a, "Lung Inflation and Alveolar Permeability to Non-Electrolytes in the Adult Sheep in Vivo," *J. Physiol.* (*London*), Vol. 260, pp. 409-424.

Egan, E. A., 1976b, "Effect of Lung Inflation on Alveolar Permeability to Solutes," In *Lung Liquids*, (ed. by C. J. Dickinson) Ciba Foundation Symposium 38 (new series): 101-114. Elsevier Excerpta Medica, North-Holland Pub. Co., Amsterdam.

Egan, E. A., Nelson, R. M., and Beale, E. F., 1980, "Lung Solute Permeability and Lung Liquid Absorption in Premature Ventilated Fetal Goats," *Pediatr. Res.*, Vol. 14, pp. 314-318.

Fishman, A. L., and Hecht, H. H. (editors), 1969, *The Pulmonary Circulation and Interstitial Space*, The University of Chicago Press, Chicago.

Frisoli, A., and Gassen, B., 1950, "A Study of Hemorrhagic Rib Markings Produced in Rats by Air Blast," *J. Aviation Med.*, Vol. 21, pp. 510-513.

Fung, Y. C., 1985, "The Application of Biomechanics to the Understanding and Analysis of Trauma," In *The Mechanics of Trauma*, edited by A. M. Nahum, and J. Melvin, Appleton-Century, Crofts, Norwalk, Conn., pp. 1-16.

Giuntini, C. (editor), 1971, *Central Hemodynamics and Gas Exchange*, Minerva Medica, Torino, Italy.

Hopping, F. G., and Hildebrandt, J., 1977, "Mechanical Properties of the Lung," In *Bioengineering Aspects of the Lung*, edited by J. B. West, Marcel Dekker, New York.

Jönsson, A., 1979, "Experimental Investigations on the Mechanisms of Lung Injury in Blast and Impact Exposure," Linköping Univ. Medical Dissertation, No. 80, Linköping.

Kooyman, G. L., 1981, *Weddell Seal: Consummate Diver*, Cambridge Univ. Press, London.

Kraman, S. S., 1983, "Speed of Low-Frequency Sound Through Lungs of Normal Man," *J. Appl. Physiol.*, Vol. 55, pp. 1862-1867.

Lau, V. K., and Viano, D. C., 1981, "Influence of Impact Velocity and Chest Compression on Experimental Pulmonary Injury Severity in Rabbits," *The Journal of Trauma*, Vol. 21, No. 12.

Nicolaysen, G., and Hauge, A., 1982, "Fluid Exchange in the Isolated Perfused Lung," In *Mechanisms of Lung Microvascular Injury, Annals of the New York Acad. of Sci.*, Vol. 384, pp. 115-125.

Rice, D. A., 1983, "Sound Speed in Pulmonary Parenchyma," *J. Appl. Physiol.*, Vol. 54, No. 1, pp. 304-308.

Staub, N., 1978, "Lung Fluid and Solute Exchange," In *Lung Water and Solute Exchange* (ed. by N. Staub), Marcel Dekker, New York, pp. 3-16.

Tao, Z. L., and Fung, Y. C., 1987, "Lungs Under Cyclic Compression and Expansion," ASME JOURNAL OF BIOMECHANICAL ENGINEERING, Vol. 109, pp. 160-162.

Viano, D.C., 1978, "Evaluation of Biomechanical Response and Potential Injury from Thoracic Impact," *Aviation Space Environ. Med.*, Vol. 49, pp. 125-135.

Viano, D. C., and Lau, V., 1983, "Role of Impact Velocity and Chest Compression in Thoracic Injury," *J. of Aviation, Space, and Environmental Medicine*, Vol. 54, No. 1, pp. 16-21.

White, C. S., Jones, R. K., Damon, E. G., Fletcher, E. R., and Richmond, D. R., 1971, "The Biodynamics of Airblast," Report No. DNA 2738T, Lovelace Foundation for Medical Education and Research, Albuquerque, NM.

Yen, R. T., and Fung, Y. C., 1985, "Thoracic Trauma Study: Rib Markings on the Lung Due to Impact are Marks of Collapsed Alveoli; But Not Hemorrhage," ASME JOURNAL OF BIOMECHANICAL ENGINEERING, Vol. 107, pp. 291-292.

Yen, R. T., Fung, Y. C., Ho, H. H., and Butterman, G., 1986, "Speed of Stress Wave Propagation in the Lung," *J. Appl. Physiol.*, Vol. 61, No. 2, pp. 701-705.

Yen, R. T., Fung, Y. C., and Artaud, C., 1987, "The Incremental Elastic Moduli of the Lungs of Rabbit, Cat, and Man," *1987 Advances in Bioengineering*, ASME, New York.

APPENDIX

Response of Pulmonary Alveoli With Trapped Gas to a Shock Wave

Experimental results show that beyond a critical compressive strain gas can be trapped in the alveoli. As a simplified model of these alveoli, we consider a cylindrical tube with spherical ends. It is a microstructure, with a diameter of the order of 1 mm and length of the order of 4 or 5 mm, much smaller than the dimension of the whole lung. Let the dynamic response of the whole lung to the shock wave be analyzed first with the lung considered as a homogeneous compressible material. Let θ (x, t) be the macroscopic volumetric strain in the lung at location x and time t. If K is the bulk modulus of the lung material, then the macroscopic stress in the lung is σ $(x, t) = K\theta(x, t)$. The microscopic cylinder is assumed to be situated at x and subjected to the external stress $\sigma(x, t)$ and strain $\theta(x, t)$. Let the wall of the shell be considered as an elastic membrane, and that the effect of the tissue and air mass surrounding the wall be described by an apparent mass m per unit area of the wall. Then the equations of motion of the wall are (Flügge, *Stresses in Shells*, Springer Verlag, 1960), for the spherical portion,

$$(\rho_0 h + m) \frac{d^2 u}{dt^2} = p - K\theta(x,t) - p_e - \frac{\tau_\theta h}{R} - \frac{\tau_\phi h}{R} , \quad (1a)$$

whereas for the cylindrical portion,

$$(\rho_0 h + m) \frac{d^2 u}{dt^2} = p - K\theta(x,t) - p_e - \frac{\tau_\theta h}{R} . \quad (1b)$$

The stress-displacement relation is:

$$\tau_\theta = \tau_\phi = \rho_0 c_0^2 \frac{u}{R} . \quad (2)$$

The initial conditions are

$$t = 0: \quad u = 0, \quad \frac{du}{dt} = 0. \quad (3)$$

Here ρ_0 is the density of wall, h is wall thickness, c_0 is the sound speed in the wall, u is the displacement in the radial direction, R is the radius of the sphere or the cylinder, p is internal pressure of gas trapped in the alveoli, p_e = pressure acting on the outside of the wall in addition to $K\theta$, τ_θ, and τ_ϕ are the circumferential and meridional stresses.

If $R = R_0$ when $t = 0$, we introduce the following dimensionless variables:

$$\bar{u} = \frac{u}{R_0}, \quad \bar{R} = \frac{R}{R_0} = 1 + \bar{u}, \quad \bar{h} = \frac{h}{R_0}, \quad \bar{p} = p/\rho_0 c_0^2,$$

$$\bar{\tau} = \tau/\rho_0 c_0^2, \quad \bar{m} = m/\rho_0 R_0, \tag{4}$$

$$\bar{x} = x/R_0, \quad \bar{t} = c_0 t/R_0, \quad \bar{\beta} = K/\rho_0 c_0^2.$$

Then, on substituting (4) into equations (1) and (3), and reducing, we obtain the following dimensionless equations:

$$(\bar{h} + \bar{m}) \frac{d^2 \bar{u}}{d\bar{t}^2} = \bar{p} - \bar{\beta}\theta(x,t) - \bar{p}_e - \frac{b\bar{u}\bar{h}}{(1+\bar{u})^2}, \tag{5a}$$

$$\bar{t} = 0: \quad \bar{u} = 0, \quad \frac{d\bar{u}}{d\bar{t}} = 0 \tag{5b}$$

where b equals 2 for a sphere, 1 for a cylinder.

If the process were isothermal, then

$$\frac{\bar{p}(t)}{\bar{p}(0)} = \frac{V(0)}{V(t)} = (1+\bar{u})^{-k} \tag{6}$$

where $p(0)$, $V(0)$ are the lung pressure and volume at time zero, and k equals 3 for a sphere, and 2 for a cylinder. If the process were adiabatic, then the two right-hand side terms in equation (6) should be raised by a power of γ, because then pV^γ = constant. Further,

$$p(0) = (1+\alpha)p_0 \tag{7}$$

$$p_0 = p_{t0} + p_e - \beta\theta \tag{8}$$

$$\alpha = \frac{\delta M}{M_0} \tag{9}$$

where M_0 and δM represent, respectively, the mass of the gas enclosed in the alveoli initially and that which was increased in the dynamic process.

If the displacement were very small, then we obtain, by linear approximation, the following solution for the circumferential stress in the wall

$$\bar{\tau} = \frac{\bar{u}}{1+\bar{u}} \cong \frac{a_0}{a_1} (1 - \cos\sqrt{a_1}\,\bar{t})$$

$$+ \frac{\beta}{(\bar{h}+\bar{m})\sqrt{a_1}} \sin\sqrt{a_1}\,\bar{t}.\bar{f}(\bar{x},\bar{t}) \tag{10}$$

where

$$a_0 = \frac{1}{\bar{h}+\bar{m}} [(1+\alpha)\bar{p}_0 - \bar{p}_e - b\bar{h}]$$

$$a_1 = -\frac{1}{\bar{h}+\bar{m}} [k(1+\alpha)\bar{p}_0 + b\bar{h}] < 0, \quad b = 2 \text{ for sphere} \tag{11}$$

$$\bar{f} = \int_0^{\bar{t}} \theta(\bar{x},\bar{t}')d\bar{t}'$$

Approximately, the stress τ will reach its maximum or minimum when

$$\bar{t} = \bar{t}_m = \frac{1}{\sqrt{a_1}} (n+1/2)\pi, \quad n = 0,1,2,\ldots \tag{12}$$

and the extremum of the circumferential stress is

$$\bar{\tau}_m = \frac{a_0}{a_1} \pm \frac{\bar{\beta}}{(\bar{h}+\bar{m})\sqrt{a_1}} f_m \tag{13}$$

where

$$\bar{f}_m = \int_0^{\bar{t}_m} \theta(\bar{x},\bar{t})d\bar{t}. \tag{14}$$

If the lung was inflated initially, i.e., $p_{t0} > 0$, then

$$p_0 = (p_{t0} + p_e) \geq \frac{p_e}{1+\alpha}, \quad \text{(because } \alpha > 0\text{)}. \tag{15}$$

Hence $a_0/a_1 > 0$. For a compression wave, $\bar{f}_m$ is negative, and we obtain from equaton (13) two peak values of $\bar{\tau}_m$. The maximum one is positive, representing tension. The minimum one is either positive or negative. For a tension wave, $\bar{f}_m$ is positive, and again we obtain from equation (13) two peak values of $\bar{\tau}_m$, the largest one represents tension.

Thus we see that when the lung is inflated initially, the maximum circumferential stress will be tensile in both compression or tension waves. The higher the initial transpulmonary pressure, the larger will be the maximum tensile stress in the walls of the alveoli with trapped gas.

CHAPTER 11

PULSATILE BLOOD FLOW IN THE LUNG STUDIED AS AN ENGINEERING SYSTEM

MICHAEL R. T. YEN

Department of Biomedical Engineering, University of Memphis, Memphis, TN 38152

WEI HUANG

Department of Bioengineering, University of California, San Diego

1. Introduction

Every part of a living organism, from molecules to the whole organ, is a marvelous engineering system, and an engineering analysis can add clarity to our understanding. The pulmonary circulation system carries blood from the right ventricle to the left atrium of the heart. Blood flows from the right ventricle to the pulmonary arterial trunk which is divided into the main right and left pulmonary arteries, entering the lungs, then they divide again and again into smaller and smaller branches like a tree, the lung finally terminating in the capillary sheet. In the capillary sheets the blood acquires oxygen and releases carbon dioxide into the air sacs called pulmonary alveoli. Then the blood flows from the capillary sheets into a venous tree, terminating into the left atrium of the heart. The blood flow is pulsatile, the system is complex, and measurement is difficult. However, the pressure-flow relations of pulmonary circulation is extremely important to the understanding of the diseases of the lung and heart, such as the pulmonary hypertension, cardiac hypertension, tissue hypertrophy, edema, various respiratory disorder, and diabetes. In this regard a thorough theoretical understanding with full experimental verification will be useful.

The pulsatile pressure-flow relations in the lungs are often expressed in the terminology of electric engineering. For example, in clinical applications it is important to estimate the pulmonary arterial input impedance, which is the ratio of the amplitude of oscillatory arterial pressure to the oscillatory inflow rate at a given frequency. The impedance as a function of frequency is said to be a spectrum (see Fig. 2 of the attached paper). Two general theoretical approaches are used to study pulsatile hemodynamics. One is the electric circuit analog, or the "lumped parameter" model, in which the anatomic details are ignored. The other is biomechanical approach, or the "continuum" model. Both approaches have been used in the studies of pulmonary pulsatile flow (Milnor, 1989). Whereas the lumped parameter approach is often convenient and effective, the continuum approach is more fundamental. To evaluate the values of the lumped parameters theoretically, for example, one has to use the continuum model.

The continuum approach is based on the principles of continuum mechanics in conjunction with detailed measurement of vascular geometry, vascular elasticity and blood rheology. The method is uniquely suitable to the study of pulmonary blood flow. It provides an analytical tool to analyze the physiological problems, and to synthesize the many components of a complex problem so that quantitative predictions are made possible. Zhuang *et al*, (1983) used this analytical approach and successfully developed a hemodynamic model for steady blood flow in cat lung. In the calculations, the sheet flow theory of Fung and Sobin (1969) is used for pulmonary capillary blood flow and an analogous "fifth power" is used for flow in the arteries and veins. Yen and Sobin (1988) performed experiments, which validates the theoretical predictions.

Recently, the biomechanical approach was used to compute the impedance for pulsatile flow in dog lung (Gan and Yen 1994) and in cat lung (Huang *et al*, 1998), and the agreement between theory and experiment was satisfactory. We shall describe this biomechanical approach below.

2.   The Biomechanical Approach Defines What Experimental Data Are Needed

Knowledge of the various factors controlling pulmonary blood flow is important in human health and disease, and today our knowledge and understanding of these factors is incomplete. For an experimenter, one should ask: what precise experimental data are required? To answer this question, one should seek theoretical guidance. For the biomechanical approach to pulmonary circulation, the answer is that we require: (1) detailed description of vascular geometry, (2) measured elasticity of blood vessels and (3) rheology of blood in blood vessels. With the pieces of information listed above, together with the basic physical laws and appropriate boundary conditions, differential equations can be written and the problems solved either analytically or numerically.

2.1.   *Morphometry of pulmonary vasculature*

Quantitative study of pulmonary was begun by Malpighi in 1661. The early history of the pulmonary vascular morphometry has been summarized by Miller (1947). Two mathematical models have been used to describe the treelike pulmonary branching systems: the generation model, and the Strahler's ordering model. The generation model was used by Weibel and Gomez (1962) for the human pulmonary arterial tree. Cumming, Horsfield and their coworkers however used Strahler's method of describing rivers and rivulets in geography to study the human pulmonary arterial and venous trees [see review by Horsfield (1991)]. Yen *et al*, (1983, 1984) used the Strahler's ordering model to study the morphometry of cat pulmonary arterial and venous trees. More recently, in order to remedy the unsatisfactory features of the generation and Strahler models, a Diameter-Defined Strahler ordering system was developed (Kassab *et al*, 1993; Jiang *et al*, 1994). The concept of the Diameter-Defined Strahler system was used by Fung and Yen for the pulmonary arteries of the dog in 1987 (unpublished). Kassab and Fung (1993) developed and used the Diameter-Defined Strahler system in the study of the coronary blood vessels in the pig. Jiang *et al*, (1994) used these refinements in their study of the rat pulmonary arterial tree. Gan *et al*, (1993) and Tian (1993) used the Diameter-Defined ordering system to describe the dog pulmonary venous tree and arterial tree, respectively. Huang *et al*, (1996) used the new system in their study of the human pulmonary vasculature.

Table 1. Morphometric and elastic data of the pulmonary arteries and veins in the cat's right lung. Data are from Yen *et al*, (1980, 1981, 1983, and 1984), and Yen (1990).

	Order	Number of Branches	D_0 (mm)	Length (mm)	Compliance $10^{-4} P_a^{-1}$	cmH_2O^{-1}
	11	1	5.080	25.000	2.6630	0.0267
	10	4	2.486	11.870	2.6630	0.0267
	9	12	1.519	15.260	2.6630	0.0267
	8	49	0.875	8.190	1.1220	0.0112
Arteries	7	202	0.533	4.600	0.7143	0.0072
	6	774	0.352	2.720	0.7959	0.0080
	5	2925	0.192	1.510	1.1220	0.0112
	4	9736	0.122	0.810	1.9280	0.0193
	3	31,662	0.073	0.433	1.9280	0.0193
	2	97,519	0.044	0.262	1.9280	0.0193
	1	300,358	0.024	0.116	1.9280	0.0193
	1	282,733	0.025	0.086	1.9280	0.0193
	2	86,241	0.046	0.247	1.9280	0.0193
	3	26,306	0.077	0.496	1.9280	0.0193
	4	8024	0.127	1.545	2.0800	0.0208
	5	2348	0.251	2.380	1.4690	0.0147
Veins	6	656	0.432	3.810	1.0920	0.0109
	7	171	0.642	4.950	1.0920	0.0109
	8	46	1.040	7.610	0.7240	0.0073
	9	13	1.727	15.120	0.7240	0.0073
	10	4	3.010	19.240	0.7240	0.0073
	11	1	4.491	25.000	0.7240	0.0073

D_0, diameters at zero transmural pressure. Compliance data are obtained with $P_A = 0$ and $P_{pl} = -10$ cmH$_2$O.

Yen *et al*, (1983, 1984) gave data on the diameters, lengths, and number of branches in the silicone elastomer casts of the cat lungs. For the smaller vessels of the cat lungs, they measured the dimensions in the thick histologic sections with injected vessels. Using the Strahler system, they found 12 orders of pulmonary arteries between the right ventricle and the capillaries (Yen *et al*, 1984), and 11 orders for pulmonary venous tree between the left atrium and the capillaries in the cat (Yen *et al*, 1983). Their data are summarized in Table 1.

Gan *et al*, (1994) and Tian (1993) gave morphometric data on the dog lung by the Diameter-Defined Strahler system (see Table 2). There are 13 orders of branches in the dog pulmonary arterial tree (including the main pulmonary artery) and 11 orders in venous tree. So the total numbers of orders of the pulmonary arterial and venous trees are not very different in dog and cat.

2.2. *Elasticity of pulmonary blood vessels*

Vascular elasticity, or distensibility, is a mechanical property of blood vessels which determines a change in the diameter of blood vessels as a result of the change in blood pressure. The mechanical properties of pulmonary blood vessels are essential factors influencing the distribution of pulmonary blood pressure, regional distribution of pulmonary blood volume, transit time distribution of blood in the lung, and pulse-wave attenuation through the lung.

Table 2. Morphometric and elastic data of the pulmonary arteries and veins in right lung of dog. Data are from Gan *et al*, (1993), Tian (1993), and Gan and Yen (1994).

	Order	Number of Branches	D_0 (mm)	Length (mm)	Compliance $10^{-4}P_a^{-1}$	cmH_2O^{-1}
	12	1	10.839	49.150	0.2486	0.0025
	11	7	6.205	20.990	0.2486	0.0025
	10	14	3.760	21.500	0.2486	0.0025
	9	40	2.339	24.220	0.4316	0.0043
	8	98	1.611	14.200	0.6827	0.0068
Arteries	7	322	1.111	9.643	0.6827	0.0068
	6	1440	0.679	6.768	1.1102	0.0111
	5	6704	0.374	3.436	1.6878	0.0169
	4	24,804	0.184	2.230	2.5694	0.0257
	3	70,020	0.106	1.498	3.2337	0.0323
	2	477,109	0.051	0.701	3.2337	0.0323
	1	3,978,476	0.021	0.341	3.2337	0.0323
	1	2,475,933	0.025	0.401	1.8561	0.0186
	2	214,642	0.066	0.683	1.8561	0.0186
	3	78,410	0.120	1.093	1.8561	0.0186
	4	10,574	0.222	1.767	1.2990	0.0130
	5	3386	0.406	2.975	0.8714	0.0087
Veins	6	989	0.704	5.190	0.4694	0.0047
	7	197	1.119	8.148	0.4286	0.0043
	8	76	1.776	12.653	0.2755	0.0028
	9	29	2.699	14.765	0.2755	0.0028
	10	12	4.384	20.516	0.2194	0.0022
	11	2	8.268	34.400	0.2194	0.0022

D_0, diameters at zero transmural pressure. Compliance data are obtained with $P_A = 0$ and $P_{pl} = -10$ cmH$_2$O.

The mechanical properties also affect the pressure-flow relationship and the change in total blood volume in response to an alteration of blood pressure.

In most cases, data on elasticity of mammalian blood vessels have been obtained from postmortem tissues. Several techniques have been used to measure the elasticity of pulmonary blood vessels. The X-ray image technique of an isolated lung is commonly used to determine the distensibility of pulmonary arteries and veins in different species. The earlier work began with Patel's (1960) direct measurement of the canine main pulmonary artery with an electronic caliper. The same vessel was measured by Frasher and Sobin (1965) by making silicone polymer casts under various controlled transmural pressures. Later Caro and Saffman (1965) used X-ray photography to record elasticity of pulmonary blood vessels of the rabbit in the range of 1 ~ 4 mm. Maloney *et al*, (1970) determined the elasticity of canine pulmonary blood vessels in the diameter range of 0.8 ~ 3.6 mm.

The elasticity of the cat pulmonary arteries in the diameter range of 100 ~ 1600 μm was measured by Yen *et al*, (1980) using X-ray photographic method. By the same method, the elasticity of pulmonary veins in the diameter range of 100 ~ 1200 μm was determined by Yen *et al*, (1981). They found that over a limited range of transmural pressures the variations of pulmonary artery and vein diameters under pressure could be expressed as a linear function of transmural pressure according to

$$D = D_0[1 + \beta(P - P_A)] \tag{1}$$

for vessels with smaller diameter than the alveolar diameter (about 117 μm), and

$$D = D_o[1 + \beta(P - P_{pl})] \tag{2}$$

for vessels with much larger diameter than the alveolar diameter. Here D stands for diameter, P_A and P_{pl} denote the alveolar gas pressure and the pleural pressure, respectively, and P represents local blood pressure, β is the compliance constant with a unit of %/cm H_2O, and D_0 is the vessel diameter when $P - P_A = 0$. The elastic values of cat pulmonary arteries and veins are summarized in Table 1.

The elasticity data listed in Table 2 were obtained from the X-ray radiographs of the dog's isolated right lung by Gan and Yen (1994). They measured the elasticity of blood vessels in the diameter range of $100 \sim 4500$ μm.

2.3. *Pulmonary capillaries*

The structure of pulmonary capillaries is very different from that of pulmonary arteries and veins. The pulmonary capillaries form a continuous network in a flat sheet which comprises the alveolar wall (Sobin *et al*, 1970; Sobin and Fung, 1992). The alveolus in the lung is the smallest unit of the air space. The dense network of the pulmonary capillary blood vessels of the cat's lung is shown in Figs. 1 and 2 of Sobin *et al*, (1970).

Fung and Sobin (1969, 1972b) described the capillary flow as a "sheet flow" taking place between two relatively flat sheets which are connected by a large number of closely spaced posts. By the sheet-flow model, the blood is not channeled in tubes, but has freedom to move in any way between the posts. This model greatly simplifies the hydrodynamic analysis of flow in the alveolar wall and the structural analysis of the elastic deformation of the sheet. The morphometric basis of the sheet-flow theory was investigated by Sobin *et al*, (1970). They measured the sheet dimensions by optical microscopy on the specimens prepared by solidifying a silicone elastomer perfused into the capillaries (Sobin *et al*, 1970). Additionally, Sobin *et al*, (1972) used a microvascular casting method to measure the elasticity of the pulmonary capillary sheet in the cat. The variation of alveolar sheet thickness in cat lung has been shown with respect to change in transmural pressure (Sobin *et al*, 1972 and 1978). When the transmural pressure is positive and less than 35 cm H_2O, the thickness h increased linearly with increasing pressure as follows (Sobin *et al*, 1972)

$$h = h_0 + \beta^*(P - P_A) = h_0 + \beta^*\Delta P \tag{3}$$

where β^* is the compliance of the sheet and h_0 is the thickness at zero transmural pressure. At higher values of ΔP, the compliance becomes very small. For negative values of ΔP, the vessel collapses and h tends to zero.

For the pulmonary capillaries of the dog, Fung and Sobin (1972) reported the dimensions and elasticity of the capillary sheet and the influence of various pressures on the capillary sheet thickness. Gan and Yen (1994), and Huang *et al*, (1998) collected the existing data of pulmonary capillary sheets in dog and cat to study the pressure-flow relationship in the lung. Table 3 summarizes the morphometrical and elastic data of the pulmonary capillary sheet in cat and dog.

Table 3. Morphometric and elastic data of the pulmonary capillary sheet.

	h_0 (μm)	β^* (μm/cmH$_2$O)	S	$\bar{L}$ (μm)
Cat	4.28[1]	0.219[1]	0.916[1]	556[2]
Dog	2.5[3]	0.122[3]	0.916[3]	447[4]

[1]data from Sobin *et al*, 1972; [2] data from Sobin *et al*, 1980; [3] data from Fung and Sobin, 1972a; [4] estimated by Gan and Yen, 1994. h_0, capillary sheet thickness at 0 transmural pressure; β^*, elasticity of capillary sheet; S, vascular space-to-tissue ratio; $\bar{L}$, average length of the capillaries.

2.4. *Blood viscosity*

When the shear strain rate of blood is smaller than 100 sec^{-1}, the coefficient of blood viscosity is inversely proportional to strain rate. When the strain rate approaches zero, the blood behaves like a plastic solid that has a small yield stress and an infinite coefficient of viscosity. However, when the shear strain rate exceeds 100 sec^{-1}, viscosity coefficient of blood is approximately a constant which is dependent upon the hematocrit. That is, at high shear rate, whole blood behaves like a Newtonian fluid. In other words, as the shear rate increases from zero to a high value, the stress-strain relation of blood changes from non-Newtonian to Newtonian behavior. The transition to the Newtonian region depends on the hematocrit. The effect of the non-Newtonian rheological behavior of blood in pulmonary arteries and veins is quite small (Fung, 1991).

Yen and Fung (1973) studied the blood viscosity on a scale model of the pulmonary alveolar sheet, with the red blood cells simulated by soft gelatin pellets and the plasma simulated by a silicone fluid. They demonstrated that the relative viscosity of blood μ_r with respect to plasma depends on hematocrit H in the following manner

$$\mu_r = 1 + aH + bH^2 \tag{4}$$

where a and b can be determined from the experiment. Applying this result to blood, they found that the viscosity of blood in the alveoli is related to the viscosity of the plasma and the hematocrit in a quadratic relationship:

$$\mu_{\text{blood in alv}} = \mu_{\text{plasma}}(1 + aH + bH^2). \tag{5}$$

The variation of the apparent viscosity of blood in the pulmonary capillary sheet with hematocrit has been determined by Yen and Fung (1973) in model experiments. The apparent viscosity is approximately 1.92 *cp* when the hematocrit is 30%. The apparent viscosity is assumed to be 4.0 *cp* in larger vessels. Then the apparent viscosity of blood in small vessels of the orders of 1–3 is obtained by linear interpolation as 2.5, 3.0, and 3.5, respectively.

3.   How to Use the Experimental Data

After collecting the experimental data of morphometry, elasticity, and blood viscosity, one may ask how to use these data. With specifying boundary conditions, we can use these experimental data to compute the pressure-flow relationship, the transit time, the wave propagation, and the stress-strain relationship in the whole lung, or anywhere in the lung.

The basic laws are the conservation of mass, momentum, and energy. A great deal of examples has been shown in Fung (1996, 1997). In the following, we present how to carry out theoretical analysis of the pulsatile blood flow in the pulmonary circulation based on the experimental data.

3.1. *Impedance of pulmonary arteries and veins*

In the model studies, the pulmonary vasculature is separated into two basic components, the arterial and venous vessels and the capillary network. The arteries and veins are treated as elastic tubes and the capillaries as two-dimensional sheets. The macro- and microcirculation is transformed into an electrical circuit analog. The pulmonary vascular input impedance and the characteristic impedance of each order in the pulmonary arterial tree are calculated under normal physiological conditions. In the model, a blood vessel is assumed as a thin-walled circular cylindrical elastic tube of uniform material and mechanical property, and blood is a homogeneous, incompressible, Newtonian fluid. The formula derivations have been detailedly presented in Gan and Yen (1994) and Huang *et al*, (1998). In applying these formulas to the pulmonary circulation, the variation of the physical properties along the vascular tree is taken into account . Each order of branches has been assigned different values for the geometrical and elastic properties of the vessels. The continuity of pressure and flow at each branch point is imposed as the boundary conditions of each vessel. The pulmonary microvascular impedance model is used to connect the arterial and venous trees.

3.2. *Microcirculation impedance*

Fung (1972, 1974) presented a one-dimensional solution of transient flow through the alveolar sheet that contains the basic feature of the more complex two-dimensional analysis. In our two-dimensional case, each alveolar sheet is supposed to stretch between parallel arterioles and venules. The derivations of formulas have been described in Fung (1972, 1974) and Huang *et al*, (1998). We do not want to repeat the derivation of formulas. When computing the impedance along the pulmonary vascular tree, the connection between the arteries and veins is by a capillary transfer matrix, A_{ij}, as

$$\left\{ \begin{array}{c} \tilde{P}_a \\ \tilde{Q}_a \end{array} \right\} = \left\{ \begin{array}{cc} A_{11} & A_{12} \\ A_{21} & A_{22} \end{array} \right\} \left\{ \begin{array}{c} \tilde{P}_v \\ \tilde{Q}_v \end{array} \right\} \tag{6}$$

where $\tilde{P}_a$, $\tilde{Q}_a$, $\tilde{P}_v$, $\tilde{Q}_v$ are the oscillatory pressure and flow at the arteriole and venule edges of the sheet respectively. A detailed analysis of pulmonary microvascular impedance based on the dimensions and elastic data of cat's pulmonary sheets has been demonstrated by Gao (1997).

4. Experimental Validation of the Theoretical Model Study

Through the biomechanical approach, a theoretical model of pulsatile flow in the pulmonary circulation can be developed based on the experimental data of vascular morphometry, elasticity, and blood viscosity. The theoretical pulsatile pressure-flow relations can be predicted. However, the predictions got to be validated in experimental studies.

To validate the theoretical predictions of dog's model, the experimental studies of pulmonary arterial pressure-flow relations in the normal and chronic pulmonary thromboembolic dogs were completed by Olman *et al*, (1994). A fluid-filled pulmonary arterial catheter was placed percutaneously for measurement of pulmonary arterial pressure. A radionuclide perfusion scan was performed for measurement of mean pulmonary blood flow. The pressure and flow of pulmonary arteries were measured in Day 0 and Day 30. The large pulmonary arteries were chronically obstructed with lysis-resistant thrombi since Day 0. Comparison of experimentally measured and model-derived pulmonary arterial pressure-flow is shown in Fig. 2 of Olman *et al*, (1994). The model-derived pulmonary arterial pressure was within 1 mmHg of the baseline (day 0) measured pulmonary arterial pressure.

Experiments on isolated perfused cat lungs were carried out to validate the theoretical model (Huang *et al*, 1998). The pulsatile blood pressure in the pulmonary arterial trunk was measured with a fluid-filled Teflon tubing. The flow in the pulmonary arterial trunk was measured with a flow probe around the vessel. The resulting data of pressure and flow were subjected to a frequency analysis. The modulus and phase of pressure and flow waves were calculated from the Fourier coefficients (Huang *et al*, 1998). Comparison of the experimental and model-derived pulmonary vascular input impedance is shown in Fig. 2 of the attached paper. At lower frequencies from 1 to 7 Hz, the input impedance of our model fits the animal data well. At frequencies higher than 7 Hz, the experimentally measured phase angle was positive, while the predicted value was negative. The possible reasons for this discrepancy have been discussed in the attached paper.

5.   Challenges for the Future

To better simulate the pulmonary vascular impedance theoretically, we need to include the information of pulmonary trunk (i.e. curvature, length, diameter), and the angles between the big branches. So far, these data are not available. Additionally, the influence from heart beat should be considered in the theoretical model.

Basic researches in biomedical science are aimed to improve the diagnosis and treatment of human diseases. After the establishment of animal models, we would like to extend our studies to the pulsatile flow in human lung. The detailed pressure-flow relations of human pulmonary circulation are unknown. A theoretical model can be done based on the morphometric and elastic data (Huang *et al*, 1996; Yen and Sobin, 1988; Yen *et al*, 1990). Additionally, experimental studies are necessary to evaluate the theoretical model. Refinement of the theoretical model is possible based on the experimental results.

The pressure-flow relation is extremely important to understand the human pulmonary circulation, and the mechanism of diseases in the human lung. As we discussed above, this relation is determined by the vascular morphometry, the vascular mechanical properties, and the rheology of the blood. Pulmonary vascular diseases may change the vascular geometry, the viscoelastic properties of the vessel walls, and the blood viscosity, such as atherosclerosis, hypertension, thromboembolism, and diabetes. Therefore, the pulmonary vascular impedances are changed in the disease lungs [see review by Nichols and O'Rourke, (1998)]. The theoretical models of pulmonary vascular impedances in human lung diseases will be useful to understand the mechanism of the diseases, and eventually helpful in the diagnosis and treatment of the human lung diseases.

References

1. Caro, C. G. and Staffman, P. G., Extensibility of blood vessels in isolated rabbit lungs, *J. Physiol.* **178**, 193–210 (1965).
2. Frasher, Jr, W. G. and Sobin, S. S., Pressure-volume response of isolated living main pulmonary artery in dogs, *J. Appl. Physiol.* **20**, 675–682 (1965).
3. Fung, Y. C., Theoretical pulmonary microvascular impedance, *Ann. Biomed. Eng.* **1**, 221–245 (1972).
4. Fung, Y. C., Fluid in the interstitial space of the pulmonary alveolar sheet, *Microvasc. Res.* **7**, 89–113 (1974).
5. Fung, Y. C., Dynamics of blood flow and pressure-flow relationship, *The Lung: Scientific Foundations*, eds. (Crystal *et al.*) Raven Press, New York (1991), pp. 1121–1134.
6. Fung, Y. C., Biomechanics: Circulation, 2nd edition, Springer-Verlag, New York (1996).
7. Fung, Y. C., Selected works on biomechanics and aeroelasticity, World Scientific, New Jersey (1997).
8. Fung, Y. C. and Sobin, S. S., Theory of sheet flow in lung alveoli, *J. Appl. Physiol.* **26**, 472–488 (1969).
9. Fung, Y. C. and Sobin, S. S., Elasticity of the pulmonary alveolar sheet, *Circ. Res.* **30**, 451–469 (1972a).
10. Fung, Y. C. and Sobin, S. S., Pulmonary alveolar blood flow, *Circ. Res.* **30**, 470–490 (1972b).
11. Gan, R. Z., Tian, Y., Yen, R. T. and Kassab, G. S., Morphometry of the dog pulmonary venous tree, *J. Appl. Physiol.* **75**, 432–440 (1993).
12. Gan, R. Z. and Yen, R. T., Vascular impedance analysis in dog lung with detailed morphometric and elasticity data, *J. Appl. Physiol.* **77**, 706–717 (1994).
13. Gao, J. Dissertation, University of Memphis, Memphis, TN (1997).
14. Horsfield, K., Pulmonary airways and blood vessels considered as confluent trees, *The Lung: Scientific Foundations*, (eds. Crystal, R. G., West, J. B., Barnes, P. H., Cherniack, N. S. and Weibel, E. R.), Raven, New York **1**, 721–727 (1991).
15. Huang, W., Yen, R. T., McLaurine, M. and Bledsoe, G., Morphometry of the human pulmonary vasculature, *J. Appl. Physiol.* **81**, 2123–2133 (1996).
16. Huang, W., Tian, Y., Gao, J. and Yen, R. T., Comparison of theory and experiment in pulsatile flow in cat lung, *Ann. Biomed. Eng.* **26**, 812–820 (1998).
17. Jiang, Z. L., Kassab, G. S. and Fung, Y. C., Diameter-Defined Strahler system and connectivity matrix of the pulmonary arterial tree, *J. Appl. Physiol.* **76**, 882–892 (1994).
18. Kassab, G. S., Rider, C. A., Tang, N. J., Fung, Y. C., Morphometry of pig coronary arterial trees, *Am. J. Physiol.* **265** (*Heart Circ. Physiol.*) **34**, H350–H356 (1993).
19. Maloney, J. E., Rooholamini, S. A. and Wexler, L., Pressure-diameter relations of small blood vessels in isolated dog lung, *Microvasc. Res.* **2**, 1–12 (1970).
20. Miller, W. S., *The Lung.* 2nd edition, Springfield, Thomas (1947).
21. Milnor, W. R., *Hemodynamics*, 2nd edition, Williams & Wikins, Baltimore (1989).
22. Nichols, W. W. and O'Rourke, M. F., McDonald's blood flow in arteries: Theoretic, experimental, and clinical principles, 4th edition, Oxford University Press, New York (1998).
23. Olman, M. A., Gan, R. Z., Yen, R. T., Villespin, I., Maxwell, R., Pedersen, C., Konopka, R., Debes, J. and Moser, K. M., Effect of chronic thromboembolism on the pulmonary artery pressure-flow relationship in dogs, *J. Appl. Physiol.* **76**, 875–881 (1994).

24. Patel, D. J., Schilder, D. P., and Mollos, A. J., Mechanical properties and dimensions of the major pulmonary arteries, *J. Appl. Physiol.* **15**, 92–96 (1960).

25. Sobin, S. S., Tremer, H. M. and Fung, Y. C., Morphometric basis of the sheet-flow concept of the pulmonary alveolar microcirculation in the cat, *Circ. Res.* **26**, 397–414 (1970).

26. Sobin, S. S., Fung, Y. C., Tremer, H. M. and Rosenquist, T. H., Elasticity of pulmonary alveolar microvascular sheet in the cat, *Circ. Res.* **30**, 440–450 (1972).

27. Sobin, S. S., Lindal, R. G., Fung, Y. C. and Tremer, H. M., Elasticity of the smallest noncapillary pulmonary blood vessels in the cat, *Microvasc. Res.* **15**, 57–68 (1978).

28. Sobin, S. S., Fung, Y. C., Lindal, R. G., Tremer, H. M. and Clark, L., Topology of pulmonary arterioles, capillaries, and venules in the cat, *Microvasc. Res.* **19**, 217–233 (1980).

29. Sobin, S. S. and Fung, Y. C., Response to challenge to the Sobin-Fung approach to the study of pulmonary microcirculation, *Chest* **101**, 1135–1143 (1992).

30. Tian, Y., Dissertation, University of Memphis, Memphis, TN (1993).

31. Weibel, E. R. and Gomez, D. M., Architecture of the human lung, *Science* **137**, 577–585 (1962).

32. Yen, R. T., Elastic properties of pulmonary blood vessels, *Respiratory Physiology: An Analytical Approach*, (eds. Chang, H. K. and Paiva, M.), Marcel Dekker, New York (1990), pp. 533–559.

33. Yen, R. T. and Fung, Y. C., Model experiment on apparent blood viscosity and hematocrit in pulmonary alveoli, *J. Appl. Physiol.* **35**, 510–517 (1973).

34. Yen, R. T., Fung, Y. C. and Bingham, N., Elasticity of small pulmonary arteries in the cat, *J. Biomech. Eng.* **102**, 170–177 (1980).

35. Yen, R. T. and Foppiano, L., Elasticity of small pulmonary veins in the cat, *J. Biomech. Eng.* **103**, 38–42 (1981).

36. Yen, R. T., Zhuang, F. Y., Fung, Y. C., Ho, H. H., Tremer, H. and Sobin, S. S., Morphometry of cat pulmonary venous tree, *J. Appl. Physiol.* **55**, 236–242 (1983).

37. Yen, R. T., Zhuang, F. Y., Fung, Y. C., Ho, H. H., Tremer, H. and Sobin, S. S., Morphometry of cat's pulmonary arterial tree, *J. Biomech. Eng.* **106**, 131–136 (1984).

38. Yen, R. T. and Sobin, S. S., Elasticity of arterioles and venules in postmortem human lungs, *J. Appl. Physiol.* **64**, 611–619 (1988).

39. Yen, R. T., Tai, D., Rong, Z. and Zhang, B., Elasticity of pulmonary blood vessels in human lungs, *Respiratory Biomechanics-Engineering Analysis of Structure and Function* (eds. Farrell Epstein, M. A., and Ligas, J. R.), Springer-Verlag, New York (1990), pp. 109–116.

40. Zhuang, F. Y., Fung, Y. C. and Yen, R. T., Analysis of blood flow in cat's lung with detailed anatomical and elasticity data, *J. Appl. Physiol.* **55**, 1341–1348 (1983).

Annals of Biomedical Engineering, Vol. 26, pp. 812–820, 1998
Printed in the USA. All rights reserved.

0090-6964/98 $10.50 + .00
Copyright © 1998 Biomedical Engineering Society

Comparison of Theory and Experiment in Pulsatile Flow in Cat Lung

W. Huang, Y. Tian, J. Gao, and R. T. Yen

Department of Biomedical Engineering, The University of Memphis, Memphis, TN

(Received 29 May 1996; accepted 27 April 1998)

Abstract—A mathematical model of pulsatile flow in cat lung based on existing morphometric and elastic data is presented and validated by experimental results. In the model, the pulmonary arteries and veins were treated as elastic tubes, whereas the pulmonary capillaries were treated as two-dimensional sheets. The macro- and microcirculatory vasculature was transformed into an analog electrical circuit. Input impedances of the pulmonary blood vessels of every order were calculated under normal physiological conditions. Pressure-flow relation of the whole lung was predicted theoretically. Experiments on isolated perfused cat lungs were carried out. The relation between pulsatile blood pressure and blood flow was measured. Comparison of the theoretically predicted input impedance spectra with those of the experimental results showed that the modulus spectra were well predicted, but significant differences existed in the phase angle spectra between the theoretical predictions and the experimental results. This latter discrepancy cannot be explained at present and needs to be further investigated. © *1998 Biomedical Engineering Society.*
[S0090-6964(98)02103-1]

Keywords—Pulmonary circulation, Pulsatile flow, Input impedance, Characteristic impedance.

INTRODUCTION

For the pulsatile flow in pulmonary circulation, the pulsatile pressure-flow relation can be expressed in terms of pulmonary vascular impedance. Engelberg and DuBois first measured the pulmonary vascular input impedance of the rabbit in 1959. Numerous investigators have since contributed to the field.[17] There are two general theoretical approaches to study pulsatile hemodynamics. The first uses continuum mechanics and morphometric and elasticity data. The other uses an electric circuit analog or "lumped parameter" model. Refinements of early models of either approach have been introduced by many authors.[17]

Weibel[23] initiated precise quantitative description of the lung anatomy using a successive bifurcation model. Cumming, Horsfield and their co-workers used Strahler's method of describing rivers and rivulets in geography to study the human pulmonary arterial and venous trees.[9] The cat pulmonary venous tree and arterial tree have been described with the Strahler system by Yen *et al.*[30,31] Recently, three innovations were introduced to improve the studies of coronary vasculature by Kassab *et al.*,[12] namely, (a) adding a new criteria based on the vessel diameters into Strahler's ordering system; (b) the concept of segment and element is used to express the series-parallel feature of blood vessels; and (c) the connectivity matrix is introduced to describe the connectivity of blood vessels among different orders. Jiang *et al.*[11] used these refinements in their study of the rat pulmonary arterial tree. The new method is called a diameter-defined Strahler's ordering system. In the study of microcirculation in the lung, investigators have found that capillaries in the lungs are not separate vessels, but form a continuous network in a flat sheet which comprises the alveolar wall. Fung and Sobin[1] have taken advantage of the geometry to describe the flow as a "sheet flow." The sheet model is especially advantageous for the description of the capillary elasticity. Data on the elasticity of pulmonary arteries and veins and the rheology of blood in these vessels have been obtained for the cat by Yen *et al.*[26,28,29] Those of arterioles, venules, and capillaries are given in Sobin *et al.*[19,20] Based on the morphometric and elastic data of cat lung, Zhuang *et al.*[33] and Fung and Yen[4] presented a detailed theoretical analysis of steady flow in zone 3 conditions and in zone 2 conditions, respectively. Experimental validation of these results was published by Yen *et al.*[32] These theoretical predictions and the experimental measurements have shown very good agreement over a wide range of conditions in steady flow.

Recently in our group, we have used the diameter-defined ordering system to describe the dog pulmonary venous[6] and arterial trees (unpublished). We have also studied the human pulmonary vasculature,[10] and developed a vascular impedance model in dog lung based on detailed morphometric and elastic data.[7] The pulmonary microvascular impedance model developed by Fung[2] was used in the vascular impedance model.[7] The model was

Address offprint requests to Michael R-T. Yen, PhD, Department of Biomedical Engineering, University of Memphis, Memphis, TN 38152.

Dr. Huang's current address is Department of Bioengineering, University of California, San Diego, La Jolla, CA 92093-0412.

validated by experimental studies of pulmonary arterial pressure-flow relations in the normal and chronic pulmonary thromboembolic dogs.[18] Additionally, a preliminary model study in cat lung was done.[22]

The objective of this paper is to present a mathematical model of pulsatile flow in cat's whole lung with detailed anatomical and elasticity data and an experimental validation of the theoretical prediction. To validate the mathematical predictions, the relation of pulsatile blood pressure and flow was measured in isolated perfused cat lungs.

MATHEMATICAL MODELS OF CAT LUNG

In the model studies, the pulmonary vasculature is separated into two basic components, the arterial and venous vessels and the capillary network. The arteries and veins are treated as elastic tubes and the capillaries as two-dimensional sheets. The macro- and microcirculation is transformed into an electrical circuit analog. The pulmonary vascular input impedance and the characteristic impedance of each order in the pulmonary arterial tree are calculated under normal physiological conditions. A brief description of mathematical models follows.

Impedance of Pulmonary Arteries and Veins

A blood vessel is modeled as a thin-walled uniform circular cylindrical elastic tube of uniform material and mechanical property. Blood is supposed to be a homogeneous, incompressible, Newtonian fluid.

The characteristic impedance Z_c is given by Womersley:[25]

$$Z_c = \frac{4}{\pi D^2} \sqrt{\frac{\rho}{2\beta(1-v^2)(M_{10}\exp(i\varepsilon_{10}))}}. \quad (1)$$

The complex wave velocity c is given by

$$c = \sqrt{\frac{M_{10}\exp(i\varepsilon_{10})}{2\rho\beta(1-v^2)}}, \quad (2)$$

where ρ is the density of blood, D is the diameter of blood vessel, M_{10} and ε_{10} are functions of Bessel functions of the Womersley's nondimensional parameter α,[25] β is the compliance of a blood vessel expressed as the percentage change in D versus change in transmural pressure,[29] and v is the Poisson's ratio. The relationship between input impedance Z_{in} and characteristic impedance Z_c depends on the structure of the vascular tree, and is put in the following form by Gan and Yen:[7]

$$Z_{in} = \frac{1-i\lambda\,\tan(\kappa l)}{\lambda - i\,\tan(\kappa l)} Z_c, \quad (3)$$

where l is the distance from the point in question to the distal end of the tube where reflection is generated, and κ is the complex propagation coefficient. The discontinuity coefficient λ is a function of the terminal impedance Z_T, which is the impedance faced by the end of the vessel, and the characteristic impedance Z_c,

$$\lambda = \frac{Z_C}{Z_T}. \quad (4)$$

In applying these formulas to the lung, the variation of the physical properties along the vascular tree is taken into account by assigning different values for the geometrical and deformational properties of the vessels in each successive order of branch. The continuity of pressure and flow at each branch point is imposed as the boundary conditions of each vessel.

Pulmonary Microvascular Impedance

Fung[2,3] presented a one-dimensional solution of transient flow through the alveolar sheet that contains the basic feature of the more complex two-dimensional analysis. In our two-dimensional case, each alveolar sheet is supposed to stretch between parallel arterioles and venules. If the permeability of water across the endothelium can be ignored and the amplitude of the thickness fluctuations is small compared with the mean pulmonary alveolar sheet thickness, the basic equation for a transient flow can be expressed as[2]

$$\frac{d^2 h^4}{dx^2} = 4\mu k f \beta^* \frac{dh}{dt}. \quad (5)$$

Here h is the thickness of capillary sheet; μ is the apparent viscosity of the blood in the capillary sheet; k is a function of the ratio of the thickness to width of the capillary sheet and has a numerical value of about 12; f is a friction factor, which is a function of the ratio of the post diameter to the sheet thickness and other flow parameters, and has a value of about 1.6; and β^* is the compliance constant of the capillary sheet.

Equation (5) is a nonlinear differential equation and does not have a harmonic solution with respect to time. Only in small perturbations can the basic equations be linearized and the concept of impedance be useful. Linearization can be justified if the amplitude of the pressure oscillation, $H(x)$, is small compared with the mean pulmonary arterial pressure. Under this condition, the solution of the governing equation is set in the following form:

$$h(x,t) = h_{SI}(x) + e^{i\omega t}H(x), \qquad (6)$$

where $h_{SI}(x)$ represents the wall thickness of the capillaries in a nonoscillatory flow, and $H(x)$ is assumed to be much smaller than $h_{SI}(x)$. According to (3), $h_{SI}(x)$ is equal to

$$h_{SI}(x) = \left[h_a{}^4 - (h_a{}^4 - h_\nu{}^4)\frac{x}{\bar{L}} \right]^{1/4}, \qquad (7)$$

where x is the distance measured from the inlet, $\bar{L}$ is the average length of blood pathway between the inlet and outlet, h_a is the steady-state sheet thickness at the arteriole inlet, and h_ν is that at the venule outlet. Similarly, the pressure and flow per unit width can also be represented as the sum of the oscillatory terms and the terms representing a nonoscillatory flow in capillaries whose walls are impervious to the solute:

$$p(x,t) = p_{SI}(x) + e^{i\omega t}P(x), \qquad (8)$$

$$q(x,t) = q_{SI}(x) + e^{i\omega t}\dot{Q}(x). \qquad (9)$$

The variables are written in dimensionless form. The nondimensional variables are related to their dimensional counterparts, denoted by circumflex ($\tilde{}$). Then the governing equation for pulsatile flow in the capillary sheet becomes

$$(d^2/d\tilde{x}^2)(\tilde{h}_{SI}^3\tilde{H}) = i\Omega\tilde{H}, \qquad (10)$$

where the dimensionless frequency parameter Ω is

$$\Omega = \mu k f \beta^* \omega \bar{L}^2 / h_0^3 \qquad (11)$$

and the pressure, flow rate and thickness relations are

$$\bar{P} = \tilde{p}_{SI} + e^{i\omega t}\tilde{H}, \qquad (12)$$

$$\tilde{Q} = -\frac{d}{d\tilde{x}}(\tilde{h}_{SI}^3\tilde{H}). \qquad (13)$$

Let the oscillatory pressure and flow at the arteriole and venule edges of the sheet be designated by $\tilde{P}_a$, $\tilde{Q}_a$, $\tilde{P}_\nu$, $\tilde{Q}_\nu$, respectively. The relation between these nondimensional quantities can be expressed in the following matrix form:

$$\begin{Bmatrix} \tilde{P}_a \\ \tilde{Q}_a \end{Bmatrix} = \begin{bmatrix} A_{11} & A_{12} \\ A_{21} & A_{22} \end{bmatrix} \begin{Bmatrix} \tilde{P}_\nu \\ \tilde{Q}_\nu \end{Bmatrix}. \qquad (14)$$

Hence the capillary network can be inserted between the arteries and veins to complete the circuit. A detailed analysis of pulmonary microvascular impedance has been reported by Gao *et al.* (submitted).

Morphometric and Elastic Data of Cat Lung

The morphometric data on the pulmonary arterial and venous trees of the cat are provided by Yen *et al.*,[30,31] on the basis of measurements of silicone elastomer casts. The elastic data for arteries and veins are given by Yen *et al.*[26,28,29] For the pulmonary artery trunk, the length (1 cm) and the diameter (0.986 cm) are from Yen *et al.*[31] The compliance β of the pulmonary arterial trunk was estimated based on the value of Moens-Korteweg wave velocity C_o, which is predicted as

$$C_o = \sqrt{\frac{1}{2\rho\beta}}, \qquad (15)$$

where ρ is the density of blood. In our modeling study, the β assigned to the pulmonary artery trunk gave a Moens-Korteweg wave velocity C_o of 120 cm/s.

The apparent viscosity values have been determined by Yen and Fung[27] in model experiments. They found that the apparent viscosity is approximately 1.92 cp when the hematocrit is 30%. The apparent viscosity is assumed to be 4.0 cp in larger vessels. The apparent viscosity of blood in small vessels on the order of 1–3 is obtained by linear interpolation as 2.5, 3.0, and 3.5, respectively. The density of blood is assumed to be 1.05 g/cm^3 in all calculations.

For pulmonary capillaries of the cat, Sobin *et al.*[19] report $h_o = 4.28$ μm, $\beta^* = 0.219$ μm/cm H$_2$O, and $S = 0.916$. Sobin *et al.*[21] give the average length of the cat capillaries to be $\bar{L} = 556$ μm. Based on the same topological maps used by Sobin *et al.*,[21] Zhuang *et al.*[33] determined, by three different stereological methods, that the total capillary area of the whole lung of the cat is 0.87, 1.27, or 1.61 m^2 when the transpulmonary pressure is 10 cm H$_2$O. Based on their results of the pressure distributions in pulmonary blood vessels for the case in which venous pressure (P_ν) is 2.1 cm H$_2$O, airway pressure (P_A) is 0, and pleural pressure (P_{pl}) is −7 cm H$_2$O, we found that the capillary inlet pressure P_{art} is 13.35 cm H$_2$O and capillary exit pressure P_{ven} is 11.29 cm H$_2$O. When the arterial pressure (P_a) is 20.22 cm H$_2$O, the dimensionless steady-sheet thickness at the arteriole inlet h_a is 1.68 and that at the venule outlet h_ν is 1.58.

The circulation of the whole lung is the sum of flow through all of its parts. To analyze blood flow in the whole lung, it is necessary to have a definite vascular circuit. Our vascular circuit construction follows the rules in the Gan–Yen model.[7] The process of obtaining statistical morphometric data from a tree is unique, given that the same set of rules is followed each time. The creation of a circuit from morphometric data, however, is not a unique process. An infinite number of circuits can be created corresponding to a set of morphometric data, but not uniquely specified by them. The topological arrangement in our theoretical calculation is based on the actual number of branches of each order. For this pattern, all vessels of the same order are identical. They have the same impedance, resistance, flow, and therefore, pressure drop.

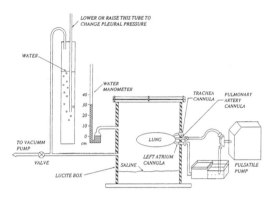

FIGURE 1. Schematic diagram of the experimental setup. The lung was placed in a sealed Lucite box with three cannulas and supported in the prone position. The lung was ventilated with negative pressure by a vacuum pump. The water column was constructed to stabilize the vacuum and provide adjustment capabilities for the negative pressure (pleural pressure).

EXPERIMENTAL STUDIES OF ISOLATED PERFUSED CAT LUNG

Specimen Preparation

Experiments were carried out on nine healthy mongrel cats, weighing between 2.6 and 3.5 kg (2.9 ± 0.6 kg). Each cat was anesthetized with pentobarbital sodium (25 mg/kg intramuscular) and heparinized intravenously (1000 IU/kg). It was then sacrificed using a minimal overdose of ketamine (10 mg/kg) and xylazine (2 mg/kg). To expose the heart and lungs, the anterior chest wall and sternum were removed, and a bandage wrapped around the cut ends of the ribs so that the jagged bone ends would not tear the lung surface. Extreme care was taken not to touch the lung surfaces, as the pleura is very thin and easily torn. The whole lung was then removed together with the heart from the chest by transection of the trachea 5 cm above the carina.

In order to prepare for the probe of an electromagnetic flowmeter, the fat pads which surround the main pulmonary artery were removed. A T-shaped polyethylene cannula was tied into the pulmonary artery trunk via a puncture wound in the wall of the right ventricular outflow tract. We ligated the aorta and placed a suture around the atrioventricular groove to occlude the ventricular lumen. Cannulation of the pulmonary artery trunk was done with the vessel submerged in saline to prevent air bubbles from entering the lungs. An identical cannula was introduced through the left atrial appendage and tied in the pulmonary vein near its junction with the left atrium. The trachea was also cannulated after gentle suction to remove retained secretion. These three cannulas were attached to vinyl tubing fitted with quick-connect fittings that in turn were appropriately connected to a perfusion circuit by mating quick-connect fittings.

Experimental Setup

The lung was placed in a Lucite box with three fittings protruding through the front. For all of the measurements, the lungs were supported in the prone position. With the lung in the box, the left atrium was connected to a venous reservoir, the pulmonary artery was attached to a Harvard pulsatile pump (Harvard Apparatus, South Natick, MA), and the trachea was left open to the atmosphere. Care was taken to see that no air bubbles were present in the arterial line. The height of the reservoir could be adjusted to set the pulmonary venous pressure at the desired level. The average time interval from death of the cat to the perfusion of the lungs was less than 1 h.

The basic features of the experimental setup and the perfusion circuit are shown schematically in Fig. 1. The lung was ventilated with negative pressure by a vacuum pump (General Electric, Wayne, IN) and connected to the box and a water column. The water column was constructed to stabilize the vacuum and provide adjustment capabilities for the negative pressure (pleural pressure). The pleural pressure was monitored from a water manometer. After full expansion with sufficiently high pleural pressure, the lung was then preconditioned by inflating and deflating eight to ten times by varying the pleural pressure from 0 to -20 cm H_2O to remove areas of superficial atelectasis. Following the preconditioning, the blood vessels from the pulmonary artery to the opposing vein were purged of blood by flushing with a small amount of Macrodex (6% dextran 70 in normal saline, a viscosity less than 200 cp) at a very low flow rate to establish vascular continuity across the lung. After the blood and dextran mixture were visible in the exit line, a Harvard pulsatile blood pump was promptly connected to the pulmonary artery, and circulated Macrodex solution through the pulmonary artery into the lung and back through the pulmonary vein into a venous reservoir. This pump provides a choice of beat frequencies, proportions of systolic–diastolic times, and stroke volumes. The flow rate could be adjusted by changing stroke volume and/or pulse rate. Pulmonary arterial pressure was adjusted by altering flow rate, whereas left atrial pressure was adjusted by the height of the venous reservoir. The zero reference level for vascular pressure was the left atrium of the heart. During perfusion, a constant pleural pressure of -10 cm H_2O approached from all inflation was maintained while the pulmonary venous pressure was fixed at 2 cm H_2O.

Pulmonary Arterial Pressure and Flow Measurements

The pulmonary arterial pressure was measured through a Teflon tubing (i.d. 0.75 mm, 5–10 cm in length) connected to physiologic pressure transducers P23XL (Spectramed Inc., Oxnard, CA). The system was filled with physiological saline solution. The Teflon tubing for pressure measurement was attached directly through a three-way pressure tap to the side arm of the polyethylene cannula which was tied into the pulmonary artery. The output of the pulmonary arterial pressure wave was displayed on a monitor. Using a similar method as described above, the pressure wave of the pulmonary vein was also recorded through another pressure transducer connected to the pulmonary vein cannula. Instantaneous blood flow was measured with a square-wave electromagnetic flowmeter (model 501D, Carolina Medical Electronics, Inc., King, North Carolina) and flow probes (type EP400, Carolina Medical Electronics, Inc., King, North Carolina). Probes were carefully matched to the diameter of the artery in each case. The probe was directly placed around the main pulmonary artery. Cautious adjustment of the probe was employed to avoid crumpling or distortion of the vessel. The flowmeter probe lead was brought out of the Lucite box and connected with the square-wave electromagnetic flowmeter. Zero flow baseline was taken to be the integrated level during diastole. The tip of the Teflon tubing for pressure measurement was positioned 2–5 mm downstream of the flow probe. Once the catheters and flow probe were placed, the pulmonary arterial pressure, the arterial flow rate, and the left atrial pressure were continuously recorded.

The frequency response of the pressure-measurement system and the flow-measurement system was checked before and after each experiment. A resonant frequency higher than 50 Hz was accepted as being satisfactory. The frequency response of the flow-measurement system was found to be flat up to 30 Hz. No correction was made for the frequency response for either pressure-measurement system or flow-measurement system. After all other experimental procedures had been completed, the flow probe was calibrated *in vitro* by excising a piece of pulmonary artery with the probe around it, connecting it to a reservoir and outflow tubing, and immersing it in a large bath of saline. The vessel was stretched to its normal length. The average output of the flowmeter was taken as the signal corresponding to the timed collections of outflow. The probes gave a linear output for the range of flows used in the experiment.

Data Analysis

An IBM PC-compatible 80386 computer was used to control the experiment and to acquire and store the data. A data acquisition board DT2821 (Data Translation, Marlboro, MA) was used for analog input and digital output. The pulmonary arterial pressures, pulmonary arterial flow, and left atrial pressure signals were simultaneously digitized by a data acquisition software package GLOBAL LAB (Data Translation, Marlboro, MA) at 500 samples per second for ten beats and displayed on a monitor. Once the data were collected and stored on the computer's hard drive, they were subsequently retrieved and analyzed with another signal processing software package DADISP (DSP Development, Cambridge, MA). Only those recorded beats with optimal pressure and flow configurations were chosen for analysis. The pressure and flow waves were smoothed by using DADISP/FILTERS. DADISP/FILTERS is a menu-driven, digital filtering module that provides full finite impulse response and infinite impulse response filtering capabilities. The filter was frequency domain, and 100 Hz filter was used. Pressure moduli less than 0.2 cm H_2O and flow moduli less than 0.1 ml/s were rejected from subsequent analysis because these values were within the noise level of measuring system.

Pulmonary Vascular Input Impedance Computations

The experimental procedure for determined pulmonary vascular input impedance was to measure the pulsatile pressure and flow simultaneously. The resulting data were then subjected to a frequency analysis, which translates the observed pressure $P(t)$ and flow $\dot{Q}(t)$ into Fourier series. The modulus and phase of pressure and flow waves of any harmonics can be calculated from the Fourier coefficients, respectively.

The pulmonary vascular input impedance at any frequency n is the ratio of pressure over flow at that frequency:

$$|Z_{\text{in}}|_n = \left| \frac{P_n}{\dot{Q}_n} \right|, \qquad (16)$$

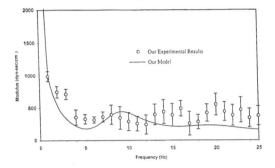

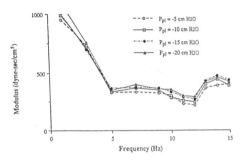

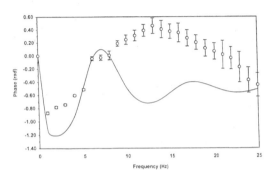

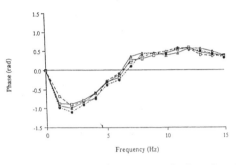

FIGURE 2. Comparison of the experimental and model derived pulmonary vascular input impedance.

FIGURE 3. Experimental pulmonary vascular input impedance in various pleural pressures.

$$\theta_n = \phi_n - \zeta_n. \qquad (17)$$

Here $|Z_{in}|_n$ is the modulus, θ_n the phase angle of the input impedance at the same harmonics frequency, $|P_n|$ and $|\dot{Q}_n|$ are the harmonic component of the pressure and flow wave, respectively, ϕ_n is the phase angle of the pressure, and ζ_n is the phase angle of the flow.

RESULTS

Hemodynamic parameters for the pulsatile control condition in the nine cats are set as follows (mean ± SD): pulmonary artery pressure (systolic/diastolic) $P_a = 35.1 \pm 0.5/11.7 \pm 0.6$ cm H_2O; mean pulmonary arterial pressure $P_{a_0} = 20.2 \pm 0.9$ cm H_2O; mean pulmonary venous pressure is $P_{v_0} = 2.1 \pm 0.2$ cm H_2O; peak arterial flow rate 18.9 ±0.6 ml/s; mean arterial flow rate 6.9 ± 0.5 ml/s; airway pressure $P_A = 0$ cm H_2O; pleural pressure $P_{pl} = -10$ cm H_2O.

Figure 2 shows our experimental and theoretical results on the pulmonary vascular input impedance of the whole lung of the cat. The input impedance of our model fits the animal data well at lower frequencies from one to seven Hz. At frequencies higher than 7 Hz, the pulmonary vascular input impedance measured experimentally

exhibited many fluctuations of the modulus compared with the input impedance spectrum generated by the mathematical model, although the order of magnitude remained comparable. At frequencies higher than 7 Hz, our experimentally measured phase angle was positive, while the predicted value was negative. The reason for this discrepancy is not clear; thus the high frequency regime needs further study.

Figure 3 shows the effects of lung inflation on the pulmonary vascular input impedance. We changed the pleural pressure to four different levels: -5, -10, -15, and -20 cm H_2O. The results show the lung inflation has little effect on pulmonary vascular input impedance; there is only a minor increase of the modulus as inflation pressure increased.

Figure 4 compares the results of our mathematical and experimental data on the cat with those of Grant and Paradowski,[8] who used both a lumped and a distributed parameter model. All predicted modulus spectra are fairly similar. At frequencies higher than 7 Hz marked fluctuations of the moduli which are not predicted by the theoretical model exist in our experimental data, whereas Grant and Paradowski presented a smooth curve. In general, our predicted impedance moduli are higher than those predicted by Grant and Paradowski's distributed parameter model. None of the theoretical models pre-

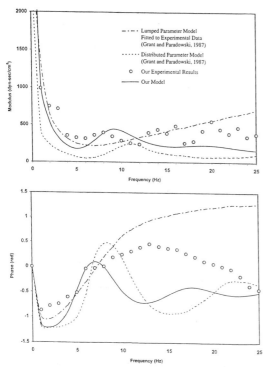

FIGURE 4. Comparison of the results of our mathematical model prediction and experimental measurement with those of a lumped parameter model fitted to experimental data and a distributed parameter model presented by Grant and Paradowski (Ref. 8).

dicted the experimental positive phase angle at frequencies higher than 8 Hz.

DISCUSSION

Experimental measurement of the input impedance of the pulmonary vasculature not only provides a concise expression of the pressure-flow relationship at the inlet to the pulmonary vascular bed, but also provides information about the elasticity and other physical characteristics of the blood vessels involved. The interpretation of the pulmonary vascular input impedance pattern has been attempted through experimental measurements and various levels of modeling. The input impedance spectra in our experiments are generally consistent with our theoretical predictions, but the prediction of the phase angle at frequencies higher than approximately 7 Hz was off as shown in Fig. 2. We may propose several reasons for this discrepancy:

(1) The pulmonary arterial trunk (artery of the order number 12) of the isolated cat lung did not simulate

the mathematical trunk model, nor did it behave as the trunk *in vivo*.

(2) The curvature and taper of the left and right pulmonary arteries (of order number 11) were not simulated in the mathematical model.

(3) The nonlinear terms in fluid mechanics and solid mechanics equations of the theory were ignored.

Looking into these possible causes would be a worthwhile objective for future research.

Part of the poor simulation of the pulmonary artery trunk *in vitro* is the uncertainty of the values of diameter, D, and the compliance, β, of the vessels. Trial computations of the impedance by varying D and β have shown that the impedance modulus and phase angle at larger frequencies are rather sensitive to the variation of D but not to that of β. Hence, in the future, a greater attention to the simulation of the diameter of the arterial trunk must be given.

The Reynolds and Womersley numbers of the flow in the larger pulmonary arteries of cats are high enough to influence the effect of curvature and the possibility of flow separation and turbulence, but nonlinear and nonstationary effects were not accounted for in the linear theory.

Nonlinearity and the effects of local geometric conditions, entry conditions, wall motion in the longitudinal axis, and viscoelasticity were neglected. Hence the agreement between the predictions of the linear theory and the experimental results on the magnitude of the frequency spectrum shown in Fig. 2 is not a justification of the linearized theory, whereas the disagreement between the linear theory and experiments definitely casts doubts on the many linearizations involved in the analysis.

In pulsatile blood flow, the paper by Ling *et al.*[15] remains as one of the most thorough. They investigated a segment of descending aorta of a dog, and showed that the flow predicted by linear and nonlinear theories for a given pressure difference may differ by a factor of 2. Li *et al.*[13] measured the propagation constant in arteries and found that the measured attenuation coefficients and phase velocities differed greatly from those predicted by linear theories. To explain the discrepancies between theory and experiments, they evaluated and compared the nonlinear and linear theories for the femoral artery. However, similar work has not been done in the pulmonary circulation. More recent work on nonlinear effects on blood flow has been focused more on flow separation, wall shear, and collapsible tubes (veins). The nonlinear collapsible flow in pulmonary capillaries has been analyzed by Fung and Yen[4] with regard to sluicing in zone 2 conditions. On the other hand, the analysis of gas flow in the airway has traditionally paid attention to the nonlinear secondary flow.

In Fig. 2, it is seen that both theory and experiment show the existence of peaks and valleys in the modulus of the input impedance and the change of sign from negative to positive in the phase angle. We believe that these features reflect the branching pattern of the vascular tree. To give a simple qualitative explanation, we note that the input impedance represents the impedance of the waves in the entire vasculature as seen at the pulmonary arterial trunk. Now, if a simple harmonic forward wave initiated at the trunk goes down a vessel of uniform diameter and elasticity, the pressure and velocity in the vessel are related by the formula $p = \pm \rho c v = \pm \rho c \dot{Q}/A$.[5] Here p is the pressure, v is the velocity, ρ is the density of the blood, c is the wave speed, A is the cross-sectional area, and $\dot{Q} = Av$ is the volume flow rate. The "+" sign applies if the wave goes in the direction of the flow, and the "−" sign applies if the wave goes the other way. When the wave reaches the bifurcation point of the blood vessel, part of the wave is reflected, the reflected wave goes in the direction opposite to the velocity of flow. The phase angle between the pressure and flow, $p/\dot{Q}$, is therefore 0° in the forward wave, and −180° in the reflected wave. This happens at every bifurcation point, back and forth. The cumulative effect is seen in the input impedance. The negative phase angle is a revelation of the reflected waves. Figure 2 thus shows that the reflected waves are significant at lower frequencies, and that the reflected–reflected waves become significant at higher frequencies. The valleys of the modulus spectrum hint at the phenomenon of resonance, which is a reverberation of forward and reflected waves.[5] The existence of shallow and multiple valleys suggests weak resonance in different parts of the vascular tree.

Grant and Paradowski[8] measured pulmonary arterial input impedance in six anesthetized cats. They adjusted eight lumped parameters in their mathematical model to fit the measured impedance spectra. They also used a distributed parameter model with geometric and distensibility data of Zhuang et al.,[33] the impedance formula by Wiener et al.,[24] and the pulmonary sheet flow model for microcirculation. Their theoretical results are quite similar to ours as shown in Fig. 4. Their experimental phase data fitted with a lumped parameter model differ significantly from ours at high frequency. This may be due to the differences of the conditions of the in vivo lung[8] and the isolated lung of our study, and the conditions of the pulmonary arterial trunks in these two different experiments.

Lucas et al.[16] used a 24-generation constrained tube model of the pulmonary vascular bed of the cat to study the lumped parameter models of the pulmonary circulation. In their model, no longitudinal changes are allowed. Their objective was to compare the spectra of their model with spectra obtained from lumped parameter

models. Lilagan et al.[14] used the two-port approach to develop a new model based on the experimental data of elasticity and topology on the pulmonary vasculature of the cat. The two-port approach treats the capillaries as small vessels. This capillary model is perhaps an unrealistic geometric representation of the capillaries, and the uncertainty about their compliance constitutes the weakest link in their model of the pulmonary vascular bed. A more realistic model can be based on the sheet-flow theory (see Addendum in Ref. 33).

We have made a thorough analysis of pulmonary vascular impedance in the dog.[7] On the basis of experimentally measured morphometric and elasticity data and model-derived mean pressure-flow conditions, we made a theoretical modeling of the pulsatile flow in the whole lung, but did not validate the model experimentally. The morphometric data applied in the dog's model were obtained from a 25-kg dog's right lung, which cannot represent the dog population. In the present model, the morphometric data were obtained from five cats with an average weight of 3.6 kg.[30,31] The pulmonary arteries and veins were treated as elastic tubes, whereas the pulmonary capillaries were treated as two-dimensional sheets. The macro- and microcirculatory vasculature was transformed into an analog electrical circuit. Input impedance of the pulmonary blood vessels of every order were calculated under normal physiological conditions. Pressure-flow relation of the whole lung was predicted theoretically. To validate the present model, a series of experiments on isolated perfused cat lungs were carried out to examine the relationship of pulsatile pressure and flow. Comparison of the theoretically predicted input impedance spectra with those of the experimental results showed that the modulus spectra were well predicted, but significant differences existed in the phase angle spectra between the theoretical predictions and the experimental results. This discrepancy cannot be explained at present. Further research should be done to understand this discrepancy.

REFERENCES

[1] Fung, Y. C., and S. S. Sobin. Theory of sheet flow in lung alveoli. J. Appl. Physiol. 26:472–488, 1969.

[2] Fung, Y. C. Theoretical pulmonary microvascular impedance. Ann. Biomed. Eng. 1:221–245, 1972.

[3] Fung, Y. C. Fluid in the interstitial space of the pulmonary alveolar sheet. Microvasc. Res. 7:89–113, 1974.

[4] Fung, Y. C., and R. T. Yen. A new theory of pulmonary blood flow in zone 2 condition. J. Appl. Physiol. 60(5):1638–1650, 1986.

[5] Fung, Y. C., Biomechanics: Circulation, 2nd ed. New York: Springer-Verlag, 1996, pp. 108-205.

[6] Gan, R. Z., Y. Tian, R. T. Yen, and G. Kassab. Morphometry of the dog pulmonary venous tree. J. Appl. Physiol. 75:432–440, 1993.

[7] Gan, R. Z., and R. T. Yen. Vascular impedance analysis in dog lung with detailed morphometric and elasticity data. *J. Appl. Physiol.* 77(2):706–717, 1994.

[8] Grant, B. J. B., and L. J. Paradowski. Characterization of pulmonary arterial input impedance with lumped parameter models. *Am. J. Physiol.* 252:H585–593, 1987.

[9] Horsfield, K. Functional morphology of the pulmonary vasculature. In: Respiratory Physiology, edited by H. K. Chang and M. Paiva. New York: Marcel Dekker, 1989, pp. 499–531.

[10] Huang, W., R. T. Yen, M. McLaurine, and G. Bledsoe. Morphometry of the human pulmonary vasculature. *J. Appl. Physiol.* 81(5):2123–2133, 1996.

[11] Jiang, Z. L., G. S. Kassab, and Y. C. Fung. Diameter-Defined Strahler system and connectivity matrix of the pulmonary arterial tree. *J. Appl. Physiol.* 76(2):882–892, 1994.

[12] Kassab, G. S., C. A. Rider, N. J. Tang, and Y. C. Fung. Morphometry of pig coronary arterial trees. *Am. J. Physiol.* 265:H350–356, 1993.

[13] Li, J. K-J., J. Melbin, and A. Noordergraaf. Pulse wave propagation. *Circ. Res.* 49:442–452, 1981.

[14] Lilagan, P. E., B. E. Marshall, and A. Noordergraaf. Analysis of deficiencies in models of the pulmonary circulation using the two-port method. *Adv. Bioeng. ASME* 26:507–510, 1993.

[15] Ling, S. C., H. B. Atabek, W. G. Letzing, and D. J. Patel. Nonlinear analysis of aortic flow in living dogs. *Circ. Res.* 33:198–212, 1973.

[16] Lucas, C. L., B. Ha, G. W. Henry, J. L. Ferreiro, and B. R. Wilcox. Toward a minimal lumped parameter model of the pulmonary input impedance spectrum. *Adv. Bioeng. ASME* 26:503–506, 1993.

[17] Milnor, M. R. Hemodynamics, 2nd ed. Baltimore: Williams & Wikins, 1989.

[18] Olman, M. A., R. Z. Gan, R. T. Yen, I. Villespin, R. Maxwell, C. Pedersen, R. Konopka, J. Debes, and K. M. Moser. Effect of chronic thromboembolism on the pulmonary artery pressure-flow relationship in dogs. *J. Appl. Physiol.* 76(2):875–881, 1994.

[19] Sobin, S. S., Y. C. Fung, H. M. Tremer, and T. H. Rosenquist. Elasticity of pulmonary alveolar microvascular sheet in the cat. *Circ. Res.* 30:440–450, 1972.

[20] Sobin, S. S., R. G. Lindal, Y. C. Fung, and H. M. Tremer. Elasticity of the smallest noncapillary pulmonary blood vessels in the cat. *Microvasc. Res.* 15:57–68, 1978.

[21] Sobin, S. S., Y. C. Fung, R. G. Lindal, H. M. Tremer, and L. Clark. Topology of pulmonary arterioles, capillaries, and venules in the cat. *Microvasc. Res.* 19:217–233, 1980.

[22] Tian, Y., and R. T. Yen. Model studies of vascular impedance of cat's lung with detailed anatomical and elasticity data (abstract). *Adv. Bioeng. ASME* 22:447–450, 1992.

[23] Weibel, E. R. Morphometry of the Human Lung. New York: Academic, 1963.

[24] Wiener, F. E., E. Morkin, R. Skalak, and A. P. Fishman. Wave propagation in the pulmonary circulation. *Circ. Res.* 19:834–850, 1966.

[25] Womersley, J. R. Oscillatory flow in arteries: The constrained elastic tube as a model of arterial flow and pulse transmission. *Phys. Med. Biol.* 2:176–187, 1957.

[26] Yen, R. T. Elastic properties of pulmonary blood vessels. In: Respiratory Physiology: An Analytical Approach, edited by H. K. Chang and M. Paiva. New York: Marcel Dekker, 1990, pp. 533-559.

[27] Yen, R. T., and Y. C. Fung. Model experiment on apparent blood viscosity and hematocrit in pulmonary alveoli. *J. Appl. Physiol.* 35:510–517, 1973.

[28] Yen, R. T., Y. C. Fung, and N. Bingham. Elasticity of small pulmonary arteries in the cat. *J. Biomech. Eng.* 102:170–177, 1980.

[29] Yen, R. T., and L. Foppiano. Elasticity of small pulmonary veins in the cat. *J. Biomech. Eng.* 103:38–42, 1981.

[30] Yen, R. T., F. Y. Zhuang, Y. C. Fung, H. H. Ho, H. Tremer, and S. S. Sobin. Morphometry of cat pulmonary venous tree. *J. Appl. Physiol.* 55:236–242, 1983.

[31] Yen, R. T., F. Y. Zhuang, Y. C. Fung, H. H. Ho, H. Tremer, and S. S. Sobin. Morphometry of cat's pulmonary arterial tree. *J. Biomech. Eng.* 106:131–136, 1984.

[32] Yen, R. T., Y. C. Fung, F. Y. Zhuang, and Y. J. Zeng. Comparison of theory and experiments of blood flow in cat's lung. In: Biomechanics in China, Japan, and USA, edited by Y. C. Fung, E. Fukada, and J. J. Wang. Beijing: Science Press, 1984, pp. 240–243.

[33] Zhuang, F. Y., Y. C. Fung, and R. T. Yen. Analysis of blood flow in cat's lung with detailed anatomical and elasticity data. *J. Appl. Physiol.* 55:1341–1348, 1983.

APPENDIX A

ABOUT THE AUTHORS

SHU CHIEN Professor of Bioengineering and Medicine; Director, Whitaker Institute for Biomedical Engineering, UCSD

M.D., National Taiwan University, 1953. Ph.D., Columbia University, 1957. Member, Academia Sinica, ROC, 1976. Fahraeus Medal in Clinical Haemorheology, 1981. Landis Award, Microcirculatory Society, 1983. NSF Special Creativity Grant Award, 1985–1988. President, American Physiological Society, 1990–1991. Melville Medal, American Society Mechanical Engineers, 1990, 1996. Zweifach Award, Fifth World Congress for Microcirculation, 1991. Founding Fellow, American Institute of Medical and Biological Engineering, 1992. President, Federation of American Societies for Experimental Biology, 1992–1993. ALZA Award, Biomedical Engineering Society, 1993. Member, Institute of Medicine, National Academy of Sciences, 1994. Joseph Mather Smith Prize for Distinguished Alumni Research, Columbia University, 1996. Member, National Academy of Engineering, 1997.

YUAN-CHENG B. FUNG Professor Emeritus of Bioengineering and Applied Mechanics, UCSD

Ph.D., California Institute of Technology, 1948. Member, Academia Sinica, 1966. Landis Award, Microcirculatory Society, 1975. von Karman Medal, American Society of Civil Engineers, 1976. Lissner Award for Bioengineering, American Society of Mechanical Engineers, 1978. Member, National Academy of Engineering of U.S.A., 1979. Centennial Medal of ASME, 1981. Worcester Reed Warner Medal, ASME, 1984. Honorary Member, Chinese Academy of Science, Institute of Mechanics, 1986. Poiseuille Medal, International Society of Biorheology, 1986. ALZA Award, Biomedical Engineering Society, 1989. Senior Member, Institute of Medicine of the NAS of U.S.A., 1991. Timoshenko Medal, ASME, 1991. Founding Fellow of the American Institute for Medical and Biological Engineering, 1992. Member, National Academy of Science of U.S.A., 1992. Borelli Award, American Society of Biomechanics, 1992. Melville Medal, ASME, 1994. Foreign member of Chinese Academy of Science, People's Republic of China, 1994. World Council for Biomechanics, Chair, 1994–1998, Honorary Chair, 1998–. Honorary Member, ASME, 1996. Founders Award, National Academy of Engineering of U.S.A., 1998. U.S. Presidential National Medal of Science Laureate, 2000.

DAVID A. GOUGH Professor of Bioengineering; Chair, Department of Bioengineering, UCSD

Ph.D., University of Utah, 1974. Founding Fellow, American Institute of Medical and Biological Engineering, 1991. M. J. Kugel Award, Juvenile Diabetes Foundation, 1996. Teacher of the Year Award, University of California, San Diego, 1996.

Research Interests: Implantable glucose sensor for diabetes. Glucose and oxygen transport through tissues, sensor biocompatibility, and glucose gradients in the bloodstream. Dynamic models of the natural pancreas, insulin islet, and beta cell-based glucose input and insulin output. Development of a scanning cytometer for time-lapsed analysis of populations of living mammalian cells attached to a substrate for studies of cell cycle phases, cell motility and attachment, concentration of labeled intracellular substances, and protein production.

MARCOS INTAGLIETTA Professor of Bioengineering, UCSD

Ph.D., California Institute of Technology, Pasadena, 1963. Humboldt Preiss-Senior Scientist Award, Federal Republic of Germany, 1982–1983. Founding Director, International Institute for Microcirculation, 1983. Abbott Distinguished Professor Lecture Series Award, Rio de Janeiro and Recife, Brazil; Caracas and Maracaibo, Venezuela; Mexico City and Guanajuato, Mexico; Bogota, Columbia, 1982–1988. Honorary Member, Italian, French, Indian, and Mexican societies for microcirculation, 1985–1989. President, American Microcirculatory Society. Honorary Founding Member, Mexican Society for Microcirculation, 1988. "Intensive Care Medicine Award 1989," 10th International Symposium on Current Problems in Emergency and Intensive Care Medicine, Münster, Germany, 1989. Malpighi Gold Medal Award, European Society for Microcirculation, 1994. Whitaker Award, Biomedical Engineering Society, 1996.

GHASSAN S. KASSAB Associate Researcher in Bioengineering, UCSD

Ph.D., University of California, San Diego, 1990. AHA Postdoctoral Fellow, 1990. NIH FIRST Award, 1997.

Research Interests: Encompass the biomechanics of the coronary circulation in health and disease. One of the goals is to develop a computational hemodynamic model of the entire coronary circulation based on a detailed set of anatomical and rheological data of the coronary blood vessels. Previously obtained the necessary set of morphological data of the entire coronary vasculature and I am currently studying the mechanical properties of the coronary blood vessels. Also studying the remodeling of the coronary circulation in disease. Specifically, focusing on the structural and mechanical remodeling of the coronary blood vessels in hypertension and flow-overload.

BERNHARD O. PALSSON Professor of Bioengineering, UCSD

Ph.D., University of Wisconsin, 1984. Assistant Professor, University of Michigan, 1984–1989. G.G. Brown Associate Professor, University of Michigan, 1990–1994. Vice President for Developmental Research, Aastrom Biosciences Inc. 1994–1995. Fellow, American Institute for Medical and Biological Engineering.

Research Interests: Cell and tissue engineering. Quantitative analysis of subcellular and cellular events focusing on the descriptions of the rates of metabolism, gene transfer, cell cycle, cell differentiation and cell movement, with the concomitant analysis of the simultaneous physico-chemical processes. Current focus is on the tissue engineering of bone marrow, enhanced gene transfer for gene therapy, genomatics, and networks underlying cell-to-cell communications.

ROBERT L. SAH Charles Lee Powell Associate Professor of Bioengineering, UCSD

Sc.D., Massachusetts Institute Of Technology, 1990. M.D., Harvard Medical School, 1991. Kappa Delta (Ann Doner Vaughan) Award, Orthopaedic Research Society, 1992. Hulda Irene Duggan Arthritis Investigator, Arthritis Foundation, 1993. National Science Foundation, Young Investigator Award, 1994. UCSD Outstanding Teacher Award, 1994, 1996, 1997.

Research Interests: Cartilage repair and tissue engineering. Function, composition, structure relationships of cartilage during growth, aging, degeneration, and repair. Biomechanics, electromechanics, and transport. Extracellular matrix metabolism.

GEERT W. SCHMID-SCHÖNBEIN Professor of Bioengineering, UCSD

Ph.D., University of California, San Diego, 1976. Malpighi Gold Medal, 1980. Abbott Award, European Societies for Microcirculation, 1984. Melville Medal, American Society of Mechanical Engineering, 1990. President, Biomedical Engineering Society, 1991–1992. Ratchow-Memorial Gold Medal, European Societies for Phlebology, 1999. President, Biomedical Engineering Society, 1991–1992. President, North American Society for Biorheology, 1998–1999. Distinguished Alumnus Lecture, Department of Bioengineering, U. C. San Diego, 1999. Teacher of the Year, Bioengineering, School of Engineering, U.C. San Diego, 1997/1998 and 1998/1999.

Research Interests: Blood flow, lymphatic transport. Molecular and cellular biomechanics. Spontaneous activation of cardiovascular cells. Mathematical model of blood flow in skeletal muscle. Hypertension, ischemia, and shock.

LANPING AMY SUNG Associate Professor of Bioengineering, UCSD

Ph.D., Columbia University, 1982.

Research Interests: Molecular basis of cell and membrane mechanics. Molecular structure and control of gene expression of membrane skeletal proteins in relation to the mechanical properties of cells and tissues in differentiation, aging, and disease. cDNA cloning, genomic organization, promoter elements and splicing isoforms of membrane skeletal proteins; molecular defects of these proteins in inherited diseases. Pseudoenzyme (protein 4.2) as a membrane skeletal protein in maintaining the stability and flexibility of erythrocyte membranes. Mechanical function of tropomodulin (a tropomyosin-binding protein) in the heart, muscles, and erythrocytes.

PIN TONG Professor of Mechanical Engineering, Hong Kong University of Science and Technology; Visiting Professor, UCSD.

Ph.D., California Institute of Technology, 1966. Von Karman Memorial Award, TRI, 1974. Secretary of Department of Transportation's Meritorious Achievement Award, 1976. Engineer of the Year of Research and Special Programs Administration, U. S. Department of Transportation, American Society of Professional Engineers, 1986. Founding head of Mechanical Engineering Department, HKUST. President of Far East Oceanic

Fracture Society, 1994–1998. President of Hong Kong Society of Theoretical and Applied Mechanics, 1998–1999. Asian subregion director, American Society of Mechanical Engineers, 1998–1999.

Research Interests: Electronic packaging reliability, crashworthiness, micromechanics, fracture, biomechanics and finite element method.

MICHAEL R. T. YEN Professor of Biomedical Engineering, University of Memphis.

Ph.D., University of California, San Diego, 1973. Founding Fellow, American Institute of Mechanical and Biological Engineering, 1990. Superior Performance in University Research, University of Memphis, 1989 and 1990. Achievement Award-Featured Engineering of the Year, University of Memphis, 1990.

Research Interests: Pulmonary circulation. Modeling of static and pulsatile blood flow in the lungs based on the experimental results of pulmonary blood vessels' morphometry, elasticity, and blood viscosity under different boundary conditions. Evaluation of blood flow in the lungs by experiments. Lung mechanics. Trauma of lung due to impact load. Mechanical properties of human heart, lung and aorta for highway safety research. Orthopedic biomechanics. Model studies on calcaneous fracture. Remodeling of long bone under pure bending.

WEI HUANG Assistant Project Scientist of Bioengineering, UCSD

M.D., Guangzhou Medical College, 1983. Cardiovascular Surgeon, 1983–1986. Biomedical engineer, 1989. *Ph.D., University of Memphis, 1994.* American Heart Association Postdoctoral Fellow, 1996–1998.

Research Interests: Vascular tissue mechanics, and vascular tissue engineering. Pulmonary vasculature, and pulmonary circulation. Tissue remodeling of pulmonary arteries in hypoxic hypertension. Measurement of blood flow and blood pressure *in vivo*. Data analysis with the nonstationary, nonlinear, and stochastic features taken into account. Measurement of mechanical properties of the tissues in the blood vessel wall as a two-layered structure *in vivo*. Development of a mathematical model of hypoxic hypertension based on the results from the *in vivo* experiments. Development of tissue engineered vascular graft in which the mechanical and biological function is compatible to the native blood vessel with a lumen diameter less than 4 mm.

APPENDIX B

ABOUT THE DEPARTMENT OF BIOENGINEERING AT THE UNIVERSITY OF CALIFORNIA, SAN DIEGO

A brief description of a typical department of bioengineering in a university may be of interest to the reader. Hence the following account.

The Bioengineering Program at the University of California, San Diego (UCSD) was founded in 1966 by Professor Y. C. Fung, the late Professor Benjamin W. Zweifach, and Professor Marcos Intaglietta as a joint effort between the Department of Aerospace and Mechanical Engineering Sciences and the School of Medicine. From its inception, UCSD Bioengineering has had a graduate program offering PhD and Master of Science degrees and an undergraduate program offering Bachelor degrees. Over the last 33 years, the program has graduated more than one hundred PhD's, a similar number of Masters, and over 500 Bachelors. Until 1999, the undergraduate program consisted of an ABET-accredited bioengineering track and a premedical track. In Fall 1999, a new undergraduate track of Biotechnology was added, as well as a new Master of Engineering degree program. Both of these programs were initiated to meet the industrial needs for bioengineering graduates. UCSD bioengineering graduates are working in academia, industry, research institutions, hospitals, and government. Many have attained leadership positions, making important contributions to the advancement of the field of bioengineering.

The UCSD Bioengineering Program began with excellence in biomechanics and micro-circulation. In 1975, UCSD Bioengineering received from the National Heart, Lung, and Blood Institute of the National Institutes of Health a predoctoral and postdoctoral training grant, which has been continuously funded to 2005 following the most recent renewal.

Dr. David Gough, with a primary interest in biosensors, joined the program in 1976, and Dr. Andrew McCulloch, with a research focus on cardiac mechanics, became a faculty member in 1987. In 1988, with the completion of Engineering Building Unit I, Dr. Richard Skalak (deceased 1997) and Dr. Shu Chien moved from Columbia University to UCSD, bringing with them molecular and cellular bioengineering as new areas of focus.

In 1990, UCSD Bioengineering received from the National Heart, Lung, and Blood Institute a Program Project Grant on "Biomechanics of Blood Cells, Vessels, and Microcirculation", which has been continuously funded till 2000, and a renewal application is being reviewed.

In 1991, the Institute for Biomedical Engineering was established at UCSD as an Organized Research Unit to foster interdisciplinary research at the interface of biology, medicine and engineering. Dr. Chien was appointed as the first Director. The Institute now has 90 members from the Schools of Engineering, Medicine, and Natural Sciences of UCSD, as well as the Salk Institute, the Scripps Research Institute, and the Burnham Institute. In 1993, the Institute won a Development Award from the Whitaker Foundation to enhance the infrastructure for bioengineering at UCSD. The Award, with a theme of "Tissue Engineering

Science", allowed the recruitment of new faculty in the Department of Bioengineering, the establishment of new core facilities, and the implementation of new education initiatives.

An Industrial Liaison Program was initiated in 1993, with active involvement of nearly twenty industrial corporations in San Diego and elsewhere. An Industrial Advisory Board was formed which has provided valuable advice and collaboration on Bioengineering education and research, including the placement of student interns in industry. The industrial internship program has been greatly enhanced by the Internship Award from the Whitaker Foundation initiated in 1996.

In July 1994, the Department of Bioengineering was established at UCSD, with Dr. Chien as the Founding Chair. It is the first such Department in the University of California system. The Departmental status has enhanced bioengineering research and education, promoted student activities, fostered industrial and other partnerships, and led to the formulation of the vision and programmatic plan for Bioengineering at UCSD. After serving for the maximum term of five years, Dr. Chien was succeeded by Dr. Gough as the Departmental Chair in July 1999.

With the aid of the Whitaker Foundation Development Award, the Department added four full-time faculty members: Dr. John Frangos (1994, mechano-chemical transduction), Dr. Lanping Amy Sung (1994, molecular bioengineering), Dr. Bernhard Palsson (1996, genetic circuits), and Dr. Sangeeta Bhatia (1998, hepatic tissue engineering). In July 1999, the Department is further strengthened by the addition of two new faculty members: Dr. Shankar Subramaniam (bioinformatics) and Dr. Gary Huber (molecular biomechanics). The Department plans to recruit six additional faculty members during the next five years.

The vision of UCSD Bioengineering is to use quantitative measurements and engineering analyses to integrate the rapidly advancing biomedical information, thus formulating the design principles of biological structure and function in health and disease. The goals are to develop biomedical engineering into a mature and premier discipline in science and engineering; to train the next generation of biomedical engineers who will continue to lead and amplify this development; to partner with scientific and industrial communities for the pursuit of interdisciplinary research and education; and to generate innovative ideas and to translate them to practical products.

Shu Chien

FAREWELL TO STUDENTS

Now you have gone through an intensive short course on design in bioengineering. Is this the first design project in your life? I trust this is the beginning of many, many to come. Many former students told me that the design experience sharpened their interest in basic science. Some told me that they would choose bioengineering as their future career. Others said they would turn to other fields. If this course helped you to make an early decision, I would consider this course a success.

This is the time to revisit the topics raised in the lectures, namely:

The quality of products and research in bioengineering
The importance of reliability of products in bioengineering
Ethical issues in bioengineering
Government and university regulations in bioengineering

A keynote to these topics is **high quality**. Anything a bioengineer does may impact somebody's health. Hence a bioengineer must be aware of his/her social responsibility.

Farewell!

Y. C. Fung